SIP SECURITY

SIP SECURITY

Dorgham Sisalem
John Floroiu
Jiri Kuthan
Ulrich Abend
Henning Schulzrinne

A John Wiley and Sons, Ltd., Publication

This edition first published 2009
© 2009, John Wiley & Sons Ltd.,

Registered office
John Wiley & Sons Ltd, The Atrium, Southern Gate, Chichester, West Sussex, PO19 8SQ, United Kingdom

For details of our global editorial offices, for customer services and for information about how to apply for permission to reuse the copyright material in this book please see our website at www.wiley.com.

Library of Congress Cataloging-in-Publication Data:

SIP security / Dorgham Sisalem ... [et al.].
    p. cm.
  Includes bibliographical references and index.
  ISBN 978-0-470-51636-2 (cloth)
   1. Computer networks–Security measures. 2.   Session Initiation Protocol (Computer network protocol)
I. Sisalem, Dorgham.
  TK5105.59.S564 2009
  005.8–dc22
                                 2008053852
A catalogue record for this book is available from the British Library.

ISBN   978-0-470-51636-2 (H/B)

Typeset in 10/12pt Times by Laserwords Private Limited, Chennai, India.
Printed and bound in Great Britain by CPI Antony Rowe, Chippenham, Wiltshire.

Contents

Foreword

In recent years I've been working on developing secure VoIP protocols to protect against wiretapping. But I'm not really a VoIP guy. I'm a crypto guy trying to learn about VoIP. And one of the first things I learned about VoIP is the lack of security. Not just security against wiretappers. VoIP can be attacked in so many ways. A call center can be targeted in a distributed denial of service attack. You can get a hundred telemarketing calls a day at home, with the calls originating where labor is cheap, out of reach of domestic laws prohibiting unwanted telemarketing calls. Or criminals can penetrate your PBX and make countless PSTN calls from your phone number, at your expense. And, of course, you can be wiretapped from criminals on the other side of the world. It's appalling how much worse VoIP is compared to the PSTN. If these problems aren't fixed, VoIP is going nowhere.

Yet VoIP is regarded by many as the manifest destiny of telephony, and for good reason. It's so much cheaper, it allows so many extra features, like video teleconferencing, and can be intelligently managed by computers under your own direct control. It puts the control back in the end user's hands, reducing the monopolistic power of the phone company. It just feels so right. It's obviously where telephony will go.

How do we reconcile these two opposing trends? Well, clearly the answer is we have to fix VoIP security. We just have to. That means a lot of engineers who work on VoIP are going to have to get up to speed on security, and start thinking like security professionals. If you want to develop VoIP applications, you need to read books like the one you're holding. This one covers a lot of the problems and solutions.

I looked at the crypto topics first. They do a good job showing the complexity in building and maintaining a PKI. They cover a number of crypto protocols in a great detail, including my own ZRTP protocol. Some of these protocols are used outside of VoIP, so this book is useful for those who want to see how crypto can be used in other applications. It's a nice crypto tutorial in its own right. Just as a source book on a number of influential crypto protocols, it's useful to have on your bookshelf. And it covers how these can be applied to VoIP. The authors have implemented the well-regarded SIP Express Router, and have run iptel.org, giving them a broad hands-on perspective on implementing SIP applications.

After treading the familiar ground of the crypto related topics, I started looking at the rest of the book. The real strength of this book lies in the vast panorama of attacks on VoIP systems, each described in meticulous detail. With their hands-on experience running a VoIP service, these guys have seen it all. I've never seen such an encyclopedic survey of real-world attacks on VoIP systems, exactly how and why the attacks work, and the known countermeasures to those attacks. I noticed that some attacks seemed to have no countermeasures, but at least you will know how they work in detail.

In any arms race, the warring parties develop an evolving body of knowledge over time, like the knowledge embedded in the genomes of cheetahs and gazelles that led to them both learning to run so fast. If you attempt to enter the field without the benefit of that hard-earned knowledge, you will become the lunch entree. This book will let you preload your arms race genome to merge into the ongoing melee in midstream, and maybe not become lunch. Sadly, so many of your colleagues in the VoIP industry will become digestible protein to the attackers, but you may be saved from that fate by your good fortune in reading this book.

Philip Zimmermann
Creator of PGP and Zfone Fellow,
Stanford Law School Center for Internet and Society

About the Authors

Dorgham Sisalem

Dr. Dorgham Sisalem received his M.Eng. and Ph.D. from the Technical University of Berlin in 1995 and 2000 respectively. He worked at the Fraunhofer Institute Fokus, Berlin, as researcher, later as head of department, and was involved in implementing and realizing the first SIP based conferencing system in 1998. He was further involved in the development of the SIP Express Router (SER) which is currently the most widely used open source SIP proxy. In 2003, he co-founded iptelorg which offered SIP-based VoIP solutions to ISPs and telecommunication providers until it was acquired by Tekelec in 2005. In the same year, Dorgham Sisalem joined Tekelec as Director of Strategic Architecture with main involvement in IMS security issues. He is a part time lecturer at the Technical University of Berlin and has more than 100 publications including international conferences and journals.

John Floroiu

Dr. John Floroiu graduated from the Polytechnic University of Bucharest, Romania in 1993 where he continued to work as a teaching assistant and received his Ph.D. in 1999. He joined the Fraunhofer Institute Fokus, Berlin in 1999 where he participated in numerous research and industry projects. His interests covered various fields including mobility, security and quality of service in IP networks, and later was involved with multimedia service architectures. Currently with Tekelec, John Floroiu works on crafting the architectures and products for the next generation of communication systems.

Jiri Kuthan

Jiri Kuthan is Assistant Vice-President for engineering with Tekelec. In this capacity, Jiri forms the company's technological strategy for all-IP-based networks, and leads two R&D teams. Jiri's career began in 1998 with a research position at Fraunhofer Institute Fokus, a renowned research institute in Berlin, Germany. His early work in the VoIP and security field began with contributing to the IETF standardization efforts and participating in EU-funded and industry-funded research projects. The most renowned result of his, by then small R&D team, was the creation of the open-sourced software for Internet telephony, known as "SIP Express Router (SER)". Jiri co-founded a company bringing the software and its concepts to the industry: iptelorg GmbH. The company deployed

Internet telephony with major Internet Service Providers, received prestigious Pulver 100 award and was acquired by Tekelec in 2005.

Ulrich Abend

Ulrich Abend graduated in computer sciences at the Technical University of Berlin in 2004. During his studies he worked as an engineer at Fraunhofer Institute Fokus where he had a major role in the development of the SIP Express Media Server (SEMS). Being part of the iptelorg team from the very beginning he was responsible for leading the development of the carrier class SIP platform SOP, based on the SIP Express Router (SER) and supporting components. SOP was successfully deployed at major customers across Europe and the United States. In early 2006 Ulrich Abend co-founded IPTEGO, an IMS service assurance company headquartered in Berlin. As CTO he is leading the team of SIP experts creating IPTEGO's next generation IMS product Palladion.

Henning Schulzrinne

Prof. Henning Schulzrinne received his undergraduate degree in economics and electrical engineering from the Darmstadt University of Technology, Germany, his MSEE degree as a Fulbright scholar from the University of Cincinnati, Ohio and his Ph.D. degree from the University of Massachusetts in Amherst, Massachusetts. He was a member of technical staff at AT&T Bell Laboratories, Murray Hill and an associate department head at GMD-Fokus (Berlin), before joining the Computer Science and Electrical Engineering departments at Columbia University, New York. He is currently chair of the Department of Computer Science. He is co-author of the Real-Time Protocol (RTP) for real-time Internet services, the signaling protocol for Internet multimedia conferences and telephony (SIP) and the stream control protocol for Internet media-on-demand (RTSP). He served as Chief Scientist for FirstHand Technologies and Chief Scientific Advisor for Ubiquity Software Corporation. He is a Fellow of the IEEE, has received the New York City Mayor's Award for Excellence in Science and Technology, the VON Pioneer Award and the TCCC service award.

Acknowledgment

We would like to express our gratitude to our employers, Tekelec, IPTEGO and Columbia University for providing us with the needed freedom and flexibility to work on this book. Further, without the support and patience of our families during the long nights and busy weekends, this book would have never been finalized. Special thanks to Jan Janak, Andrei Pelinescu-Onciul, Cristian Constantin and Robert Sparks from Tekelec for their support, for reviewing the book and their invaluable feedback. We are further obliged to Alan Johnston and Philip Zimmerman for their diligent review of the ZRTP section. Finally, we are grateful to the Internet Engineering Task Force contributors working on SIP and security issues for providing a nearly endless number of RFCs and drafts, as well as feedback and clarifications provided to the authors.

1

Introduction

Such a thing could never be more than a scientific toy (Casson 1910). This sentence did not describe Intel's surf board with embedded tablet PC,[1] but was written in the late nineteenth century and described the telephone built by Graham Bell. If anything, this shows how difficult it is to anticipate the success or failure of a technology and even more the desires and needs of users. More than a hundred years later, telephones are a practical necessity, as well as an occasional annoyance. We all have at least two phones and even children of elementary school age firmly believe that a mobile phone is one of the most important gadgets they need to survive everyday life.

The vast majority of the telephony services used today are still broadly based on the system envisioned by Graham Bell and use the concept of circuit switching. In circuit-switched networks, also known as public-switched telephone network (PSTN), the communicating parties are connected through a circuit or channel with fixed bandwidth for the entire duration of the call. The first experiments with transmitting voice over IP networks were conducted in the early 1970s (Cohen 1977). The first commercial applications and devices appeared in the mid 1990s based on proprietary protocols. H.323 (ITU-T Rec. H.323 2006) was first published in 1996 and was the first widely deployed Voice over Internet Protocol (VoIP) standard. The Session Initiation Protocol (SIP) was first published in 1999 (Handley *et al.* 1999) and then updated in 2002 (Rosenberg *et al.* 2002b). In recent years, SIP has increasingly gained in popularity and has become the de-facto standard for public VoIP offerings. It was adopted under the name of IP Multimedia Subsystem (IMS) by the mobile telephony networks as the signaling protocol for next-generation networks.

When compared with the PSTN, the VoIP market is still small in terms of number of subscribers and revenue. However, with more than 25 million subscribers and a revenue of US$3 billion in 2007,[2] the VoIP market now has considerable size. To ensure that the VoIP market can continue to grow, VoIP cannot compete only on the basis of price and features offered. For VoIP to succeed in the long term, VoIP services must offer similar security and protection levels to what is available today in the PSTN. A headline in the newspapers about the telephony service of a VoIP provider not being reachable for

---

[1] http://www.intel.com/cd/corporate/pressroom/emea/eng/150308.htm
[2] http://www.telegeography.com/products/euro_voip

SIP Security   Dorgham Sisalem, John Floroiu, Jiri Kuthan, Ulrich Abend and Henning Schulzrinne
© 2009 John Wiley & Sons, Ltd

a couple of hours due to a denial of service will certainly make a lot of people think whether they should really replace their current PSTN phone with a VoIP one. Rumors spread in blogs and Internet forums that VoIP services do not provide the same level of privacy protection as PSTN or about their vulnerability to fraud and identity theft can have tremendously negative effects on the reputation of VoIP. Finally, the proliferation of spam calls over VoIP services to levels similar to what we see today with email spam could further contribute to users yearning for the closed and protected PSTN service.

Security threats on VoIP services can be roughly categorized as follows (VoIPSA 2005):

- Social threats–because of the similarity to email services, VoIP services are expected to have the same social threats as email. This might include unsolicited calls, intrusion on the user's privacy, fraud, identity theft and misrepresentation of identity or content. To overcome these threats, we discuss in Chapter 6 various approaches for providing authenticated identities and combating identity theft and fraud as well as for ensuring the user's privacy. Chapter 9 presents different technologies that can be used for reducing the threats of unsolicited calls.
- Eavesdropping–by monitoring signaling and media information sent and received by a user, an attacker can collect various pieces of information about the user such as her identity, the identities of her communication partners and the content exchanged by the user. To reduce the possibility of eavesdropping on the user's audio or video calls, Chapter 7 describes different protocols used for securing the media communication. The security of the signaling data is discussed in Chapters 3, 5 and 6.
- Interception and modification–by intercepting exchanged signaling and media information or by getting access to the components providing the VoIP service, an intruder can reroute calls to malicious destinations, block calls from or to certain users or degrade the quality of the calls. Chapter 8 discusses possible attack scenarios that can be used for intercepting and modifying VoIP calls and approaches for defending against these attacks. Chapter 7 presents different protocols for supporting the encryption of media data and exchanging the necessary keys for encrypting and decrypting the media traffic.
- Service abuse–service abuse describes improper use of services by bypassing a provider's authentication mechanisms, stealing the service of other users or misusing the service provider's components for launching attacks on other users or service providers. Abuse scenarios and defense strategies are described in Chapters 6 and 8.
- Interruption of service–attackers launch denial of service attacks with the goal of interrupting a service and making it unavailable to legitimate users. In Chapter 8, different threats and attack scenarios and possible defense mechanisms are described.

In this book, we explore the security aspects of SIP and IMS services, with the goal of providing the reader with an insight into possible scenarios for launching denial of service attacks on SIP-based services, conducting fraud and identity theft and misusing the SIP service for distributing spam, as well as defending against such threats.

Chapter 2 lays out the basic technologies and mechanisms used for securing and encrypting traffic. Chapter 3 is an overview of SIP. In this context, the different usage scenarios of SIP and the different methods and authentication schemes supported by SIP are described. Further we touch on different deployment issues. Chapter 4 describes the usage of SIP in mobile and next-generation networks. The differences from the basic SIP specifications are highlighted and the call model in these networks is explained. Security aspects of user authentication and internetwork communication in next-generation networks are described in Chapter 5. Chapter 6 presents an overview of the different mechanisms suggested for securing the user identity in SIP and preventing fraud. The mechanisms used for securing the media traffic in VoIP environments are described in Chapter 7. Chapter 8 presents the different possibilities for launching denial of service attacks on SIP-based services as well as the possible solutions for reducing the risks of attacks. Finally, possible misuse of spam services for distributing spam is described in Chapter 9, along with the legal aspects of SIP and the different measures for protecting against such misuse.

The SIP specifications especially in the security related areas are still evolving. Further, even after an extensive reviewing process we still suspect that there are a number of errors that we could not catch and would appreciate if our readers would point us to. For a central place listing the latest updates to the SIP specifications and security related issues as well as collecting reader feedback and keeping an up-to-date errata please visit the www.sipsecurity.org web site.

2

Introduction to Cryptographic Mechanisms

This chapter introduces the basic concepts, algorithms and protocols in cryptography that are used to secure the signaling and data of multimedia sessions. Cryptography is a vast field and has a large number of applications, which the present chapter does not and cannot cover in an exhaustive manner. For further details the reader may refer to a number of comprehensive works in this area, including (Menezes *et al.* 1996) and (Schneier 1996).

Cryptography is the science of protecting messages exchanged over public channels. Four basic security services are required for this purpose:

- **Confidentiality** enables a sender to encrypt a plaintext message into ciphertext using a key so that only the recipients who possess the appropriate key can retrieve the original plaintext message by decrypting the ciphertext.
- **Integrity protection** enables the recipient to ensure that a received message has not been tampered with along the transmission path.
- **Authentication** may apply to both an entity or a message. In the former case we talk about **entity authentication** (or *identity authentication*), which is the assurance of one party about the identity of the other party involved. *Mutual authentication* is achieved when two parties involved in a communication (e.g. a key exchange) each succeed in authenticating the identity of the peer. The latter case is referred to as **data origin authentication** and enables the recipient to verify that the received message originates from the entity that claims to have produced it. In point-to-point communication integrity protection and data origin authentication are provided together.
- **Nonrepudiation** prevents a sender from denying the ownership of a message that he previously produced and sent to a recipient.

Section 2.1 introduces the basic symmetric and asymmetric cryptographic algorithms used for key establishment and data protection and also provides a brief incursion into PKI systems. Section 2.2 describes the key establishment protocols capable of negotiating the cryptographic algorithms and other security parameters required to establish secure channels of communication between applications. Section 2.3 addresses the legacy cryptographic scheme that makes the foundation of the mutual authentication and

SIP Security   Dorgham Sisalem, John Floroiu, Jiri Kuthan, Ulrich Abend and Henning Schulzrinne
© 2009 John Wiley & Sons, Ltd

key agreement mechanisms used in the next-generation mobile networks and the IP Multimedia Subsystem (IMS).

2.1 Cryptographic Algorithms

A cryptographic algorithm receives as input the data on to whom security services are to be provided. The output of the cryptographic algorithm is the "protected data", without elaborating here on what "protected" means beyond the services just enumerated. Most of the cryptographic algorithms also require a key as an input parameter. The key influences the output of a cryptographic algorithm in that only an entity that possesses the same key or a key that is in a particular relationship with the key used to produce the protected data is able to recover the original input or produce identical output from the same input. Some sort of feedback is also common for many cryptographic algorithms.

Based on the type of the keys used, one may distinguish between:

- Symmetric key cryptography–addresses the class of cryptographic algorithms where the encryption key and decryption key can be easily calculated from one an other or are (as in most cases) identical. In order to be able to communicate securely, two parties need to agree first on a shared key.
- Public (or asymmetric) key cryptography–addresses the class of cryptographic algorithms where the encryption and decryption are each performed with a different key out of a ⟨private key, public key⟩ pair: whatever is encrypted with the private key can be decrypted with the public key or vice versa. The private and the public keys are interrelated in that the public key can be easily derived from the private key, whereas the private key is practically impossible to derive from the public one. Each participant must keep his private key secret, while the public one must be made publicly available.

There are also keyless cryptographic functions, such as random number generators and cryptographic hash functions (see section 2.1.3.2).

2.1.1 Symmetric Key Cryptography

The symmetric key cryptography is mainly concerned with ciphers. A cipher is an algorithm that encrypts plaintext into ciphertext (encryption) and decrypts the ciphertext into the original plaintext (decryption), based on the assumption that the encryption and the decryption ends share the same secret key. The main property of the ciphertext must be that it looks like a sequence of random bits to any other entity that does not possess the encryption key.

Ciphers are used in conjunction with a **cryptographic mode**, which combines a basic cipher with some simple operations (usually Boolean XOR operations) and feedback (previous sequences of encrypted output are fed into the encryption process of subsequent plaintext sequences) that enable a number of attacks to be mitigated. The most basic requirement is to force identical plaintext sequences to appear differently each time in the generated ciphertext, so that an attacker cannot correlate the ciphertext with specific patterns in the plaintext. The use of a specific cryptographic mode has implications for error propagation and computational efficiency (in particular, related to the possibility of parallelizing the encryption and decryption process).

In (Dworkin 2001) some of the most commonly used cryptographic modes are described. The are two classes of ciphers: block ciphers and stream ciphers.

2.1.1.1 Block Ciphers

A block cipher encrypts one block of plaintext (P) into one block of ciphertext (C) of the same size, by applying the same transformation to every block of input data and using the same key. A block cipher also provides a decryption function that performs the inverse operation. The encryption and decryption operations may be described as: $C_i = E(P_i)$ and respectively $P_i = D(C_i)$, where E and D denote the encryptions and the decryption functions, respectively and i represents the block index.

Two of the best-known block ciphers are AES (AES 2001) and DES (DES 1999). AES, or Rijndael to give it its original name, is the current encryption standard selected by the National Institute of Standards and Technology (NIST) from among 15 candidate proposals (Roback and Dworkin 1999) to replace its predecessor DES. No theoretical weakness of the AES algorithm is known so far and AES is also fast in hardware and software implementations on a large variety of platforms (including for instance smart cards), while requiring little memory.

As already indicated, the input data when encrypting with a block cipher is not the plaintext itself. In Cipher Block Chaining (CBC) mode, for instance, every block of plaintext is XOR-ed with the previous block of ciphertext before being encrypted, while the first block of plaintext is XOR-ed with a random Initialization Vector (IV), which usually precedes the ciphertext in the encrypted message. This mechanism is called a feedback function and is essential to ensure that each occurrence of repeating sequences of plaintext is encrypted to a different ciphertext. The CBC mode may be described in the following way:

$$C_i = E(P_i \oplus C_{i-1}), \text{ for } i = 1 \ldots n, \text{ where } C_0 = \text{IV and } P_i = C_{i-1} \oplus D(C_i)$$

As a result of block ciphers working on blocks of data, a payload needs to be first padded to a size which is a multiple of the block size before being encrypted with a block cipher.

2.1.1.2 Stream Ciphers

Stream ciphers generate the stream of ciphertext by performing a simple operation on the stream of plaintext, which usually consists of bit-XOR-ing the stream of plaintext with a keystream of the same length as the plaintext stream. Such ciphers are called additive stream ciphers. This may be represented as: $C_i = P_i \oplus K_i$, where C_i is the ith bit of the ciphertext, P_i is the ith bit of the plaintext and K_i is the ith bit in the keystream.

The strength of a stream cipher lays in the key generator that generates the keystream, which must be a pseudorandom stream of bits. At the decryption end, the keystream is regenerated and the stream of plaintext is retrieved by performing a bit-XOR between the stream of ciphertext and the same keystream: $P_i = C_i \oplus K_i$. Stream ciphers may be implemented using block ciphers and appropriate cryptographic modes. In this combination, the block cipher is used to generate the keystream.

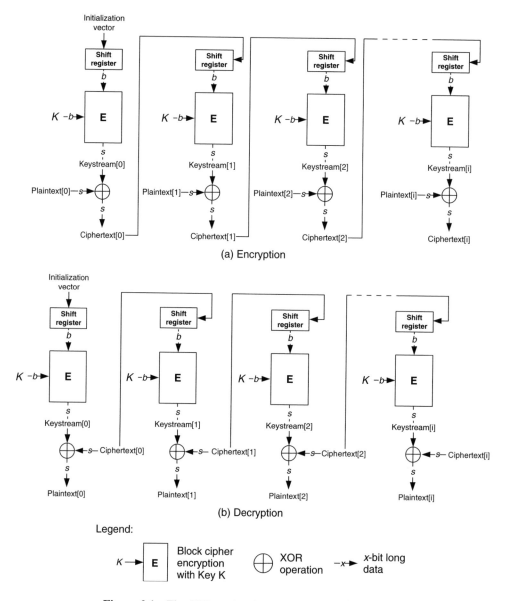

Figure 2.1   The CFB mode of operation for stream ciphers

Figure 2.1 illustrates the Cipher Feedback mode (CFB). CFB implements a self-synchronizing stream cipher, which is a class of stream cipher for which each bit of the keystream depends on a fixed number of previous ciphertext bits.

Each round of the CFB encryption produces s bits of ciphertext, with $1 \leq s \leq b$, where b is the block size. After each round the s bits of ciphertext are shifted into the shift register and used as input into the next round, while the remaining $b - s$ are thrown away.

A one-bit error in the ciphertext produced using CFB results in a larger number of plaintext bit errors, until the errorenous bit is shifted out of the shift register. The advantage, however, is that the re-synchronization happens automatically, without the need for additional synchronization mechanisms.

Figure 2.2 illustrates the Counter (CTR) mode and the Output Feedback (OFB) mode. They implement synchronous stream ciphers, which are stream ciphers where the keystream is generated independently from both the plaintext and the ciphertext. The

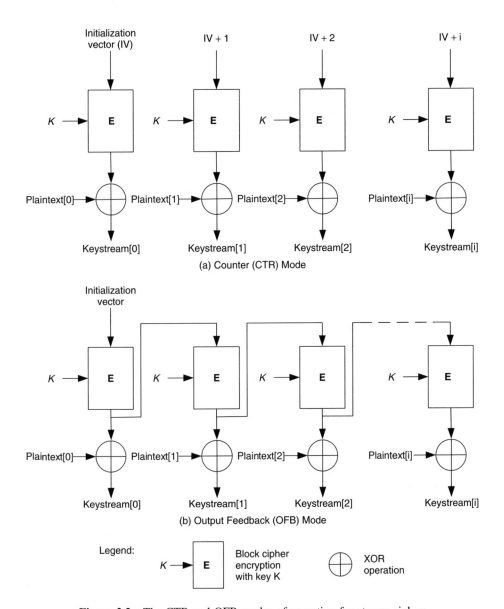

Figure 2.2   The CTR and OFB modes of operation for stream ciphers

synchronization of the keystream at the encryption and decryption end is essential for the correct operations of such stream ciphers, which cannot recover from the physical loss of ciphertext bits. Only the encryption operation is depicted in Figure 2.2; for decryption the roles of the plaintext and ciphertext are swapped.

Another common cryptographic mode used for building stream ciphers is the f8-mode. It has been standardized by 3GPP (35.201 TS 2007) and is a variant of the OFB mode that uses more complex initialization and feedback function (the IV is encrypted before being fed into the first encryption round and the feedback also includes the value of a counter).

Using a block cipher to generate the keystream involves the keystream being a multiple of the block cipher block size, so that whatever is in excess of the plaintext is discarded. When used for keystream generation, only the encryption function of the block ciphers is used for both encrypting and decrypting the data stream, which has the advantage that it reduces the implementation's footprint.

Stream ciphers have a number of properties that make them suitable for encrypting real-time media and preferred over block ciphers:

- Stream ciphers do not require any padding.
- In case of synchronous stream ciphers, the sender can precompute the keystream because the keystream is independent of both the plaintext and the ciphertext. Whenever the data becomes available, it is XOR-ed with the keystream. This has the potential of reducing packet latency.
- The CTR mode is similar to the OFB mode with the notable difference that the keystream is generated by encrypting the successive values of a counter. This enables the parallelization of both encryption and decryption operations because the ith keystream block can be generated independently of the previous ones so that any number of blocks can be processed in parallel.

On the downside, stream ciphers have tight requirements for the keystream generation. The most important aspect is that it is essential for a stream cipher not to use the same keystream more than once. The reason for this is that XOR-ing two ciphertexts produced with the same keystream yields the result of XOR-ing the corresponding plaintexts. In this way the plaintext becomes easy to break and then XOR-ing it with the corresponding ciphertext also yields the keystream.

Also, an attacker that knows both the plaintext and the ciphertext can XOR them and retrieve the keystream. Even guessing sequences of plaintext by an attacker is not a non-negligible risk because in particular the headers of most protocols contain fields whose values are easy to determine. Another example is the comfort noise used in real-time media transmissions. Keystream reuse is denoted as "two-time pad".

Synchronous stream ciphers are also easy to manipulate because flipping a bit in the ciphertext will result in that bit being flipped in the plaintext. In this way, an attacker that knows the plaintext can alter the ciphertext so that it decrypts to whatever plaintext he wants. For this reason synchronous stream ciphers must be used in conjunction with a Message Authentication Code (MAC, see section 2.1.3.2).

More sophisticated attacks consist of looking for collisions between the ciphertext and a large database of precomputed values obtained by encrypting a fixed plaintext with many distinct keys (McGrew and Fluhrer 2000). The attack takes advantage of the "birthday paradox" to detect collisions, which reduces the effective key size as compared with the case when an exhaustive search of the key is performed.

In case of additive stream ciphers, the attacks are also effective when a linear relationship (known to the attacker) between the plaintext bits holds with some probability. The use of a **salting key** in this context is necessary to increase the effective length of the key and hence robustness against these attacks by introducing additional randomness in the keystream generation. The salting key must be random and may be public.

2.1.1.3 Pseudorandom Functions

The main property of a pseudorandom function (PRF) is that it produces, in a computationally efficient way, output that cannot be distinguished from random data. We refer here to one particular application of the PRFs, which is generating additional keying material from the shared secret that two parties have derived as result of running a key establishment protocol. This is necessary because cryptographic applications usually require a relatively large number of session keys or long keystreams, which exceed the length of the derived shared secret. Session keys and keystreams may be used to protect protocol signaling and data traffic. They may need to be distinct in the send and receive direction and they may need to be different for different data streams exchanged during the same communication session.

PRFs make use of a HMAC function (see section 2.1.3.2) or a cipher, which is applied recursively in order to produce an arbitrary length of keying material. As input, the PRF uses the derived shared secret (which may need to be padded or truncated to fit the size required by the cryptographic algorithm), random numbers (exchanged by the peers during the key establishment) and some constant values to ensure that the different sets of session keys are distinct.

2.1.2 Public Key Cryptography

The public key cryptography is based on the very large difference in terms of computational complexity that exists between calculating a mathematical function as compared with calculating its inverse. For example, factoring a product of two large primes and computing discrete logarithms are computationally complex functions, while multiplication of two large primes and exponentiation are not. Therefore the latter two mathematical functions may be used in the process of generating the ⟨private key, public key⟩ pair, while the computational complexity of the former two guarantees that it is practically unfeasable to retrieve the private key from the public one.

The main applications of the public key cryptography are:

- key agreement;
- public key encryption;
- digital signatures.

2.1.2.1 Key Agreement

The Diffie–Hellman (DH) exchange (Rescorla 1999) is the most notable example of a key agreement method that makes use of public cryptography. It enables two parties to independently derive a shared secret from each party's own private DH key and the peer's public DH key. The DH key agreement is used by many protocols that negotiate the establishment of "secure channels" between two endpoints, such as IKE (see section 2.2.1.2) and TLS (see section 2.2.2.1). The DH keys cannot be used for encryption or signing.

Figure 2.3a illustrates a DH exchange. The parties, denoted as Alice and Bob, need to have previously agreed on "DH group" consisting of a generator g and a large prime number P. They then independently choose the private keys, A and B, which are random numbers between 1 and $P - 2$, and calculate and exchange the public keys, $g^A \bmod P$ and respectively $g^B \bmod P$. Each party exponentiates the peer's public key by its own private key, which yields in both cases $g^{AB} \bmod P$, the shared secret. An eavesdropper who obtains the DH public keys will not be able to find out Alice's and Bob's shared secret unless he solves the discrete logarithm problem, which is computationally impractical in the case of large prime numbers.

It may be observed that a DH exchange in its basic form is vulnerable to man-in-the-middle attacks, as illustrated in Figure 2.3b. The attack consists of the man in the middle replacing the public DH keys of the participants with his own public DH key without the participants noticing. This enables the man in the middle to decrypt and re-encrypt all the communication–presumed to be secure–between the two participants. The attack is possible because there is one missing element in this scheme, namely

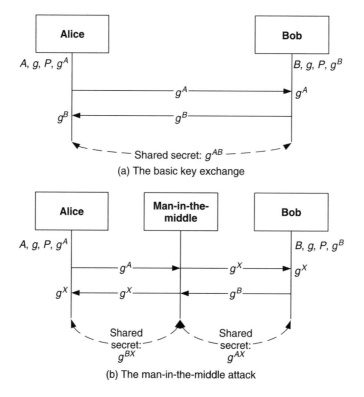

Figure 2.3 A Diffie–Hellman exchange

a provable relationship between a DH public key and the identity of its owner. This problem is in general solved by means of digital certificates (see section 2.1.2.4) in two ways:

- by including the public DH keys in digital certificates and reusing them across many key agreement sessions (also known as "static" DH keys);
- by exchanging "ephemeral" DH keys (that is, one-time DH keys) and using digital certificates to authenticate the key exchange messages.

2.1.2.2 Public Key Encryption

Public key encryption is used to provide confidentiality and consists of the sender encrypting a message with the receiver's public key. This ensures that only the intended recipient, who holds the corresponding private key, can retrieve the original message. The most widely used cryptosystem that provides public key encryption is the RSA cryptosystem, named after the names of the inventors, R. Rivest, A. Shamir and L. Adleman (Jonsson and Kaliski 2003).

The public key encryption algorithms involve operations of exponentiation, which makes them a couple of orders of magnitude slower than the symmetric key encryption algorithms. As a result, public key encryption is not employed to encrypt large amounts of real-time data.

A frequent use of the public key encryption is rather for the key transport. Key transport together with the key agreement represent the two major **key establishment** alternatives. While in the case of a key agreement both participants contribute to the generation of the shared secret (as in case of the DH exchange), key transport consists of one participant (say Alice) choosing a secret key and securely passing it to the other party (say Bob) by encrypting it with the recipient's (Bob's) public encryption key. The secret key is subsequently used to encrypt and decrypt the communication between the two parties using symmetric key algorithms.

The key transport scheme is vulnerable to a man-in-the-middle attack that consists of the man in the middle managing to substitute the public key of the legitimate peer (which is Bob, in our example) with his one. In doing so, the man in the middle will be able to decrypt the secret key (sent by Alice) and re-encrypt it using the public key of the legitimate recipient (Bob). In this way the attacker will have unlimited access to the entire communication between the two participants. As in the case of DH public keys, a mechanism is required to ensure the authenticity of the public encryption key (see section 2.1.2.4).

2.1.2.3 Digital Signatures

Digital signatures are used to provide integrity protection, authentication and nonrepudiation. They enable a sender to sign a message using her own private key so that any receiver that obtains the sender's public key can verify the digital signature and, if the verification is successful, conclude that the message is genuine.

Assuming the attacker did not manage to break or otherwise obtain the victim's private key (a situation denoted as a "total break"), specific measures need still to be taken with regard to the way a digital signature is calculated in order to ensure that it is robust against forgery. A successful forgery attack would enable an attacker to produce a valid signature for: (i) an apparently valid message over the content of which, however, the attacker has little or no control (existential forgery) or, even worse, (ii) a chosen message (selective forgery).

The digital signature schemes may be classified into:

- Digital signature schemes with appendix–for the purpose of signature verification the original message is required. In the basic form, the digital signature is produced by encrypting a hash value (see section 2.1.3.2) calculated over the original message using the sender's private key. The signature is validated by decrypting the hash value using the sender's public key and then comparing it with the result of the hash function applied to the signed message.
- Digital signature schemes with message recovery–the original message can be recovered from the digital signature itself. In the basic form, the digital signature is generated by encrypting the message itself using the sender's private key. The message is recovered by decrypting it using the sender's public key.

However, in the basic forms presented above, the digital signatures are vulnerable to selective forgery attacks and therefore a number of transformations (consisting of padding and including randomness) are performed on the hash value and on the message before encrypting them. Relevant standards are PKCS#1, ANSI X9.31 and ISO 9796.

Two of the best-known cryptosystems that provide digital signatures are RSA and the Digital Signature Algorithm (DSA; DSS 2000). The DSA is a digital signature scheme with appendix, while RSA supports both modes of operation. Also, DSA keys cannot be used for encryption.

2.1.2.4 Digital Certificates

As shown in sections 2.1.2.1 and 2.1.2.2, the ability to verify the authenticity of a public key is crucial to key establishment protocols. This is exactly what digital certificates provide: a digital certificate establishes a connection between a public key and the identity of its owner (or "subject"), digitally signed by a trusted third party, called an Certification Authority (CA) or "issuer".

A Public Key Infrastructure (PKI) is the management system that provides the framework necessary for requesting, issuing, distributing, validating, revoking and archiving digital certificates, as well as for establishing trust relationships inside and across PKIs.

Basic certificate management operations   In order to obtain a certificate, a subject must send a certificate request message to a Registration Authority (RA), which validates the binding between the subject and the private–public key pair. If this step is successful, the RA forwards the certificate request message to a CA, which issues the certificate.

The main components of a certificate request message are the certificate template, some optional attributes (which in this context are called "controls") and a Proof of Possession (POP; Schaad 2005). The certificate template contains the values of a number of digital certificate fields that are filled in by the subject, like for instance the public key and the suggested subject name. Following that, the RA/CA provides the values of the remaining ones (e.g. the serial number, the name of the issuer, the signature, etc.). The controls contain information to support the identity verification of the subject, among others.

The POP has the role of proving that the subject indeed possesses the matching private key. The POP mechanism employed is specific to the type of public key contained in the certificate request message (signing, encrypting or key agreement). For instance, if the public key is a signing key, the subject is required to sign its certificate request message; if the public key is an encrypting key, the certificate is returned to the subject in encrypted form. Once issued, the digital certificates are published in a certificate repository that may be accessible by means of the Lightweight Directory Access Protocol (LDAP; Zeilenga 2006).

Digital certificates are valid for a specific time interval, which is indicated in the certificate itself. During this time interval, a digital certificate may, however, be:

- updated–this involves a new private–public key pair being generated, which results in the public key contained in the certificate being updated (as well as the signature);
- invalidated, or revoked–this results in the digital certificate being included in a Certificate Revocation List (CRL).

These situations, and particularly the latter one, requires that, for the proper operation of a PKI, the status of the unexpired certificates needs to be checked before relying on them.

A CRL lists all unexpired certificates (by their serial numbers) that have been revoked, together with the revocation date. The revocation reasons may be that the private key has been compromised, the CA has been compromised, the user affiliation has changed, the issuing CA has been decommissioned, etc., and may be optionally specified for each CRL entry. CRLs are issued periodically or as soon as a certificate has been revoked, and they are digitally signed by the CRL issuer.

A CRL is characterized by:

- the CRL issuer, which may be a CA or another entity delegated by the CA;
- the date of issue;
- the list of revoked certificates;
- the scope, identifying the class of users or entities to which the revoked digital certificates belong or the reason why the digital certificates listed have been revoked;
- the signature algorithm and the digital signature over the CRL.

A CRL may take the form of a "delta CRL", which refers to a previously issued "complete CRL" for a given scope, and only lists the differences.

Even from this brief description it may be observed that managing a PKI involves a number of operations that include requesting a digital certificate, POP verification, updating or revoking a digital certificate, looking up a certificate in the CRL, etc. Some of them may be performed manually, but at the same time a number of protocols have been specified to support doing them automatically. For details about these protocols the reader is referred to their respective specifications: Certificate Management Protocol (CMP; Adams *et al.* 2005), Certificate Management over CMS (CMC; Schaad and Myers 2008) and Simple Certificate Enrollment Protocol (SCEP; Nourse *et al.* 2008).

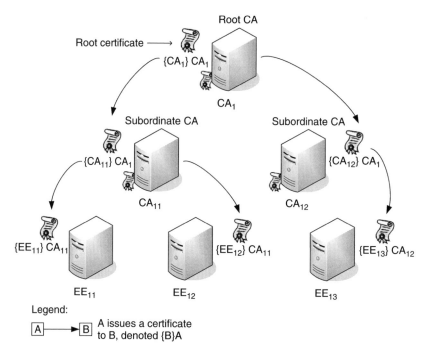

Figure 2.4   A hierarchical PKI organization

Certificate authorities and PKI organization   Figure 2.4 illustrates the structure of a hierarchical PKI organized on two levels. A CA may issue certificates either to subordinated CAs or to End Entities (e.g. a user or a host, denoted as EE).

The top-level CA is called the **root CA** and its certificate is a self-signed certificate, that is, it contains its public signing key signed with its own private signing key. The notation {E}A denotes a digital certificate issued by the CA "A" (i.e. signed by A using its private signing key) to entity "E" and may contain a public key of any type (used for encryption, signing or DH) that belongs to an EE, or a public signing key that belongs to a subordinate CA.

Returning to our example in Figure 2.4, EE_{11}, EE_{12} and EE_{13} trust the root CA; therefore they will be configured with the self-signed certificate of the root CA, which enables them to validate any digital certificate signed by the root CA.

Assuming EE_{11} needs to validate a public key provided by EE_{13} in the $\{EE_{13}\}CA_{12}$ certificate, EE_{11} needs to first obtain and validate the public signing key of CA_{12}, which is contained in the $\{CA_{12}\}CA_1$ certificate. This chaining constitutes the **certification path**, which may be visualized as the path along the directed graph starting from the verifier's trust anchor (CA_1, in our case) and ending at the certificate's owner (EE_{13}).

Figure 2.5 illustrates a multirooted hierarchical PKI. In this configuration, EE_{11}, EE_{12} and EE_{13}, as well as EE_{21}, are configured with the certificates of several root CAs, CA_1 and CA_2 in our example. This enables, for instance, EE_{11} to validate a public key provided by EE_{21} in the $\{EE_{21}\}CA_1$ certificate.

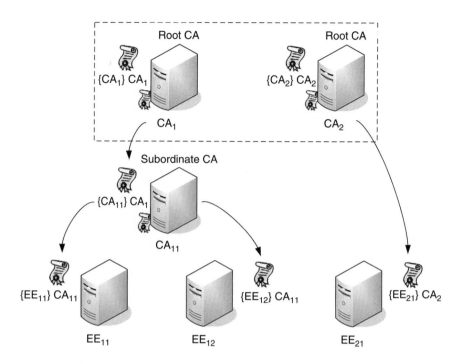

Figure 2.5   A multirooted PKI organization

The trust relationship may be extended across PKIs through cross-certification, which is the process by which two root CAs sign each other's certificate, as illustrated in Figure 2.6. CA_1 issues the certificate $\{CA_2\}CA_1$ by signing the public signing key of CA_2, and similarly CA_2 issues $\{CA_1\}CA_2$. In order to validate the certificate $\{EE_{21}\}CA_2$ provided by EE_{21}, EE_{11} builds a certification path up to the trust anchor CA_1 by including $\{CA_2\}CA_1$ in the chain.

Another possible PKI configuration is the one involving a Bridge CA (see Figure 2.7), which facilitates a more scalable way of interconnecting PKIs. The Bridge CA acts basically as a trust relay, so that any PKI that cross-certifies with the Bridge CA will be automatically trusted by and will automatically trust all other PKIs that are cross-certified with the Bridge CA.

For EE_{11} to validate the certificate $\{EE_{21}\}CA_2$, the following certification path needs to be established: $\{EE_{21}\}CA_2$, $\{CA_2\}BCA$, $\{BCA\}CA_1$, $\{CA_1\}CA_1$. More details about the practical aspects of using cross-certification and Bridge CA configurations are discussed in section 5.2.

As PKI structures can become quite complex, particularly when trust relationships are extended across PKIs through cross-certification, building an optimal certification path becomes a nontrivial task. The problem is complicated by the policies that may be associated with the use of individual certificates and various other constraints (e.g. basic constraints, name constraints). A comprehensive description of the procedures involved in building certification paths is provided in (Cooper *et al.* 2005).

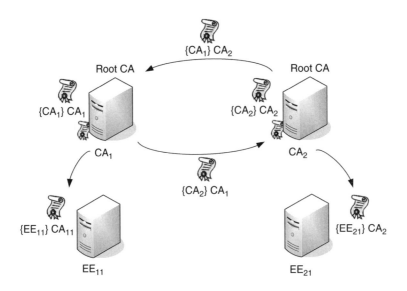

Figure 2.6   Cross-certification between PKIs

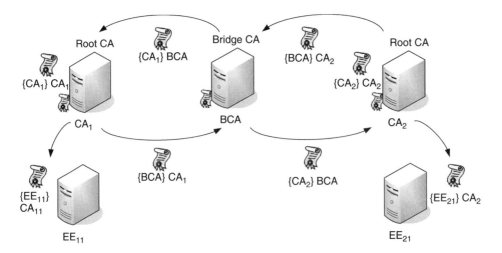

Figure 2.7   Cross-certification through a Bridge CA

The structure of a digital certificate   The format of a digital certificate is specified
in the X.509 standard (X.509 1997) (see also Cooper *et al.* 2008). A digital certificate
contains certain information:

- version number–the version of the encoded certificate;
- serial number–a unique number for each certificate issued by a CA;
- issuer–identifies the entity that signed and issued the certificate;
- validity–specifies the validity period of the digital certificate;

- subject–identifies the entity associated with the public key stored in the certificate;
- subject public key information–contains the public key and its type (RSA, DSA, DH, etc.);
- extensions–additional information associated to the certificate that facilitate the building of certification paths, indicate what purposes the contained public key should serve and what are the restrictions on using it;
- signature algorithm–identifies the cryptographic algorithm used by the CA to sign the digital certificate;
- signature–contains the signature.

A number of standard extensions may be used in conjunction with a digital certificate. They include:

- Authority key identifier–contains an identifier (e.g. the hash value) of the public signing key that corresponds to the private signing key used to sign the certificate. This extension is useful in facilitating the certification path construction.
- Subject key identifier–enables the identification of the certificate that contains a particular public key. This extension complements the Authority key identifier in the process of building the certification paths.
- Key usage–indicates, in the form of a bit string, what purpose the public key should be used for, which may be one (or more, if applicable) of:
  - digitalSignature–the public key is meant for signing purposes, other than signing certificates or CRLs. These purposes may be entity authentication or message integrity protection with data origin authentication, as used by the "digital signature authentication" exchange modes of many key establishment protocols (e.g. IKE, MIKEY, etc.);
  - nonRepudiation–indicates that the public key is used to verify digital signatures and provides nonrepudiation;
  - keyEncipherment–the public key is used for key transport;
  - dataEncipherment–the public key is used for encrypting user data other than secret keys;
  - keyAgreement–the public key is used for key agreement (for instance, as in the case of the public DH keys);
  - keyCertSign–the public key is used for signing certificates;
  - cRLSign–the public key is used for signing CRLs;
  - encipherOnly–if the keyAgreement bit is set, this indicates that the public key may be used only for enciphering data while performing key agreement;
  - decipherOnly–if the keyAgreement bit is set, this indicates that the public key may be used only for deciphering data while performing key agreement;
- Extended key usage–is typically used in end entity certificates and enables a more specific (but consistent) usage than that specified in the key usage extension to be indicated. Examples include: TLS WWW server/client authentication and email protection.
- Certificate policies–for an end entity certificate they provide a list of one or more identifiers that specify the formal terms under which the certificate can be used. Each identifier may be accompanied by a user notice (to be displayed to the user) or a URI

where a Certification Practice Statement (CPS) that defines the terms of use is located.
- Policy mappings–are used in cross-certificates to provide the mapping between the certificate policies. They take the form of a list of ⟨issuerDomainPolicy, subjectDomainPolicy⟩ mappings, indicating that the subject's certificate policy is equivalent to the respective issuer's certificate policy.
- Subject alternative name–allows identities that do not fit into the format of the Subject certificate field to be bound to the subject of the certificate. It may contain an email address, a DNS name, an IP address or a URI.
- Issuer alternative name–is an alternative name for the issuer.
- Basic constraint–indicates whether the subject is a CA (and must be correlated with the key usage keyCertSign bit) and if yes it also indicates the maximum depth a certification path that includes this certificate must have.
- Name constraint–indicates a name space within which all subject names and subject alternative names in subsequent certificates in a certification path must be located. The name constraints are specified in terms of permitted subtrees and excluded subtrees.
- Policy constraints–enable constraints to be imposed on the certificate policies when certifying outside the trusted domain. They indicate the number of certificates that may follow in a certification path before either policy mapping is no longer permitted, or a specific certificate policy is required to appear.
- CRL distribution points–contains a list of locations (e.g. HTTP, FTP or LDAP URIs) from where CRL information can be obtained.

2.1.3   Key-less Cryptographic Functions

2.1.3.1   Random Number Generation

Random numbers are obtained by generating a correspondingly long string of random bits. A random bit generator is a hardware device or software algorithm that generates a sequence of bits that are statistically independent (the probability of the source emitting a 1 does not depend on previously emitted bits) and unbiased (the probability of the source emitting a 1 is equal to the probability of emitting a 0). A random bit generator exploits the randomness of physical processes such as thermal noise from a semiconductor, the frequency instability of a free running oscillator, mouse movement, etc.

Random numbers have many applications in the cryptographic algorithms. Random numbers may be: (i) used as secrets, such as the private DH, RSA or DSA keys, or the secret key sent to a peer using a key transport protocol; or (ii) public, such as nonce values transported during the key exchange between two peers, or Initialization Vectors sent along with the encrypted data.

2.1.3.2   Cryptographic Hash Functions

The cryptographic hash functions are transformations that take a message as input and generate a fixed size "message digest", also referred to as a "hash value" or "digital fingerprint". The main properties of a cryptographic hash function are:

- It is computationally inexpensive to produce the message digest.
- It is computationally impractical to find out the message that has produced a given message digest.

- It is hard to find two different messages that yield the same message digest. This property means that, given a message, it is hard to find another message that yields the same message digest.

The most widely used cryptographic functions are the SHA algorithms (SHS 2007) and the MD5 algorithm (Rivest 1992).

Cryptographic hash functions have a large number of applications in the context of both symmetric and asymmetric cryptographic algorithms:

- Integrity protection–enables the recipient of a message to verify that a message has not been tampered with in transit by calculating the message digest and comparing it with the message digest provided by the sender, which is usually sent along with the message. In order to defend against the message digest being itself altered in transit, the hash value is calculated over a concatenation of the messsage itself and a secret key known only to the sender and the receiver. This construct is called a keyed-Hashed Message Authentication Code (HMAC; Krawczyk *et al.* 1997). HMAC provides **data origin authentication**, which enables the recipient to verify not only the message integrity but also the fact that the message was produced by the entity that claims to be the originator. The HMAC is a particular case of a Message Authentication Code (MAC), which makes use of a cryptographic hash function. Other MAC algorithms employ block ciphers, like for instance the CBC-MAC block cipher cryptographic mode (Dworkin 2005).
- Commitment schemes–enable a participant (e.g. in a key exchange) to first commit itself to some piece of data by providing the hash value of it before presenting the data itself. This may be used, for instance, to make it more difficult for an attacker to "sneak in" and provide its piece of data (e.g. public Diffie–Helman key) instead of the original one. One example is the ZRTP hash commitment scheme (see section 7.3.3.7).
- Password verification–enables a client to prove that it knows a password without having to send it in cleartext, but rather only a hash value of the password (which is usually combined with other random data). One example is the SIP Digest authentication scheme (see section 5.1.3);
- Digital signatures–digitally signing a message (see section 2.1.2.3) involves in many cases calculating a digital signature over a piece of information that includes among other elements a hash value calculated over the message.

2.2 Secure Channel Establishment

Establishing a secure channel aiming to protect the communication between two, or in some cases more, entites requires the participants not only to perform an authenticated key exchange (that is a key exchange between mutually authenticated entities), but also to negotiate the type of protection and the cryptographic algorithms used. We describe in this section two largely deployed generic mechanisms (in the sense that they are not dedicated to any particular type of application), one of them operating at the IP layer and the other one operating at the application layer.

2.2.1 IP Layer Security

2.2.1.1 The IP Security Architecture

The IP security architecture (IPsec; Kent and Seo 2005) provides security services to IP data flows and consists of three major functional components:

- A data plane component–enables the data traffic to be transported in protected form. This function is achieved by means of the Authentication Header (AH) and/or the Encapsulation Security Payload (ESP) security protocols. AH (Kent 2005a) can offer integrity protection and data origin authentication, with optional replay protection features. ESP (Kent 2005b) can offer the same set of services and also confidentiality (Manral 2007). When ESP is used with confidentiality enabled, there are provisions for limited traffic flow confidentiality (concealing packet length, generation and discarding of dummy packets, encryption of the inner IP header when operating in "tunnel mode"). The AH and ESP themselves make use of various cryptographic algorithms and cryptographic modes to provide confidentiality and integrity protection.
- A policy component–defines a set of IPsec policy rules against which the IP data flows are being matched. The disposition of these rules is either to discard a datagram, to allow it to bypass the IPsec processing or to offer to it IPsec processing. In the latter case, the matching IPsec policy rule defines which security protocol(s), cryptographic algorithms and keys must be used in conjunction with each specific IP data flow. These parameters make up an IPsec "security association" (IPsec SA). The IP data flows are identified by traffic selectors. A Traffic Selector (TS) defines a matching criterion for the IP datagrams, which includes source and/or destination IP addresses or ranges, source and/or destination port numbers and transport protocols. Depending on that, a traffic selector may have a finer granularity (e.g. by specifying a transport protocol and port numbers) or a coarser granularity (e.g. by only specifying subnet prefixes).
- A signaling component:
    - performs the mutual authentication of the IPsec peers and establishes a "signaling security association" between them to protect the negotiation of subsequent parameters;
    - negotiates the parameters of the IPsec security associations and the traffic selectors;
    - performs the key exchange that produces the keys necessary to the cryptographic algorithms used by the security protocol(s).

The signaling protocol is the Internet Key Exchange (IKE) protocol, specified in (Harkins and Carrel 1998) and updated by (Kaufman 2005).

The AH and ESP security protocols define new headers, which are inserted, depending on the operation mode, in:

- transport mode–the AH/ESP header is inserted between the IP header and the transport header, which may be a User Datagarm Protocol (UDP; Postel 1980), Internet Control Message Protocol (ICMP; Postel 1981a) or Transmission Control Protocol (TCP; Postel 1981a); or

- tunnel mode–the AH/ESP header encapsulates the entire original IP datagram and an outer IP header is prepended for the AH/ESP datagram.

Note, however, that in IPv6, when operating in transport mode, the AH and ESP headers are inserted before the Destination Options IP extension headers. The AH and ESP headers contain a sequence number used for replay protection and a Security Parameter Index (SPI) used to identify the corresponding IPsec SA. The AH security protocol protects the IP datagram payload (transport mode) or the entire IP datagram (tunnel mode) and most of the (outer) IP header. AH does not protect mutable fields in the (outer) IP header, that is those fields the value of which change between the sender and the receiver in an unpredictable way. Time-To-Live, Flags, Fragment Offset, Type-Of-Service and Header Checksum are mutable fields in an IP header. The Message Authentication Code (MAC) that results from applying the authentication algorithm is carried in the AH header.

The ESP security protocol encrypts the IP datagram payload (transport mode) or the entire IP datagram (tunnel mode) and authenticates it and the ESP header. Before being encrypted, the original data is padded up to a size which is a multiple of the block size of the encryption algorithm, if the cryptographic mode of operation requires it, or at least up to a 32-bit boundary. Additional padding, known as Traffic Flow Confidentiality (TFC) padding, may be used to conceal the size of the real datagram in order to deny an eavesdropper the possibility of discovering any patterns in the data traffic. After encryption, an Initialization Vector, as required by the encryption algorithm, is inserted at the beginning of the encrypted data. Finally, the MAC calculated over the ESP header and the encrypted data including the IV, denoted the ESP Integrity Check Value (ICV), are appended to the ESP datagram.

The replay protection mechanism protects against replay attacks that consist of an attacker storing datagrams and re-injecting them into the network at a later time. While the information contained in these datagrams is authentic, the fact that it is being repeated more than once[1] or the fact that it is obsolete may have a negative impact on the application using it. The replay protection works by adding a sequence number to the datagrams (or, in a more general context, to any kind of message) and maintaining a sliding window on the receiver side. Datagrams that are successfully authenticated, have a sequence number within the antireplay window and are not duplicates, are accepted. Datagrams carrying lower sequence numbers are discarded. The antireplay window is advanced when a datagram carrying a sequence number larger than the upper antireplay window limit is successfully authenticated. The recommended size of the antireplay window is 64 datagrams, but other values larger than 32 datagrams may be used as well.

The tunnel mode is mainly intended to be used when one or both of the communicating hosts are located behind a security gateway. This is a typical configuration of two hosts that reside in two different intranets and communicate securely over a public network. In this scenario the security gateways, the IP addresses of which appear in the outer IP header, perform all the IPsec processing. The ultimate source and destination of the IP datagram (contained in the inner IP header) are the intranet hosts and, if ESP is used, they are encrypted as the ESP datagrams travel between the security gateways.

Another common scenario is that of a "road warrior" that communicates with hosts in an intranet while it is physically located outside of it. A "road warrior" is essentially an

---

[1] A typical example here is that of a financial transaction that deposits a certain amount into an account.

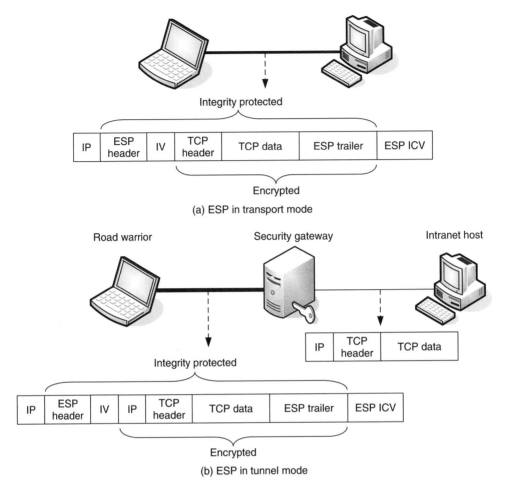

Figure 2.8 ESP in transport and tunnel modes

IPsec client accessing the intranet through a security gateway. The IPsec tunnel established between the IPsec client and the security gateway is also known as a Virtual Private Network (VPN). In some configurations the source IP address in the outer IP header is the IP address allocated to the "road warrior" in the local IP network where it currently resides, while the source IP address in the inner IP header is an IP address that belongs to the addressing space of the intranet.

Figure 2.8 illustrates two examples of using ESP in transport and tunnel mode. In addition to security-related functionality, the IKE/IPsec stack offers support for the negotiation and establishment of datagram compression at the IP layer, by means of the IP Payload Compression Protocol (IPComp; Shacham *et al.* 2001).

The two central concepts of the IP security architecture are the IPsec policies and the IPsec Security Associations. An IPsec SA defines the security protocol, operation mode (transport/tunnel) and cryptographic algorithms used to protect the traffic carried by it. A single IPsec SA protects data in one direction; two IPsec SAs must be set up in order to secure traffic in both directions between two IPsec peers.

An IPsec SA is uniquely identified by the datagram's destination IP address, the security protocol (AH or ESP) and the Security Parameter Index contained in the AH/ESP header. The IPsec SAs are stored in the Security Associations Database (SAD). An entry in the SAD also contains a sequence number (used by the AH/ESP headers), the antireply window parameters and the IPsec SA lifetime.

The IPsec policies are stored in a Security Policy Database (SPD) in the form of an ordered list. Each IPsec policy is identified by one traffic selector and traffic direction (inbound/outbound). An IPsec policy defines the security protocols and operation modes to be used for protecting the respective traffic flow and it also provides an IPsec SA specification. An IPsec SA specification lists the cryptographic algorithms supported by the local IPsec subsystem in the order of preference, as well as other parameters related to the IPsec SA lifetime and the generation of the keys. This list of attributes may be specific to an IPsec policy or a generic one. A traffic selector describes the pattern of the IP source and destination addresses, transport protocol and TCP/UDP port numbers that a traffic flow must match. Traffic selectors must allow wildcards and ranges to be used in describing the traffic flows so that the IPsec policies can be applied at different levels of granularity. If the disposition of an IPsec policy is to apply IPsec processing to the corresponding traffic flow, then the IPsec policy must link to an IPsec SA (or IPsec SA bundle). If no IPsec SA is in place, the appropriate ones are created on demand.

The creation of an IPsec SA represents the process of instantiating an IPsec SA specification and involves the peers negotiating and agreeing on the set of cryptographic algorithms, keys and SPI to use. This task is usually accomplished by:

- A key exchange protocol, such as IKE (see section 2.2.1.2);
- A protocol bundle able to provide authenticated key agreement and negotiation capabilities, such as AKA and SIP extensions (see section 5.1.1.7);
- Other out-of-bound mechanisms, such as command line interface.

2.2.1.2 The Internet Key Exchange Protocol

The IKE protocol is specified in (Harkins and Carrel 1998) and updated by (Kaufman 2005). The two protocol versions are denoted as IKEv1 and IKEv2. IKE implements an authenticated key exchange mechanism and operates in two phases:

- During Phase 1 the two peers exchange ephemeral Diffie–Hellman keys and their identities, and mutually authenticate and negotiate an IKE Security Association (IKE SA, denoted ISAKMP SA in IKEv1) that protects the subsequent IKE signaling. In IKEv2 the following authentication mechanisms may be employed:
  - digital signatures–certificates may be transported when necessary in the IKE messages;
  - pre-shared key;
  - EAP methods–leverages the use of client–server authentication schemes based on username and password.

  Support for EAP methods has been added in IKEv2, while two IKEv1 authentication mechanisms based on public key encryption have been deprecated.

- In Phase 2 IKE negotiates IPsec SAs and their corresponding traffic selectors on behalf of the applications that require security services. These security associations are denoted as "child SAs" in IKEv2. The keying material for the child SAs may be derived from the secret keys produced during the IKE SA setup. This enables fast generation of the IPsec SAs keying material, but it comes at the expense of not providing Perfect Forward Secrecy (PFS). Alternatively, an ephemeral DH key exchange may be performed for each child SA, conferring it PFS.

A security association negotiation involves the initiator of the IKE exchange providing a list of supported cryptographic algorithms (encryption and/or HMAC) that match the type of the security protocol (AH, ESP), from which the responder selects one combination. ESP requires in general one encryption and one HMAC algorithm; however, "combined mode" cryptographic algorithms, e.g. AES-CCM, (Housley 2005), which offer both confidentiality and integrity protection, do not negotiate a separate HMAC algorithm. AH requires only a HMAC algorithm. Besides the cryptographic algorithms, a security association negotiation also enables the peers to determine the lifetime of the security association as well as security associations identifiers (such as the SPI).

Perfect Forward Secrecy Perfect Forward Secrecy (PFS) is the property that ensures that the compromise of a long-term key does not allow past session keys to be compromised. Taking the IKE Phase 2 as an example, if the IPsec SAs shared secret keys are solely based on the secret keys generated during the IKE Phase 1, then once the latter ones are compromised, the IPsec SAs keys are compromised as well and hence all the data protected by them. If a separate ephemeral DH key exchange is performed for each IPsec SA, then compromising the keys of one SA does not result in the other SAs being compromised. Also, if the private signing key used by some IKE endpoint to digitally sign the IKE exchange is compromised, then the respective IKE endpoint can be impersonated (that is, the attacker can subsequently establish IKE sessions with other IKE endpoints posing as the victim); however, the past data exchange is not compromised as a result.

2.2.1.3 The Interworking of IKE and IPsec

Figure 2.9 provides an overview of the IKE/IPsec stack operation.

The processing of the outbound traffic is triggered when a datagram is sent to an external destination. First, the entries in the SPD are checked in strict order for a matching traffic selector (see arrow numbered 1 in Figure 2.9). The following results are possible.

- The traffic is IKE signaling. In this case the traffic is processed by a default IPsec policy that allows the signaling to pass through (if no IKE/ISAKMP SA exists) or applies the security services defined in the appropriate IKE/ISAKMP SA.
- A matching outbound IPsec policy is not found. A default policy is applied in this case, which either enables the traffic to bypass the IPsec processing or drops the datagrams.
- A matching outbound IPsec policy is found. The next step is to determine whether a corresponding IPsec SA or SA bundle is already in place:
  - If the IPsec policy points to a SAD entry then the respective security services are applied to the traffic (as illustrated by the arrow numbered 2b);

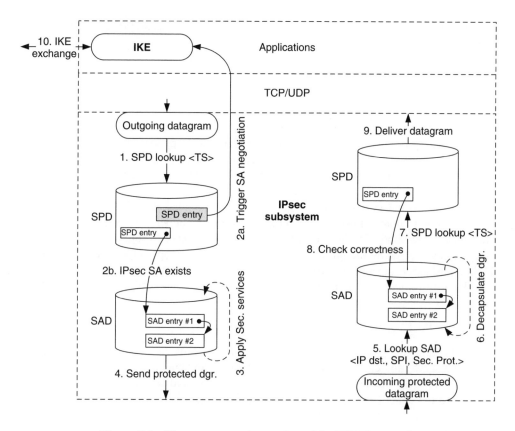

Figure 2.9   The structure and operation of the IKE/IPsec stack

– If an appropriate IPsec SA or SA bundle has not been established yet, the IKE
   daemon is invoked (see 2a). If the local host and the intended destination of the
   traffic (indicated by the IP destination address in the outer IP header) already share
   an IKE/ISAKMP SA, then a Phase 2 exchange is triggered; otherwise, a Phase 1
   exchange will take place first (see 10). Until the IPsec SAs are established, the
   outgoing datagrams are dropped.

Figure 2.9 illustrates an IPsec SA bundle composed of two security services, which
may be transport mode ESP (SAD entry #1) followed by tunnel mode AH (SAD entry
#2), just to give an example.

When inbound AH or ESP encapsulated traffic is received, the appropriate SAD entry
is looked up based on the IP destination address, SPI value and security protocol (see 5).
If the datagram is protected by an SA bundle, the SAD lookup is repeated until all the
ESP and/or AH headers have been processed (see 6). The next step is to look up the SPD
(see 7) in order to verify that the processing applied to the inbound datagram complies
with the IPsec policies. Assuming a matching IPsec policy is found, it is verified that
the correct SAD entries (and in the correct order) have indeed been used (see 8). If an

appropriate IPsec policy is not found or the IPsec SAs do not match, the inbound datagram is discarded.

2.2.2 Application Layer Security

2.2.2.1 Transport Layer Security

The Transport Layer Security (TLS; Dierks and Rescorla 2008) offers confidentiality, integrity protection and data compression to client/server applications. TLS runs in user space in the form of a library. Applications requiring security services link with the TLS library and use a generic API to send and receive datagrams and to manage the TLS sessions.

TLS is structured in two layers. The bottom layer, called the TLS Record Protocol, lays on top of a reliable transport protocol, which may be either a TCP or Stream Control Transmission Protocol (SCTP; Stewart 2007). The TLS Record Protocol encapsulates the messages exchanged by the protocol entities located on the upper layer, which are: (i) the TLS Handshake Protocols and (ii) the application to whom security services are offered. The Record Protocol assigns to the two of them distinct content type values.

The TLS Handshake Protocols are:

- The Handshake Protocol–used to:
    - Authenticate the peers. Only the server (which is also the most common use case), both server and client, or none of them may be authenticated.
    - Exchange the keying material used to derive the shared secret. The key exchange is protected against eavesdropping and if at least the server is authenticated then it is secure also against man-in-the-middle attacks.
- The Alert Protocol–used to signal error conditions.
- The Change Cipher Protocol–used by both client and server to notify the receiving party that subsequent records will be protected by the newly negotiated cipher suite and keys

The Record Protocol takes the messages sent by the upper layer protocol entities and fragments them into blocks. It then applies the desired processing, which may include data compression, MAC calculation and encryption. Finally it transmits the resulting records over the TLS connection (in fact TCP or SCTP). Received records are decrypted, verified, decompressed and reassembled by the Record Protocol operating at the other end of the TLS connection, and delivered to upper layer protocol entities.

Records containing data that belong to different upper layer protocol entities may be interleaved by the Record Protocol, the TLS handshake protocols data taking precedence over the application data. TLS records must be delivered to the transport protocol (TCP or SCTP) in the same order as they are processed by the Record Protocol.

A TLS connection is characterized by a set of security parameters that include:

- A cipher suite that consists of the hash algorithms used by the PRF and respectively the HMAC functions, together with a block cipher, its block size and cryptographic mode.
- A compression algorithm.

- The keying material that consists of the master secret (that results from the key exchange) and the random values exchanged by the client and the server. The keying material is used by the PRF to generate the following keys:
  - client write HMAC secret–used by the client to generate the MAC and by the server to check the integrity of the messages;
  - server write HMAC secret–used by the server to generate the MAC and by the client to check the integrity of the messages;
  - client write encryption key–used by the client to encrypt and by the server to decrypt the messages;
  - server write encryption key–used by the server to encrypt and by the client to decrypt the messages.

A TLS exchange starts with the client sending a Client Hello message, which contains the list of supported cipher suites, compression algorithms and key exchange method (to be described next). The Client Hello also contains a Session ID that uniquely identifies a TLS connection with a certain server and enables a client to resume a previous TLS session. In case of TLS session resumption, new cipher suites are negotiated and random numbers are exchanged between the client and the server, while the master secret of the resumed TLS session is reused and hence the authentication and key exchange phases are skipped altogether; this results in a much quicker TLS session setup between the client and the server.

The server responds with a Server Hello that contains the selected cipher suites, compression algorithms and key exchange method and concludes the negotiation phase.

The authenticated key exchange phase follows. TLS implements four methods:

1. RSA encryption (illustrated in Figure 2.10): for simplicity, only a description of the content of the relevant messages has been included, rather than the detailed payload formats. When using the RSA encryption method, the server sends a digital certificate containing a public RSA encrypting key (message 3); the client validates the certificate and uses the public RSA encrypting key to encrypt and send back a secret key (message 7). The server may optionally require the client to authenticate itself. In this case the client sends a certificate containing a public signing key (message 6) followed by a Certificate Verify message that confirms the possesion of the corresponding private key and digitally signs all messages starting with the Client Hello (message 1) and including the client Key Exchange (message 7).
2. Diffie–Hellman (illustrated in Figure 2.11): the server and the client use the Certificate messages to exchange digital certificates containing public Diffie–Hellman keys (messages 3 and 5). Both the server and the client derive then the DH shared secret.
3. Ephemeral RSA (illustrated in Figure 2.12): the server sends a digital certificate containing a public signing key (message 3), which it uses to sign an ephemeral public RSA encrypting key (message 4). The client responds with a secret key encrypted with the received public RSA encrypting key (message 8). The server may optionally require the client to authenticate itself, using a similar procedure as described in point 1.
4. Ephemeral Diffie–Hellman (illustrated in Figure 2.13): the server sends a digital certificate containing a public signing key (message 3), which he uses to sign an ephemeral public DH key (message 4). The client responds with an ephemeral public

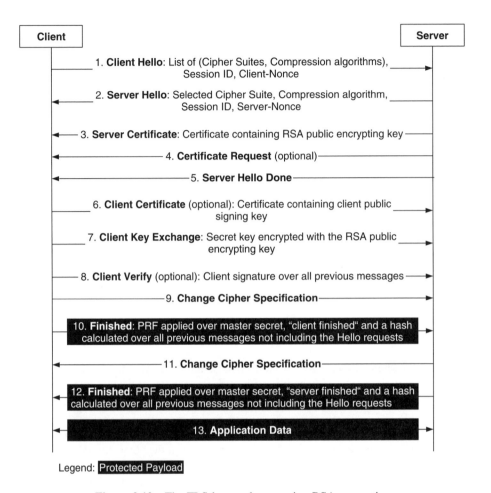

Figure 2.10 The TLS key exchange using RSA encryption

Diffie–Hellman key (message 8). Both the server and the client then derive the DH shared secret. The server may optionally require the client to authenticate itself, using a similar procedure as described in point 1.

After obtaining the shared secret, the client and the server use the Change Cipher Specification message to synchronize with each other the setup of the protected channel. The Finished messages are the first ones sent over the protected channel. They provide key confirmation and authentication of the message exchange. After successfully validating the Finished message the receiving party may start sending and receiving protected data.

2.2.2.2 Datagram TLS

TLS is used by a large number of applications requiring security services. However, it relies on running over a reliable transport protocol. Lost and reordered datagrams break both the TLS handshaking protocol as well as the TLS Record layer.

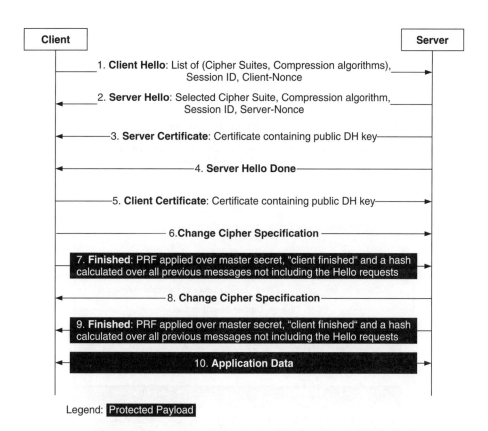

Figure 2.11 The TLS key exchange using Diffie–Hellman

On the other hand many other protocols operate over UDP (e.g. real-time media, gaming) or there are situations when UDP is preferable for protocols able to run over both TCP and UDP (e.g. SIP). In order to address this class of applications, Datagram TLS (DTLS) has been designed to be "a datagram-compatible variant" of TLS.

DTLS is specified in (Rescorla and Modadugu 2006) and is similar to TLS with regard to the security services offered to the applications. In addition to that, DTLS solves the following aspects related to the operation on top of an unreliable transport.

- DTLS record sizes: DTLS records must fit into one UDP datagram in order to avoid UDP fragmentation. On the other hand, multiple DTLS records may be carried in one UDP datagram.
- DTLS handshaking: as a result of point 1 above, the handshake message headers are enhanced to contain a distinct sequence number for each message as well as a fragment offset and fragment length. This enables the handshake messages to be exchanged in multiple fragments. A retransmission mechanism is also provided in order to cope with handshake messages being lost. Finally, a stateless cookie mechanism (Karn and Simpson 1999) is used to avoid denial of service (DoS) attacks using spoofed IP addresses;

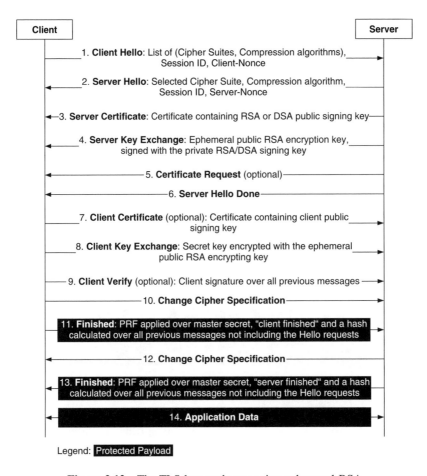

Figure 2.12   The TLS key exchange using ephemeral RSA

- Message ordering: TLS uses the last cipher text block in the previous TLS record as an Initialization Vector for the current record. The HMAC calculation also includes an implicit sequence number. TLS 1.1 already introduced explicit per-record IVs in order to protect against a security attack based on knowing the IV for a record. In addition to that, DTLS also introduces an explicit sequence number so that DTLS records can be individually verified and decrypted. The replay protection is implemented using a sliding window, in a similar way to AH (Kent 2005a).

2.3   Authentication in 3GPP Networks

AKA (33.102 TS 2008) is a challenge–response protocol that uses symmetric cryptography and was initially designed to handle the secure access in the UMTS networks. AKA achieves mutual authentication between the user and the network as well as session key generation.

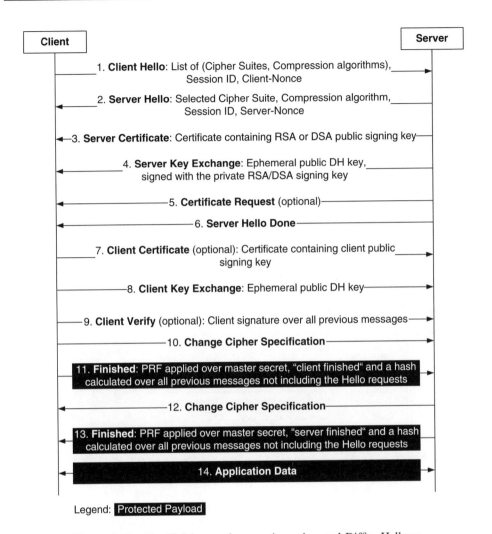

Figure 2.13 The TLS key exchange using ephemeral Diffie–Hellman

Figure 2.14 depicts a generic scenario that uses the AKA protocol, where a **supplicant** requests access to some resource or service from a server. The server acts as an **authenticator** and authenticates the **supplicant** with the help of an **authentication server**. The **authentication server** is the entity that stores the **supplicant**'s credentials and provides the **authenticator** with the necessary data that enables the latter to verify that the identity of the **supplicant** is genuine.

The authentication procedure consists of the following steps (as identified by the message numbers in Figure 2.14:

1. Whenever the authenticator wants to authenticate the supplicant, it sends an authentication request containing the supplicant's identity (described in more detail in Chapter 4), to the authentication server.

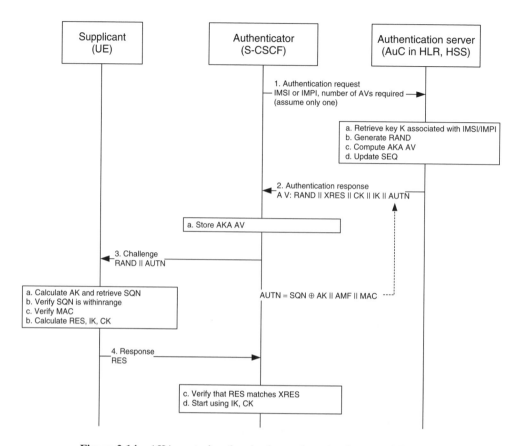

Figure 2.14 AKA mutual authentication and session keys establishment

2. The authentication server uses the private user identity to retrieve the corresponding secret key *K* and generate one or more AKA Authentication Vectors (AV), which it sends back to the authenticator (for simplicity, our example shows only one AKA AV). The AKA AV provides the authenticator with all the necessary parameters to challenge the supplicant (RAND, AUTN) and verify the response (XRES), as well as session keys (IK, CK) (see section 2.3.1). One AKA AV is valid for one authentication session and therefore the authenticator invokes the authentication server each time new AKA AVs are necessary.

3. The authenticator retains the response received (XRES) in the AKA AV and uses RAND and AUTN to challenge the supplicant.

4. AKA enables the supplicant to verify the authenticity of the challenge (see section 2.3.2) and in case this step is successful, the supplicant generates a response (RES) and sends it back to the authenticator. The authenticator uses XRES to verify that the response provided by the supplicant is correct, which ultimately proves that the supplicant is genuine.

2.3.1 AKA Authentication Vectors

Besides the secret key K, the authentication server and the supplicant share the following information, which is relevant to the management of the AKA AVs:

- A per-AKA AV sequence number (SQN), which is an always-growing value that ensures the freshness of the AVs. A re-synchronization procedure is executed between the authentication server and the supplicant each time the supplicant detects that the SQN in the AV provided by the authentication server is obsolete.
- An Authentication Management Field (AMF), which may be used to encode miscellaneous information such as the algorithm and key used to generate a particular AV, in case multiple authentication and key agreement algorithms are supported, or a limit L that indicates the range of validity for the SQN values.

An AKA AV is a concatenation of the following elements (see Figure 2):

$$AV = RAND\|XRES\|CK\|IK\|AUTN$$

with AUTN itself having the following structure:

$$AUTN = SQN \oplus AK\|AMF\|MAC$$

where:

- RAND is a random number generated for each AV.
- AUTN is an authentication token that basically represents the challenge. It encodes the SQN_{HE}, which is concealed by XOR-ing it with an Anonymity Key (AK). AUTN also has the property that it allows the supplicant to authenticate the network by validating the Message Authentication Code (MAC). The Authentication and Management Field (AMF) encodes some miscellaneous information related to the authentication mechanism and for simplicity it may be regarded as a constant value configured identically on both the supplicant and the authentication server.
- XRES is the expected response.
- CK, IK are a cipher and an integrity key. CK and IK are generated as a side effect to the authentication process and may be used to protect subsequent communication between the supplicant and the authenticator.

A number of cryptographic functions denoted f1, f1*, f2, f3, f4, f5 and f5* (35.206 TS 2007) are used in the computation of the different elements of the AKA AV (see Figure 2.15). These cryptographic functions make use of the AES block cipher in encryption mode, using the secret key K shared by the supplicant and the authentication server as cipher key, and some simple XOR and rotation operations.

A sequence number (SQN) is itself a concatenation of a counter (SEQ) and an array index (IND):

$$SQN = SEQ\|IND$$

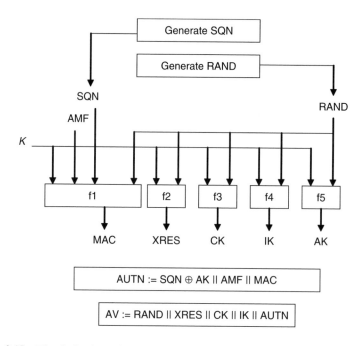

Figure 2.15 The derivation of the various components of an AKA AV (source: 3GPP)

All the AKA AVs that are generated by the authentication server as result of one request constitute an AV set (or batch), and share the same index value. The authentication server circularly increments the index value for each new generated batch and, whenever the identity of the entity requesting the AVs can be determined, the same index value is used for the same such entity. The counter is generated either by incrementing it for each new AV or by reading the value of a global timer.

In order to defend against replay attacks, both the authentication server and the supplicant maintain their own instance of SQN: the authentication server maintains one SQN_{HE} per user, while the supplicant maintains one SQN_{MS} per index.

The current value of SQN_{HE} is encoded into the AUTN produced by the authentication server as part of the AKA AV that is used to challenge the supplicant. In order for a challenge to be considered valid, the supplicant checks that the SEQ_{HE} value received in the challenge does not lag behind his locally stored SEQ_{MS} for the respective index by more than a certain limit L, and that the SQN value has not been used yet. This latter requirement involves that the supplicant must maintain a list of the most recently used SQN values. The limit L is by default 1 but a different value may be encoded into the AMF. If $SEQ_{HE} > SEQ_{MS}$, the SEQ_{MS} for the respective index is advanced to the new value. If $SEQ_{MS} - SEQ_{HE} > L$, the supplicant triggers a resynchronization procedure, the purpose of which is to allow the HSS to advance the SEQ_{HE} so that it becomes in sync with the SEQ_{MS} (see section 2.3.3).

The index value (IND) is not directly involved in the process of verifying the freshness and validity of the SQNs. It has, however, the important role of enabling multiple batches of AKA AVs to be concurrently stored on different entities and the validity of the contained

AVs to be independently verified without generating SQN re-synchronization requests when moving back and forth across the different batches.

2.3.2 AKA Mutual Authentication

As soon as the authenticator has obtained the AKA AVs, it challenges the supplicant with RAND‖AUTN (message 3 in Figure 2.14). The supplicant first authenticates the network by verifying that the AUTN has indeed been generated by a party that knows the secret key K and is using a fresh SQN. In order to do so, the supplicant calculates the AK based on the secret key K and the received RAND. It then retrieves SEQ_{HE} by XOR-ing the SQN $\oplus$ AK term contained in the AUTN with AK. Using the secret key K, the configured AMF, the received RAND and the SQN value retrieved from the AUTN, the supplicant can now calculate the expected MAC value (XMAC, see Figure 2.16), which it compares with the MAC contained in the AUTN. If the two values match, then the AV is authentic and the supplicant proceeds next to verify that the AV is also fresh. This latter test involves comparing the received SEQ_{HE} with the local SEQ_{MS} using the procedure described in section 2.3.1.

As a result to receiving an authentic and fresh challenge, the supplicant calculates RES, IK and CK and sends back an appropriate response (message 4 in Figure 2.14). The supplicant and the authenticator can then make use of IK and CK to protect the communication between them.

2.3.3 AKA Resynchronization

If the supplicant determines that the challenge received is stale (see section 2.3.1), it triggers a re-synchronization procedure. In order to do so, it generates the AUTS parameter

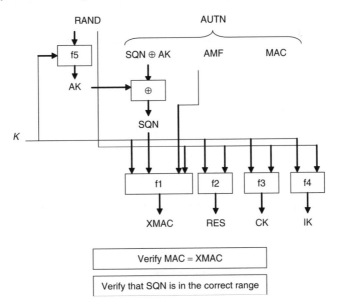

Figure 2.16 The user authenticates the network (source: 3GPP)

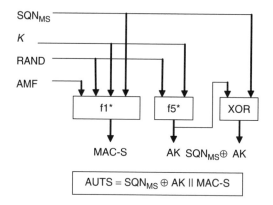

Figure 2.17   The structure of the AUTS (source: 3GPP)

as $SQN_{MS} \oplus AK\|MAC\text{-}S$ (see Figure 2.17) and sends it back to the authenticator instead of responding to the challenge. MAC-S is a Message Authentication Code similar to MAC but using a different cryptographic function. The authenticator must forward the AUTS to the authentication server, which calculates AK and retrieves SQN_{MS} by XOR-ing the received $SQN_{MS} \oplus AK$ parameter with the calculated AK. The authentication server then calculates and validates MAC-S and if the AUTS is authentic it sets SQN_{HE} to the value of SQN_{MS}.

A re-synchronization procedure is also initiated if $SEQ_{HE} - SEQ_{MS} > \Delta$, where Δ is typically 2^{28}. This puts an upper limit on how much SEQ_{MS} can jump and protects against SEQ_{MS} rolling over during the key validity lifetime. When this condition occurs, the supplicant generates a resynchronization failure message using SQN_{MS}.

2.4   Security Mechanisms Threats and Vulnerabilities

In order to secure the communication between two parties the following basic components are required:

- Mutual authentication–enables each participant to determine that the credentials presented by the peer belong to the legitimate entity.
- Key exchange–enables the participants to obtain a shared secret. When used in conjunction with the mutual authentication of the participants, we can talk about an **authenticated key exchange**, which ensures that the message exchange occurs with the intended party.
- Symmetric cryptographic algorithms–enable the data communication to be integrity-protected and, if desired, also confidential.

The goal of an attacker is to be able to decipher confidential communication, to inject faked information into the communication channel (sometimes by just replaying a genuine message) or to otherwise disrupt the communication between two parties. In order to achieve the first objective the attacker will have to obtain the secret used to encrypt the data. The attacker has for this purpose a number of alternatives:

- He may try to break the cryptographic algorithm itself–the attacker can try to use the ciphertext to mount a **ciphertext-only attack**. In its basic form this is a **brute force attack**, where the attacker exhaustively tries all possible keys. This type of attack is, however, practical only if the key space is relatively small. Alternatively, the attacker may exploit weaknesses of the cryptographic algorithm in order to break it more efficiently than using the brute force approach (Scott Fluhrer 2001). In some cases he may also know the pieces of plaintext that correspond to the ciphertext, resulting in a **known-plaintext attack**; message headers and padding are the primary sources of known (or at least easy to guess) plaintext. Moreover, he may be able to choose the plaintext that is encrypted, which represents a much more effective type of attack, known as **chosen-plaintext attack**. A particular case of a chosen-plaintext attack is the **differential cryptanalysis** (Biham and Shamir 1991). The differential cryptanalysis is based on the observation that pairs of plaintexts that XOR to one particular value produce pairs of ciphertexts that when XOR-ed produce a highly nonuniform distribution of values (with a few values having a high probability of occurrence). This weakness is exploited to determine which of the possible keys is the most probable to be the encryption key.
- The attacker may exploit weaknesses in the way a certain cryptographic algorithm is used. Keystream reuse is the most eloquent example of this category of attacks.
- The implementation of the cryptographic algorithm may also reveal useful information to an attacker. A typical example is the **timing attack**, which consists of an attacker who knows the hardware characteristics of the victim's machine being able to measure the time required by the victim's machine to decrypt a number of known ciphertexts. This provides information that can ultimately facilitate breaking the key (Bernstein 2005). Concealing the timing characteristics of an algorithm may be used as a counter-measure against such attacks, at the expense of decreased performance.

Protection against replay attacks can be achieved using a replay protection mechanism (for instance in case of data exchange) or by introducing randomness in each new exchange (for instance in case of credential exchange, such as the server and client nonces in the HTTP digest authentication, see section 3.9.1).

DoS attacks (see Chapter 8) fall into the third category, of the attacker disrupting the communication between two parties without necessarily getting hold of sensitive information. Another vector of attack, is the key exchange protocol itself. A typical example here is the **man-in-the-middle attack**, which enables an attacker to substitute the keying material exchanged between the legitimate participants with his own. If this is not possible, a man-in-the-middle can sometimes mount a **bid-down attack** that consists of the attacker interfering in the process of the victim's negotiation of the security mechanism in such a way that the victims choose a "weaker" security mechanism that the attacker can break more easily. These types of attacks can be easily eliminated if the participants mutually authenticate each other and integrity protect the key exchange.

This, however, opens in many key exchange protocols another opportunity for an attacker, maybe less dangerous in terms of access to sensitive information, but still annoying. Whenever protecting the key exchange requires computationally expensive operations, such as digital signature verifications and Diffie–Hellman secret key derivation, an attacker may mount a DoS attack by opening a large number of key exchange sessions. In order to do so, the attacker does not even need to effectively be a man in the middle, therefore the first line of defense of a key exchange protocol against DoS attacks should be checking the ability of the peer entity to receive messages at the indicated IP address and return a piece of data previously sent by the local entity (a "cookie").

Finally the attacker may choose to bypass all security mechanisms and attack directly the places where credentials such as private keys, secret keys or user passwords are stored. In this way he can obtain unlimited access to confidential communications and can impersonate legitimate entities in further communications. Key exchange protocols can help limit the impact of such an attack by providing perfect forward secrecy (PFS).

So far it looks as though the cryptography provides reasonable solutions to all known security problems; why then is security such a big problem? The answer is that security mechanisms are sometimes deployed not fully or not at all. Also, complex architectures may use multiple components, each of them using their own security mechanisms, which are not always able to fully interoperate (but rather they "complement" each other, in some cases resulting in security gaps). Yet, the way some protocols work, in particular those that require the involvement of intermediate entities, may result in a complex trust model that requires the participants involved in the communication to trust third parties (e.g. service providers, network operators, etc.). Finally, some agencies may require a **key escrow** to be in place to enable them to recover in some legal, special circumstances the communications of a certain person.

To this end, a number of limitations can be identified in practice:

- Lack of a security infrastructure–while digital certificates may be used to achieve mutual authentication, a PKI must be deployed first. Cross-certification is necessary to extend the trust relationship across different security domains. This involves a great deal of complexity and scalability issues, which have resulted in the lack of PKI availability at a large scale.
- Terminal capabilities–on the one hand, this involves the terminals being able to perform complex cryptographic calculations (like for instance public key cryptographic operations), which may be a challenge for battery-powered mobile devices (particularly if we also consider the potential DoS threats). On the other hand, terminals must have the means to securely store the credentials. If this cannot be guaranteed, the security of the whole system is compromised. Smart cards may be used to solve this problem and to this end it is worth mentioning the deployment of AKA-capable smart cards on the mobile phones; AKA, however, only enables a terminal to establish a security relationship with its service provider (rather than end-to-end).
- Perceived threat–users may simply not consider the data that they exchange as sensitive and they may simply not bother to secure it. Actually this is the gate to "user awareness" realm which has more to do with educating users about the threats facing them (such as identity theft, phishing, etc.) and the measures they should take to defend themselves (see Chapters 6, 9).

- Architectural complexity–this frequently results in security mechanisms being back-fitted into protocols, security mechanisms "complementing" each other or protocols operating concurrently. There are plenty of examples in both categories; we will mention three of them that we will encounter later on in the book. The first example is provided by the key exchange protocol messages being piggy-backed into another signaling protocol. Examples are SDES and MIKEY (see sections 7.3.1 and 7.3.2.5). In this way the security of the key exchange protocol depends on the security of the transporting signaling, which needs to be at least integrity protected (encrypted, in the case of SDES) between the communicating parties. An example in the second category is the HTTP digest over TLS mechanism, which uses the HTTP digest for user authentication and TLS for securing the communication. This results in a **tunneled authentication** protocol, which is highly vulnerable to **tunnel attacks** (see section 6.7). In the last category a typical example is provided by multimedia communications, where the signaling plane and the media plane operate independently. Securing each level independently is not enough, because it may not provide sufficient proof that the peer in the data plane is the same as the peer with whom the signaling has taken place. A cryptographic binding between the two planes is required in such a case. Examples in this category are the media plane key exchange protocols, such as ZRTP and DTLS-SRTP (see sections 7.3.3 and 7.3.4).

3

Introduction to SIP

By the mid 1990s, the Internet Engineering Task Force (IETF) had specifications for media transport (Schulzrinne *et al.* 1996), multimedia session description (Handley and Jacobson 1998) and multimedia streaming (Schulzrinne *et al.* 1998). With these capabilities it was possible to exchange packetized speech over the Internet. However, it was still not easy to establish a multimedia session without exchanging an email first with the IP addresses to use. This gap in the "puzzle architecture" of the Internet protocols has been filled in by Session Initiation Protocol (SIP).

SIP in its technical definition is a protocol, i.e. "language", which different pieces of equipment use to speak to each other over a network. The SIP protocol has been designed to allow participating devices to set up and tear down calls over the Internet. Moreover, it has been defined to be sufficiently general and is not limited to audio calls. For example, one can use SIP to set up video calls or gaming sessions, one can run instant messaging and presence, and potentially other applications that need multiparty sessions to be managed.

The following sections provide a brief overview of the functionality of SIP, the components used by SIP and its different usage scenarios. Section 3.1 provides a short summary of the history of SIP and the different competing VoIP technologies. Section 3.2 discusses the most widely used scenarios for SIP today. In section 3.3 we show the basics of SIP operations. In Section 3.4 we describe the different components used in a SIP network and the different kinds of user devices and servers used to provide a SIP-based service. Identities used for addressing and identifying users of SIP services are described in section 3.5. The way that SIP signaling messages are structured in transactions for session management is explained in section 3.7. Techniques for routing SIP signaling to appropriate destination are described in section 3.8. Then we discuss interaction of SIP signaling with other systems: authentication, authorization and accounting in section 3.9, middlebox traversal in section 3.10 and other important components in section 3.11. Eventually, we review the design decisions that led to the SIP protocol as we know it in section 3.12.

SIP Security   Dorgham Sisalem, John Floroiu, Jiri Kuthan, Ulrich Abend and Henning Schulzrinne
© 2009 John Wiley & Sons, Ltd

3.1   What is SIP, Why Should we Bother About it and What are Competing Technologies?

The work on the SIP specifications (Rosenberg *et al.* 2002b) started in 1995 in the IETF[1] and describes the SIP architecture and its components, the protocol they use to communicate with each other and security considerations. Frequently the term "SIP" is used to refer to a whole protocol family, which includes SIP itself as well as many other supporting protocols that are additionally needed to build Internet telephony applications. These include but are not limited to Real-time Transport Protocol (Schulzrinne *et al.* 2003) for conveying voice and other real-time traffic and DNS for finding other party's servers.

It is also important to know what the SIP specification does not specify. Briefly it does not prescribe how to build actual applications and networks that are based on the SIP protocol. This follows the IETF tradition of specifying protocols and leaving innovation freedom to those who actually engineer the applications and networks. For example, the SIP standard does not describe how to build a SIP softphone and combine the SIP protocol stack with other stacks, namely Real-Time Transport Protocol (RTP), voice codecs, user-interface, configuration and management, and other features. Interestingly, from the SIP specification viewpoint, a SIP desktop phone is viewed same as a large gateway that translates millions of SIP calls into the Public Switched Network. Neither does the SIP specification mention how to integrate all the SIP equipment in a consistent network, in which SIP telephones traverse Network Address Translators (NATs), find each other, terminate calls in the PSTN and are easily configured. To some extent, we try to address these practical aspects in this book.

It is also important to know that SIP has had its competitors that are listed here for sake of completeness. H.323 (ITU-T Rec. H.323 2006) used to be the greatest head-to-head rival in the 1990s. Its standardization began actually earlier than SIP's and thus gathered a greater momentum. The protocol is in many design respects similar to SIP with the key difference being its binary encoding, which allows for better performance. We believe, however, that this performance-driven feature was also the reason why many eventually chose SIP. Text encoding is simpler to debug, modify and integrate with all kinds of UN*X applications. A great push for SIP was its endorsement by the mobile-phone standardization body, 3GPP, as part of its strategy for deploying the Internet in mobile networks at the beginning of the millennium. Ironically, while this endorsement pushed SIP forward, it is not really clear what part of the 3GPP plans will be eventually deployed. There are still live H.323 deployments, but SIP appears to have the final word.

Another competitor was MGCP/Megaco (Arango *et al.* 1999; Cuervo *et al.* 2000). This twin protocol family was based on the traditional telephony approach in which network equipment controls telephones with limited intelligence in them. Innovators have been concerned that this assumption makes the introduction of new services too difficult. Support for each new feature has to be introduced both to the telephones and to the network. With SIP it is sufficient that the telephones agree on the same services and changes to

---

[1] The Internet Engineering Task Force (IETF) is a large open international community of network designers, operators, vendors and researchers concerned with the evolution of the Internet architecture and the smooth operation of the Internet.

the network are rarely needed. Nevertheless, there are still some MGCP/Megaco-based VoIP deployments today, frequently for some reason in France.

We believe that the most significant competitor to SIP is in fact Skype. Skype has taken a different approach to most of other VoIP players did. Skype invented its own protocol, which is highly proprietary. It is secured against common security threats as well as reverse engineering. Most attempts to reverse engineer as of the time of writing this book have provided rather modest results (Baset and Schulzrinne 2006; Biondi and Desclaux 2006; Ehlert *et al.* 2006; Guha *et al.* 2008). Skype's software client is a brilliant piece of engineering: it is built using very well-chosen technologies to deal with voice encoding and firewall and NAT traversal, and its user interface is very convenient. The proprietary mode in which Skype has obtained traction is, however, its greatest weakness too. Many technological companies are reluctant to support closed walled-garden environments and prefer open standards instead. It remains to be seen how the success is split between open SIP and proprietary Skype over the coming years (Dryburgh 2008).

SIP is to be understood in its evolution too. While research publications about VoIP began to appear as early as in the 1970s (Cohen 1977), work on the SIP standard began much later, in the mid 1990s, at about same time as the first proprietary solutions began to be deployed by companies like Vocaltec. SIP received additional interest with 3GPP's endorsement and publication of Microsoft's Windows Messenger in early years of this Millennium. Mass deployments began to appear in 2004 when the technology began to be affordable: the price of many consumer SIP terminals sank to below $100 and open-source SIP servers became available and mature. We think that the current SIP status as of 2008 could be compared to the Web in early 1990s: the core HTTP/HTML technology was in place and allowed basic hypertexting scenarios; however extensions to accommodate applications such as forms had just begun to appear and ground-breaking applications such as Amazon, ebay, google, not speaking of peer-to-peer and "Web 2.0", were as yet not in sight. Similarly the SIP technology is today deployed, and basic calls work on large basis. For most of the 'childhood illnesses' of the 1990s and early 2000s, solutions have been found even though they are sometimes not yet perfect. Successful SIP service providers are known to have passed the one-million-subscriber mark. SIP-based Private Branch Exchanges (PBXs) ship worldwide. There is a wide choice of SIP equipment: SIP softphones, hardphones, servers, Integrated Access Devices (IAD), dual-mode mobile telephones, gateways, media servers and other devices. Integration of all the equipment is feasible even though frequently quite laborious. Lagging are the added-value services built on top of SIP: in the PBX environment "legacy features" remain popular, SIP conferencing has been sighted, but still the focus has remained on affordable telephony in the past. In summary the SIP technology has taken off and continues to seek appealing applications on top of itself.

We think that in assessing the SIP development trends SIP itself is not alone the key factor—it is really the Internet underneath it. The Internet allows the SIP applications to be meshed up with other applications like calendars, phonebooks, email and web. The Internet allows the same service to be offered globally. Last but not least, the Internet environment has been traditionally highly open, which helps to keep prices competitive for the technology's users. In the next sections we describe the usage scenarios of SIP, the different components of a SIP-based service and the messages and addresses used by SIP.

3.2   SIP: the Common Scenarios

The purpose of this section is to familiarize readers with the commonly deployed SIP scenarios. This is important for two reasons. Firstly, each use-case relies on slightly different technologies and puts them together in a slightly different way. Secondly, focus on the real scenarios helps readers to get better sense of difference between what is state-of-the-art today and what is yet to become real. Whereas the SIP protocol specification remains the same in all scenarios, dealing with operational aspects, such as firewalls and NATs, differs on a case-by-case basis. In this section we will consider the following use-cases: SIP trunking, public ASP service and PBX deployment. We are not going to study emerging use-cases in this book. Such would include rich applications and SIP use in converged fixed–mobile networks. These are very interesting and intensively marketed; however we have not yet seen developed usage patterns that we could document.

SIP trunking has been from historical and volume points of view the most successful use case so far. With SIP trunking, Figure 3.1, the actual calls are both originated and terminated in the PSTN. Only in middle of the path does the call traverse the Internet (or sometimes a private IP network) through SIP-to-PSTN gateways. Call participants frequently remain unaware of the fact that they are using SIP. A particular reason for the success of this scenario is twofold. First, it is economically viable: with large traffic volume the VoIP saving can return the investment quite quickly. Secondly, it is easy to integrate. In many cases it takes PSTN gateways from a single vendor, all configured very similarly and interoperating with each other smoothly. Proprietary solutions began

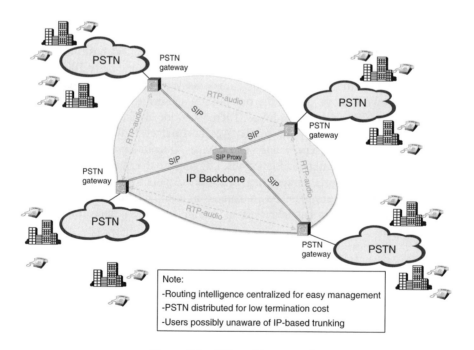

Figure 3.1   SIP trunking scenario

to be in operation from about 1995 on, to be succeeded by SIP-based implementations years later.

Technically, this scenario is characterized by star topology, use of PSTN tunneling and careful use of bandwidth-saving codecs. The gateways are distributed so that they are close to users and the longest possible segments of the calls traverse the Internet. A SIP server in the middle of the star reconnects the gateways with its centrally managed routing scheme implementing a kind of least-cost-routing policy. With PSTN tunneling original PSTN signals are attached to SIP messages and preserved across the Internet haul. The SIP-based tunneling is known as "SIP-T" (Vemuri and Peterson 2002; Zimmerer *et al.* 2001), and its ITU-originating version as "SIP-I" (ITU-T Rec. Q.1912.5 2004). Choice of bandwidth-saving codec is important, since with trunking volume bandwidth cost is considerable. Competing solutions to SIP-based trunking are based on transporting PSTN signaling using the Sigtran (Ong *et al.* 1999) or BICC (ITU-T Rec. Q.1902.3 2003) protocols, but are beyond the scope of this book.

In about 2004, VoIP began reaching consumers on a wider basis when Internet Service Providers (ISP) and Applications Service Providers (ASP) started offering consumer services (see Figure 3.2). Their offerings have usually included discounted Internet telephony with additional features like voicemail and call forwarding. Even in this case, the VoIP experience has usually been kept transparent to subscribers: subscribers received a SIP terminal adapter that allowed SIP calls to be made with their legacy telephones. As of

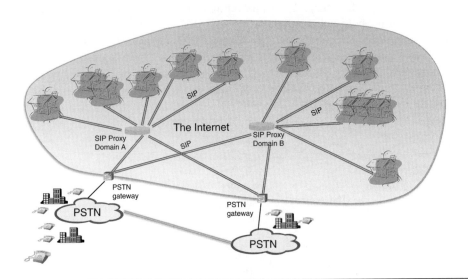

Note:

-World split in administrative domains

-The domains frequently form islands that don't speak to each other but use PSTN instead (due to lack of shared addressing)

-PSTN termination and origination frequently a separate business

Figure 3.2 SIP residential scenario

the time of writing this book, successful deployments have achieved a size of low multi-million subscribers. It is worth mentioning that frequently VoIP deployments use PSTN to interconnect to each other and remain thus isolated islands. This is caused by the fact that there is no easy way of letting the calling party know if the destination number is linked to a SIP account in the Internet or if it is only available in the PSTN. This leads to the use of PSTN as the interconnecting element, despite its sometimes prohibitive cost. The ENUM (Faltstrom and Mealling 2004) DNS-based distributed phone number directory has been envisioned to address this problem, but so far deployments have not really taken off. The apparent reason for this is that it just takes the consensus of too many competitors. Instead, many operators have begun to consider dips into "legacy databases" such as Local Number Portability databases.

Yet another frequently deployed use-case is SIP-based Private Branch Exchanges (PBXs), as illustrated in Figure 3.3. Motivation for deployment has been multifold. One of them is cost. As an example, the authors of this book have been able to install an open-source SIP-based PBX with conference bridge on it on a sub-$100 box. The installation reused existing data wiring and was able to power tens of telephones. Even commercialized solutions are likely to make a significant price point. Also programmability has been making SIP-based PBXs popular. Software solutions frequently based on open-source are easy to customize by skilled programmers for the exact needs of the users. Last but not least, SIP telephony's design is decoupled from sub-IP technologies. The separation from access technologies allows SIP telephones to be moved in and out of the office. The same infrastructure (Ethernet typically in office uses) can be shared for both voice and data.

Trunking, residential telephony and IP-based PBX replacements have in fact been the most widely deployed scenarios in the past years. The industry has been

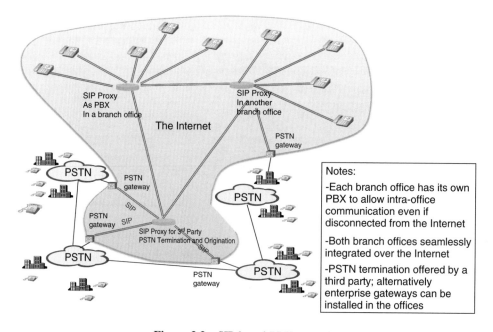

Figure 3.3   SIP-based PBX scenario

trying to find more cases, primarily in combination with other Internet services: computer–telephony–integration, messaging and presence, remote collaboration. It still remains to be seen which scenarios are actually adopted. In the next sections we are going to walk through different aspects of building SIP networks. We will look at SIP equipment, SIP protocol elements, the role of SIP servers and some specific aspects of real SIP networks.

3.3   Introduction to SIP Operation: the SIP Trapezoid

In this section, we give a preview of how SIP works and show how two SIP telephones set up a voice call between them. The actual protocol details are explained in subsequent sections: Section 3.4 gives an overview of all components in a SIP network and their behavior, section 3.5 explains how these components, services and users are addressed using email-like addresses, and section 3.6 describes details of how components communicate with each other using SIP messages.

In a nutshell SIP is a protocol that creates, updates and terminates multimedia sessions, such as voice calls, between two or more participants. Its operational model has been largely inspired by and inherited from the email world. In the email world, users are unambiguously identified by their addresses, e.g. jiri@iptel.org. Each user belongs to an administrative domain. Every user has a software application or an appliance that communicates with other users inside and outside their administrative domains. The domains are identified by DNS names, which appear in users' addresses. All of these statements are true of SIP as well. They lead to abstract network topology which is frequently referred to as the "SIP Trapezoid" (Figure 3.4).

The bottom corners of the trapezoid are represented by end-devices: SIP telephones that human users use to talk to each other. With each phone there is an email-like SIP address associated as well as its IP address. The actual packetized duplex voice transmitted using the RTP protocol is represented by the bottom dotted-line. However, before media exchange begins, the two phones must find each other and agree to have a voice session using encoding both telephones understand. This is in a nutshell the task of the signaling protocol: SIP. SIP signaling requests propagate through the network to discover their final destination. The signaling path is represented by the remaining edges of the trapezoid: each SIP phone communicates with its domain's SIP server (left and right edge) and the domains communicate with each other (upper edge).

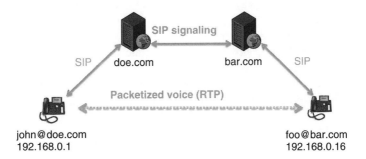

Figure 3.4   The SIP trapezoid

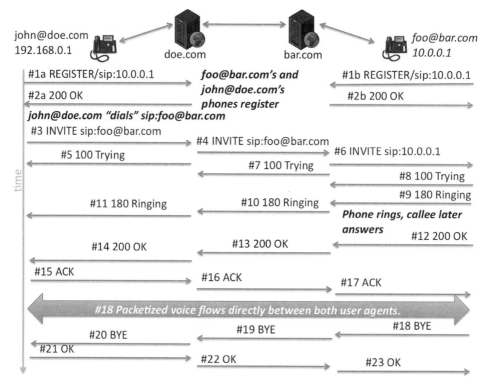

Figure 3.5   Life-cycle of a SIP phone call

The decomposition of signaling and media is a key difference from email: SIP only creates, modifies and terminates sessions, and leaves transport of the actual application data to another protocol, RTP. Generally the physical path of the voice and media packets differs as suggested by the trapezoid abstraction: whereas packetized media are sent using the shortest path, signaling packets traverse the Internet via domains' SIP servers. Even if two travelers staying in a European hotel are using SIP services from Asia and the USA, only signaling will visit all the continents; the media packets will not leave the European hotel.

Figure 3.5 shows the life-cycle of a phone call in a SIP trapezoid as an example. Initially, both SIP phones make themselves known to their respective servers. This is achieved by sending a REGISTER request (#1a and #1b in the figure). The SIP servers confirm the requests positively using a *"200 OK"* response and then know the phones' whereabouts. When later the user of the left-hand telephone, john@doe.com, "dials" foo@bar.com, his telephone sends a call setup request to its domain's server (#3). In SIP parlance the request is called "INVITE", and it includes information about who is calling whom, what capabilities the telephone features and more useful information. The INVITE request then propagates along the SIP path in the trapezoid towards the callee's SIP telephone. The doe.com server finds out that the destination address is located in another domain, bar.com, and forwards the request there (#4). The bar.com server uses the registration information to forward the request to the actual SIP telephone (#6). Each of the hops also confirms receipt of the INVITE request using a provisional *"100 trying"* response (#5, #7 and #8)–if this response was not sent, the senders would continue to retransmit the requests.

The call setup in our example is finished when the called party answers the call. Then the callee's telephone chooses codecs to be used in the call and sends a final *"200 OK"* response back (#12). The answer propagates along the same SIP path to the caller (#13, #14), who acknowledges it for reliability purposes (#15, #16, #17). At this moment the session is established and both phones begin to exchange packetized voice using the RTP protocol.

When later either party decides to hang up, it sends a BYE request. The BYE request follows the same trapezoid path (#18, #19, and #20), as does the answer to it in the reverse direction. Media packets are no longer sent and both parties have terminated the session.

In the real world, there are many alternatives to this trapezoid scheme. First of all, advanced routing schemes, described in section 3.8, can result in SIP paths shaped differently than a trapezoid. The media path may be more complex due to interaction with devices known as Middleboxes–see more about these in section 3.10. Last but not least, signaling may simply fail. There are many reasons for a possible failure: the caller cancels the pending call, the network connectivity is broken, the SIP server configuration is broken, the call setup request times out before the callee answers the call, race conditions occur as described in (Hasebe *et al.* 2008), and so on and so forth.

3.4   SIP Components

For a SIP session to take place, several components are necessary that we are going to study in this section. In the most trivial case, a pair of SIP telephones will allow the participants to have a phone call. In the complex case of public SIP services, there are also SIP servers keeping track of all users, PSTN gateways to translate SIP calls to the PSTN and vice versa, conferencing servers, voicemail servers, servers to process Call Detail Records, and more. Such equipment typically includes a complex mix of software, including media processing and communication, user-interface, protocols for configuration and monitoring, support for traversal of Network Address Translators, and more. In the heart of it, there is a signaling SIP server that implements session management for linking SIP equipment together. The SIP components are normatively described in (Rosenberg *et al.* 2002b). The specification defined logical components, which means that actual SIP-compliant products can choose to implement an arbitrary combination of these, as long as each conforms to its specification. In particular, the SIP standard defines the following SIP logical components: SIP User Agents, SIP proxy servers, SIP redirect servers and SIP registrars.

3.4.1   User Agent

The SIP User Agent (UA) is the basic element–it represents the SIP signaling functionality installed in every pieced equipment connected to a SIP network, most prominently in SIP telephones. The SIP User Agent is a peer-to-peer element in that it can both initiate SIP requests and receive them. The SIP specification draws an even finer distinction on a transaction-by-transaction basis: the part responsible for sending a request and processing replies to it is called the User Agent Client (UAC), whereas the part responsible for receiving, processing and answering incoming requests is called the User Agent Server (UAS). In the trivial case of two SIP telephones setting up a call between them, both phones include a built-in SIP User Agent. The calling telephone acts for this particular call setup as the User Agent Client, whereas the called phone acts as a User Agent Server. When later the called telephone's user decides to hang up, the roles are swapped. His SIP telephone's User Agent will be acting as Client for the sake of terminating the current SIP session.

The SIP User Agent is typically, though not always, just one piece of many forming a whole complex product. The functions of SIP telephones include negotiation of capabilities using SDP (Handley *et al.* 2006), media processing using a variety of codecs, communicating packetized media using RTP (Schulzrinne *et al.* 2003) and remote configuration and management. A block diagram, Figure 3.6, shows two SIP phones, consisting of multiple components. When a call is made, the components interact as follows: when the left-hand user dials using the user interface, the user interface triggers signaling using SIP. The signaling message describes who is calling whom, includes SDP description of available codecs and is sent to the right-hand side recipient. The recipient's phone receives it and answers it; its answer also includes a description of available codecs. Then media exchange between the two SIP phones using RTP conveying the packetized voice begins.

Of course the type of equipment building upon a SIP User Agent may vary. Users frequently see a SIP hard-phone as resembling traditional desktop phones and visually

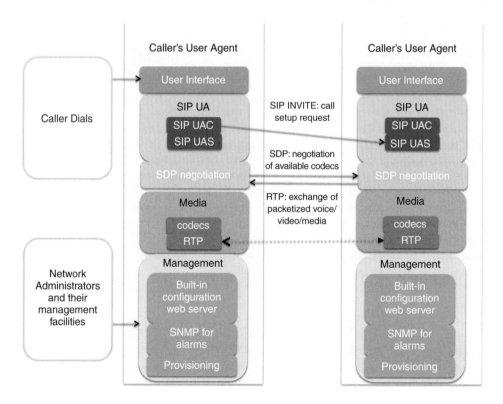

Figure 3.6   Example block diagram of communication between User Agents

Figure 3.7 SIP consumer equipment

differing only in different cabling connecting to them. Some users like to use soft-phones, i.e. a "virtual telephone" running on a PC equipped with a headset. Conservative users may wish to plug in their traditional analog telephone in a "terminal adapter" that makes the telephone appear to be a regular SIP telephone. And if the terminal adapter is one box too many, the user can use an Integrated Access Device instead: IADs include an Internet router, terminal adapter and frequently also a WiFi transmitter and Ethernet hub. Also "dual wireless phones" are available that support both traditional GSM telephony and VoIP telephony over WiFi. Examples of such consumer devices from various manufacturers are shown in Figure 3.7.

A variety of SIP User Agents, however, remains invisible to the end-users. In a SIP network, there are often several types of equipment for bulk processing: gateways for interconnection with the PSTN, conference bridges and media servers for playing announcements, storing voicemail messages and other media processing. Such equipment is based on the notion of the SIP User Agent as well–the only difference from the SIP point of view is the high number of User Agents running in parallel on a single host.

3.4.2 Registrar

There is a major problem with the simplest-possible two-party call-flow as depicted in Figure 3.8: it assumes callers' knowledge of callee's whereabouts. The caller must know the IP address of his peer's equipment. The IP addresses can be hard to learn and remember, and can quickly become useless because they frequently change. In short, they are temporary. The solution to this problem in SIP is the concept of permanent SIP addresses, called the Address of Record (AoR).

The AoRs are permanently associated with a user such as sip:jiri@iptel.org or an abstract service such as sip:conferencebridge@iptel.org, and advertised for example on business cards. The AoR are then temporarily bound to current contacts of the user or service. The bindings are stored in a specialized User Agent Server, called the registrar, or more accurately in the registrar's user location database. The user location database is updated by User Agents using REGISTER requests through the registrar. The REGISTER

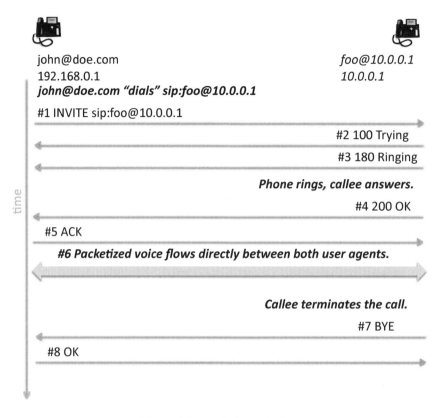

john@doe.com
192.168.0.1
john@doe.com "dials" sip:foo@10.0.0.1

#1 INVITE sip:foo@10.0.0.1

foo@10.0.0.1
10.0.0.1

#2 100 Trying

#3 180 Ringing

Phone rings, callee answers.

#4 200 OK

#5 ACK

#6 Packetized voice flows directly between both user agents.

Callee terminates the call.

#7 BYE

#8 OK

Figure 3.8   Basic SIP call flow

requests include most importantly the AoR in the *To* header-field and the contacts bound to it in the *Contact* header field. The *Contact* header field value can suggest a period of time, after which the contact expires if not refreshed. If the interval is set to zero, the binding is removed instead.

The registrar answers REGISTER requests with a reply that includes a complete list of bindings associated with the given AoR. The SIP standard puts no limitation on the number of contacts that can be bound to an AoR. This facilitates a feature known as forking: call invitations for an AoR can be "forked" to multiple destinations, such as office and home. An example of register transaction is shown in Figure 3.9. In the example, a User Agent Client registers itself, something that typically happens when a SIP telephone is powered up and later when it periodically refreshes its registration. The registrar's answer reveals that, in addition to the newly registered contact, there is another one associated with the AoR. Note that the call-flow is simplified as it does not include verification of client's identity. If a registrar did not demand an authentication, a malicious user could register his SIP phone with someone else's Address of Record to receive incoming calls for the victim. SIP authentication is explained with in detail in section 3.9.1.

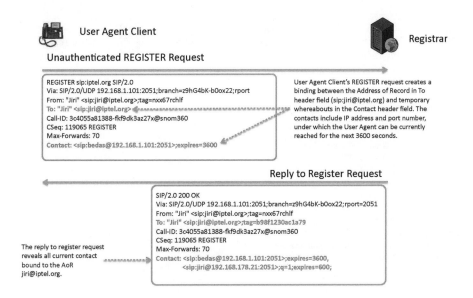

Figure 3.9   SIP registration transaction

3.4.3   Redirect Server

Another specialized instance of a User Agent Server is the so-called redirect server. A redirect server does a very simple job: it answers the incoming request with a suggestion to the User Agent Client to try again at some other destination(s). This can be useful, for example, if a user leaves a company and wishes incoming calls to be redirected to her new place. Another use may be load-balancing: a redirect server can redirect clients to some other place that will handle the actual SIP processing load. This scales well since processing load for SIP redirection is very low. In fact, a redirect server does not even need to parse the whole SIP message.

Note that a well-designed User Agent Client should notify its user of a redirection and request consent–otherwise the caller could be redirected to a destination that is subject to high charges. An example call-flow in Figure 3.10 shows a redirect transaction. The initial attempt to set up a call to user sip:jiri@jkuthan.org is redirected to the address "sip:voicemail@iptel.org".

3.4.4   Proxy

A proxy server is a central element residing in the network and "gluing" SIP User Agents together. The "gluing job" is more accurately called routing: a SIP proxy between User Agents routes requests from User Agent Clients to User Agent Servers and responses on the way back. The most obvious way to make the routing decision is using registrar's user location database. However the proxy server is free to make its routing decision using any logic it chooses. This may, for example, include lookup in telephone directories or

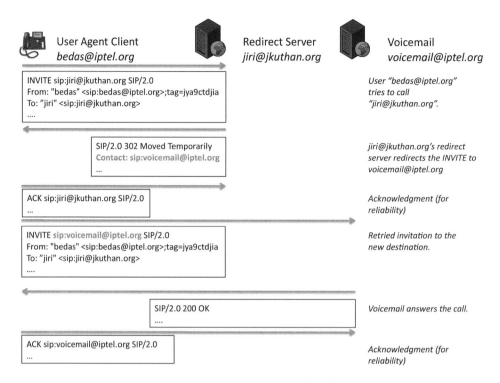

Figure 3.10   Redirected SIP INVITE

a least-cost routing policy. Section 3.8 provides more details on various types of routing logic and how these can be combined with each other.

The centralized position of a proxy server in the network is not only useful for using the central user location database for finding callees. It also a place to perform security policy and apply user preferences to signaling. Typically the tasks performed by a proxy server on receipt of a request include:

- Authentication of a request, i.e. verifying that the request originator really is who she claims to be and not an attacker pretending someone else's identity. The user's identity is important as input to authorization decisions and for producing Call Detail Records (CDR). Also, receivers of a request forwarded by a server responsible for a domain certainly appreciate if the user's identity, as asserted in the SIP request, is verified by the server. If request authentication fails, the request is declined.
- Routing decision, i.e. finding where to forward the request; this routing decision may be governed by a combination administrative policy and user's preferences. For example, the administrative policy may suggest that INVITEs for a user are routed to where his SIP telephones were registered from, unless the user overrides this behavior by setting up permanent call forwarding to voicemail.
- Authorization decision, i.e. verifying if the request is eligible for processing. For example an authorization policy may only permit off-net calls from the Internet to the PSTN for privileges users. If the authorization fails, the request is declined.

- Appending additional pieces of information that can be helpful for downstream SIP components. For example, SIP servers frequently add telephone numbers owned by the caller to the request, so that the callers' phone numbers can be displayed in the PSTN.
- Modifying the request in conformance with the definition of SIP protocol. This includes updating the request URI to the new destination, as determined by the routing decision, and leaving the server's "signature" in the request to remember the SIP path for subsequent SIP messages. In particular, the SIP server adds its address in a *Via* header field for routing of subsequent replies, and in *Record-Route* header-field for routing of subsequent requests.
- Forwarding the request to the next hop. Note that if routing logic has identified multiple destinations, the request can be forked to all of them. If that is the case, it is the proxy server's responsibility to pick the "best choice" from returning responses based on the rules specified in RFC3261 (Rosenberg *et al.* 2002b).

Figure 3.11 shows the changes that a proxy server makes to forwarded SIP requests: its "signature" appears in *Via* and *Record-Route* header fields, the request URI is rewritten to the new destination and a header-field with a telephone number to be displayed in PSTN is appended.

Some server implementations offer additional possibilities to administrators and subscribers to define processing behavior for their SIP requests. The standardized ways include Call Processing Language (Lennox *et al.* 2004), SIP CGI (Lennox *et al.* 2001) and Java servlets (Kristerser 2003).

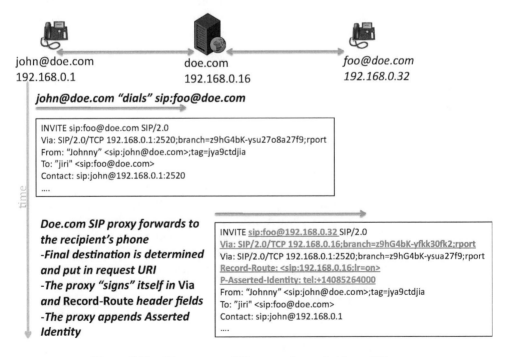

Figure 3.11   Changes to a SIP request forwarded by a SIP proxy

3.4.5   Real-world Servers

There are several important aspects of a SIP server that are not defined in the SIP specification: integration of the logical components and the underlying data model. Both are
important to understand how an actual server deployment works. Throughout this chapter,
we are providing examples based on an existing open-source SIP server, SIP Express
Router (SER) (Pelinesu-Onciul *et al.* 2003).

3.4.5.1   Servers Integrate Logical Components

The SIP standard specifies multiple logical components introduced in the previous
sections. The purpose of the specification is a comprehensive explanation of the concepts
and a clean definition. Such separation in individual components however does not appear
meaningful for building actual server implementations. In fact, in most conceivable
deployments all of the logical components are required: registrar for running the user
location database, proxy server for implementing routing and security policies, and user
agent server for declining unroutable requests. Operating those separately would keep
maintenance overhead high and require yet more components to dispatch among them.
Instead server implementations combine the components in a single piece of software
and evoke the components on a transaction-by-transaction basis. The decision on how to
process a request is driven by content of the request and administrative policy further
refined by subscriber preferences.

The decision on how to handle a request can be quite complex. For example, the
live iptel.org SIP service powered by SER is configured to act as registrar for incoming
REGISTER requests, redirect server for speed-dials (requests with short numerical code in
destination that represents a longer SIP address), proxy server for most other requests and
User Agent Server for requests whose processing failed or was declined for policy reasons.
Further, in-dialog requests obtain a different treatment than dialog-initiating requests,
instant SIP messages get forwarded to recipients if they are online and stored in a database
otherwise and requests from one of served domains are authenticated, whereas the others
are not. Simply said, formulating a sound routing and security policy for processing SIP
requests by a SIP server is non-trivial.

3.4.5.2   Back-to-back User Agent

There are situations in which the proxy server model is considered too "passive" or
"dumb": a proxy server compliant to RFC3261 only relays SIP messages and introduces
only minor modifications to them. Once a SIP request is forwarded, it is forgotten again
after some period of time needed to collect responses, absorb possible retransmissions
and produce CDRs. A SIP proxy server is aware of calls being set up but keeps no track
of active calls and initiates no transactions on its own. This allows for good performance
and resilience against failures.

The proxy behavior is not sufficient for network applications that require active call
control from inside a network. For example, third-party call control applications (see
Rosenberg *et al.* 2004) require a component that initiates calls from network. Another
argument for placing more signaling functionality in a network is interoperability: service
logic located in the network allows its operators to interoperate even with very simple

SIP phones. These desires for more control in the network led to an emerging concept known as the "Back-to-back User Agent" (B2BUA).

A B2BUA is a proxy-like server that splits a SIP transaction into two pieces: on the side facing the User Agent Client, it acts as the server; on the side facing User Agent Server it acts as the client. It keeps the dialog state and is thus in position to initiate and manipulate calls in addition to routing them. With keeping all the session information, the B2BUA is also in shape to provide services based on mangling: anonymization of call parties and topology hiding.

The most direct disadvantage of B2BUA is that it breaks message integrity. If it chooses to modify part of SIP message whose integrity is protected, for example, using (Peterson and Jennings 2006), verification of such message will fail and the message will be rejected. Another disadvantage is somewhat lower scalability: a SIP proxy server only needs to keep track of transactional state for the duration of transaction. A B2BUA must keep dialog state for the duration of a whole session.

Nowadays, B2BUA-based servers frequently find their place in a feature-rich PBX environment, sometimes even combined with media-processing features such as voicemail. Session Border Controllers (SBC) are frequently implemented using the B2BUAs for a higher degree of control. While SBC is a marketing term with no technical meaning associated with it, it is frequently a centralized component that relays both SIP and media. A detailed attempt to define SBCs may be found in (Hautakorpi *et al.* 2008).

3.4.5.3 SIP Server Data Model

The other important aspect of a SIP server is the data model upon which it is operating. While the SIP standard does not model its "view of the world" formally, a data-model is to a large extent implied by the specification and varies among server implementations rather marginally. In the following text, we are providing readers with details of data model as implemented in the SIP Express Router (Pelinescu-Onciul *et al.* 2003).

The central notion in the data model is that of a subscriber. Subscriber is a user of service identified by an internal unique identifier (UID), authentication username and security credentials for proving her identity. SER keeps this information in the "credentials" table. Each subscriber owns one or more identities that represent her to the outside world. Multiple identities may be useful, for example, to differentiate between calls to and from office and home. SER stores all identities owned by their respective subscribers in "uri" tables. All other information characterizing the subscriber is stored in subscriber profile represented by the "user_attrs" table. It typically contains call-forwarding preferences, access privileges, PSTN numbers to be used as caller-id for calls to the PSTN, as well as SIP-unrelated information: name, credit-card number, web application preferences and in fact almost anything. The profile is flexibly structured in the Attribute-Value form, so that new characteristics can be easily introduced "on-the-fly" without changing database schema. The whole relational model centered around the subscribers is depicted in Figure 3.12.

The whole database is typically used as follows: when a SIP INVITE request comes in, the proxy authenticates its originator against the credentials table using digest authentication. If the client successfully proved knowledge of shared secret to the server, the server continues with authorization. The authorization is typically about verifying that the request originator is using an displaying a legitimate identity, and has sufficient privileges

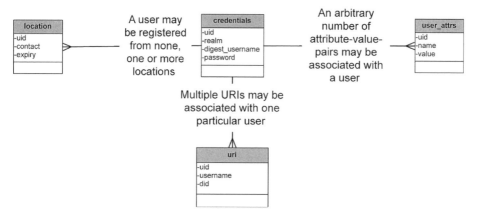

Figure 3.12   SER 2.0 data model

for this particular request. Identity verification is done by a dip in the URI table; privileges are checked by a lookup in the user profiles. In the next stage the server proceeds to processing data associated with the request recipient. The server translates recipient's address in her unique internal ID, and checks the user's profile for special preferences such as "do not disturb mode" or "call forwarding". If none such applies, the server looks up the destination in the user location database. We can observe an interesting fact: one SIP INVITE transaction has caused about five database reads. In fact, in real-world deployments, optimizing databases is at least equally important to optimizing the SIP servers.

3.5   Addressing in SIP

Addressing is an extremely important part of the SIP protocol specification. SIP addresses uniquely identify users, services and equipment. SIP applications use the same addressing scheme to name users and services as email, web and other Internet applications do: Unified Resource Identifiers, URIs (Berners-Lee *et al.* 1998). A user Joe with email address of mailto:joe@iptel.org is likely to have SIP address of sip:joe@iptel.org.

SIP addresses are generally used to uniquely identify a sender or receiver of a message, be it a user, an IP phone, a service or a telephone number. While not frequently used, a URI may even be unrelated to any kind of telephony: for example a SIP server may redirect a caller to a web URI for a textual announcement to be rendered on her screen.

The following examples illustrate some uses of URI addresses within the context of SIP applications.

1. Public SIP user address, formally known as *address of record. sip:joe@iptel.org* is an example. This address is printed on business cards, stored in phonebooks and

displayed on web pages for use with click-to-dial clients. A client attempting to send
to this address will use DNS extensions NAPTR and SRV to find out the IP address,
transport protocol (TCP or UDP) and port number to communicate with the machine
"iptel.org". A request sent to this address usually reaches a SIP server. The server then
resolves it to a SIP telephone address using its database of phone registrations.

2. The same SIP address can be accompanied by a list of parameters dictating in detail
   how a message for this address should be sent. These parameters then override the
   DNS resolution. For example, the parameters can dictate the use of a specific transport protocol, TCP in this example: *sip:joe@iptel.org;transport=tcp*. The full list of
   parameters can be found in (Camarillo 2004). The most frequently used parameter is
   ";lr", which is used to indicate an RFC3261-compliant proxy server in *Record-Route*
   header-fields.

3. A device address: *sip:10.0.0.1:16348;transport=udp*. This URI simply specifies the IP
   address, transport protocol and port number of a SIP telephone. This form is commonly
   used during phone registrations to associate a telephone's address with a user's address
   of record.

4. Public SIP address reachable via a secured channel: *sips:joe@iptel.org*. The address
   with "sips" in protocol discriminator mandates that the client and subsequent hops
   use privacy protocols to convey SIP signaling and prevent eavesdropping. The actual
   behavior of the "sips" specifier was somewhat underspecified in RFC3261, which is
   being currently addressed in the IETF (Audet 2008). The refined specification draft
   suggests that "sips" implies hop-by-hop TLS encryption all the way down from client
   to the final recipient. Hops that cannot guarantee secure forwarding are supposed
   to decline such an SIP request. Unlike in its web counterpart, "https", sips is more
   complex in that the SIP path may include multiple hops. Then transitive trust must be
   assumed. Users with high confidentiality desires may not like it as there is no way to
   detect a server in the SIP path that cheats and forwards the SIP traffic in plaintext.

5. PSTN telephone number: *tel:+1-972-353-4554*. The "tel URIs" (Schulzrinne 2004)
   specify that the sender of a SIP message shall find a way to break out to the PSTN.
   Unlike SIP URIs, it leaves the routing choice open in that there is no server name.
   Usually an outbound SIP server resolves a tel URI into a SIP URI using some
   Least-Cost-Routing algorithm or a number database like ENUM (Faltstrom and
   Mealling 2004). It is recommended that the tel URIs include a full international E.164
   (ITU-T Rec. E.164 2005) number format. However, numbers following a private
   dialing plan are allowed as well.

6. A service address such as *sip:chatroom@conference.org*. This is exactly the same as a
   user's address except that a machine is associated with it. Such a machine can offer a
   variety of services: weather announcements, conference bridges, automated voicemail,
   voicemail-to-email, etc.

7. Textual announcement: *http://iptel.org/joe/away*. This is an example of a nontelephony
   URI used to create "meshed" Internet applications. A SIP server may redirect a call to
   an unavailable user to a web page. The web page may include information like "I'm
   unavailable by SIP, but you can click my email address to send me an email, or find
   me using my GPS coordinates".

The SIP-based URI in the examples above follow the scheme in Figure 3.13.

```
protocol:[user@]host[:port][;parameter=value]
```

Figure 3.13 URI scheme

A protocol discriminator identifies the protocol to be used to communicate with the recipient of the traffic; in our context the commonly used ones are SIP, SIPS and TEL. The user name identifies a user within a domain specified in the host part of the URI. Additional parameters specify details of how to send messages to the destination described in the URI. The characters that are valid within a URI are specified in (Rosenberg *et al.* 2002b).

3.6 SIP Message Elements

Key properties of the SIP protocol are that it is a client–server, request–response, text-based, transactional protocol. Every transaction consists of client sending a textual request to a server, some computation by the server and eventually a final response from the server. Sometimes the server can also send one or more provisional responses back before issuing the final one. Figure 3.14 shows example messages forming a simple transaction: a REGISTER request, followed by a positive answer to it.

Every single SIP message has the same three-part structure: first line, message header and message body. The first, most important line states the purpose of the message. For requests it identifies its type and the destination in form of a URI. For replies the first line

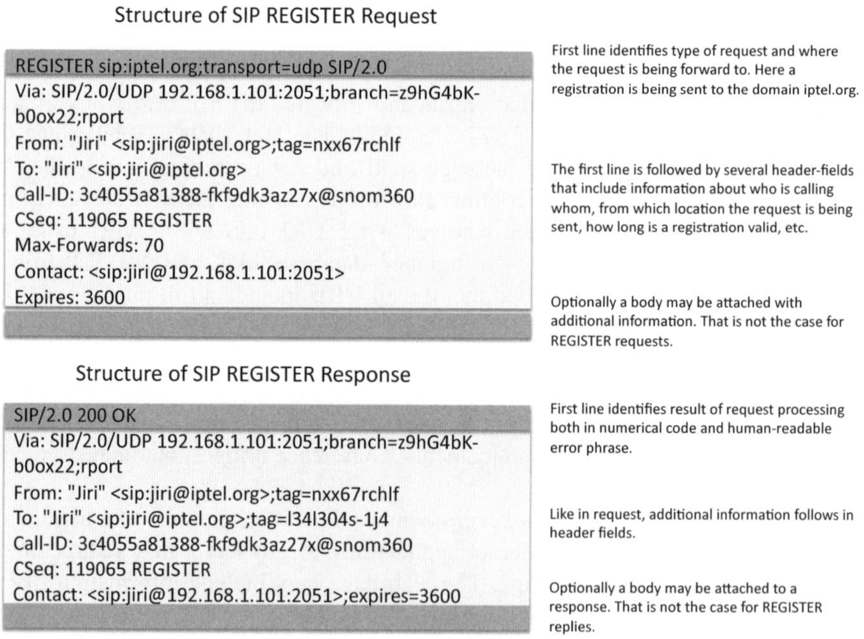

Structure of SIP REGISTER Request

```
REGISTER sip:iptel.org;transport=udp SIP/2.0
Via: SIP/2.0/UDP 192.168.1.101:2051;branch=z9hG4bK-
b0ox22;rport
From: "Jiri" <sip:jiri@iptel.org>;tag=nxx67rchlf
To: "Jiri" <sip:jiri@iptel.org>
Call-ID: 3c4055a81388-fkf9dk3az27x@snom360
CSeq: 119065 REGISTER
Max-Forwards: 70
Contact: <sip:jiri@192.168.1.101:2051>
Expires: 3600
```

First line identifies type of request and where the request is being forward to. Here a registration is being sent to the domain iptel.org.

The first line is followed by several header-fields that include information about who is calling whom, from which location the request is being sent, how long is a registration valid, etc.

Optionally a body may be attached with additional information. That is not the case for REGISTER requests.

Structure of SIP REGISTER Response

```
SIP/2.0 200 OK
Via: SIP/2.0/UDP 192.168.1.101:2051;branch=z9hG4bK-
b0ox22;rport
From: "Jiri" <sip:jiri@iptel.org>;tag=nxx67rchlf
To: "Jiri" <sip:jiri@iptel.org>;tag=l34l304s-1j4
Call-ID: 3c4055a81388-fkf9dk3az27x@snom360
CSeq: 119065 REGISTER
Contact: <sip:jiri@192.168.1.101:2051>;expires=3600
```

First line identifies result of request processing both in numerical code and human-readable error phrase.

Like in request, additional information follows in header fields.

Optionally a body may be attached to a response. That is not the case for REGISTER replies.

Figure 3.14 Registration process in SIP

states the result both in numerical three-digit status code and textual human-readable form. In this example *200* stands for successful completion which is also textually described as *OK*. The SIP standard formally describes only the success codes; the text phrase is not standardized and may vary, for example, in languages used: *200 bravo* would also be a perfectly valid answer. The second part of the message, header, includes a variety of useful information such as identification of the User Agent Client and the SIP path taken by the request. The third part includes a message body–a payload that includes application specific information, such as information about codecs available for use with voice calls.

The *Request-URI*, sip:iptel.org;transport=udp, in the first line of the SIP request is of key importance to SIP's routing mechanism: it identifies the target the request is trying to reach. The request URI may change as the request is routed through one or more SIP servers who may update the target as result of its routing policy or services. For example, if a proxy server decides to forward a request for user sip:jiri@iptel.org to a registered phone location, the request URI will be updated to *sip:jiri@10.0.0.1;transport=tcp*. Similarly if the same user has set up call forwarding, the URI may be updated to sip:jiri@voicemail.com before the request is forwarded.

Other frequently used SIP message elements are described in the following sections. An exhaustive list of SIP header-fields defined in (Rosenberg *et al.* 2002b) and extensions to it may be found at http://www.iana.org/assignments/sip-parameters.

3.6.1 Who are you Calling?

How do you find who a request was initially targeted at? That is what the *To* header field stands for, which consists of a human-readable display name and a URI, see Figure 3.15. The *To* header field does not change en route. That is different from *Request-URI*, which describes next-hop and changes as intermediate nodes en route find the next destination. The *To* header field refers to the initial destination and remains unchanged even during call-forwarding and other routing scenarios.

3.6.2 Who is Calling You?

For a variety of purposes, it is useful to know who is calling. First of all, human callees like to know. Servers in the network like to know too for the sake of access control and accounting. Without the notion of identity, they would not be in position to determine if a user may call expensive PSTN 900 numbers, and produce appropriate Call Details Records. Identity is also a prerequisite for all kinds of black-listing and white-listing. Unfortunately there is a confusing mix of possibilities today that have accumulated over the course of time.

The SIP message element used to convey caller's identity is the *From* header field. The *From* header field is used to display human readable identification of message originator and originator's Address for Record, which is useful for example for answering missed

```
To:  "Donald Duck" <sip:don33@iptel.org>
```

Figure 3.15 *To* header

calls or storing in a phonebook. The SIP specification puts no other constraints than syntactical ones on the content of the header. A user may thus configure his SIP telephone to display a surprising *From* header field such as *"Barack Obama"* <*sip:bo@* whitehouse.gov>. Therefore most SIP servers put additional restrictions on the content of the header field in requests they forward. Typically they only permit URIs in the *From* header field that belong to an authenticated user. See Figure 3.16 for an example; note that the tag parameter included does not constitute part of the address; it is merely put there by the sender as a part of the "session identifier".

Frequently one identity is not sufficient–that is particularly the case with SIP calls terminated in the PSTN. PSTN equipment cannot display non-numerical caller ID. Thus proxy servers are frequently configured to append a telephone number owned by the caller to signaling, so that it can be displayed in the PSTN. The particular header field used for this purpose is *P-Asserted-Identity* and is defined in (Jennings *et al.* 2002), which specifies the header-field, and its proper use. A SIP request fragment then includes both *From* and *P-Asserted-Identity* header fields, as shown in Figure 3.17.

From a security viewpoint, the *From* header and *P-Asserted-Identity* header fields are completely insecure. They are self-asserted and can be easily changed by a man in the middle. Use of underlying secured transport for relaying the requests, namely TLS (Dierks and Allen 1999), can be used to provide resistance against tampering.

To make sure that a request originator is who he claims to be, SIP uses the digest authentication protocol (Franks *et al.* 1999) based on a secret shared between client and server. The protocol allows the server to securely verify that the client is in possession of a valid password using cryptographic hash function MD5 (Best and Walsh 2001) without disclosing the password in cleartext. Details of the authentication protocol are explained in section 3.9.1, and an example of an authentication header field is shown in Figure 3.18. The header field includes username, realm (domain name for sake of simplicity) and hashed password. The header is usually used only on the way from client to proxy server, which deletes it afterwards before forwarding.

```
From: "john doe" <sip:joe@iptel.org >;tag=9tcxmf7
```

Figure 3.16 *From* header

```
From: "john doe" <sip:joe@iptel.org >;tag=9tcxmf7
P-Asserted-Identity: <tel:+19191234567>
```

Figure 3.17 *From* header in combination with *P-Asserted-Identity* header

```
Proxy-Authorization: Digest username="bedas",realm="iptel.org",
nonce="SP5E9Uj+ROsknmWP7fP6E4/uYc6iUoML",
uri="sip:1000@iptel.org",
response="8ad9f13f7b584367d84d5df15ec5ec08",
algorithm=md5
```

Figure 3.18 *From* header in combination with *Proxy-Authorization* header

```
From: Alice <sip:alice@atlanta.example.com>;tag=1928301774
Identity:
   "ZYNBbHC00VMZr2kZt6VmCvPonWJMGvQTBDqghoWeLxJfzB2a1pxAr3VgrB0SsSAa
    ifsRdiOPoQZYOy2wrVghuhcsMbHWUSFxI6p6q5TOQXHMmz6uEo3svJsSH49thyGn
    FVcnyaZ++yRlBYYQTLqWzJ+KVhPKbfU/pryhVn9Yc6U="
Identity-Info: <https://atlanta.example.com/atlanta.cer>;
   alg=rsa-sha1
```

Figure 3.19 *From* header in combination with *Identity* header

Level of digest security is generally considered good enough but definitely not bullet-proof. It does not provide message integrity, i.e. we know certainly who the message came from but we are not quite sure that content of the message is not bogus. MD5 features weaknesses that raise further concerns, even though these are not known to be exploitable in authentication use. In (Abdelnur *et al.* 2008) a relay attack is revealed, that allows a SIP caller to obtain his peer's credentials and reuse them for another call. Last but not least, digest authentication requires storage of plaintext passwords (or weakly encoded passwords) in the server's database, which creates too appealing an attack target for malicious users.

Typically all these identity-related header fields are used together as follows: User Agent Client states its identity in the *From* header field of a request sent to his domain's proxy. The SIP proxy uses digest authentication to cryptographically verify the sender's identity. After that it verifies if the URI in the *From* header field is owned by that particular owner, deletes the authentication header field and optionally appends a *P-Asserted-Identity* with another URI owned by the user. Unfortunately this practice inherits all of the security weaknesses from above. The really big limitation is that identity is established securely only to the server in possession of user's secret. Any other elements in the SIP path, including the User Agent server, cannot verify request sender's identity securely.

To establish verifiable identity even for SIP elements without knowledge of the client's secret password, a signature-based mechanism has been designed and specified in (Peterson and Jennings 2006). The idea is to create trust between proxy servers by certificates, and have the proxy servers vouched for identify of users within their respective domains. Such a proxy verifies the user's identity as displayed in *From*, for example using digest authentication, and cryptographically signs it. The signature is appended to the request along with a link to server's certificate, as shown in the Figure 3.19 taken from the specification. Any SIP element with possession of or access to the originating domain's certificate through a trusted certification authority can securely verify the request is coming from the domain. Similarly identity can be disclosed securely in the reverse direction from callee to callers using the procedures specified in (Elwell 2007). That is particularly useful if the caller does not know who the actual caller is due to redirection or forwarding of caller's INVITE or a call transfer.

Known critique of the RFC4474 mechanism is limited to the "bait attack" (Kaplan and Wing 2008) and the additional message integrity it provides that sometimes conflicts with the need to change parts of SIP requests (Wing and Kaplan 2008). Further details of authentication and authorization are explained in section 3.9.

3.6.3   How to Route SIP Traffic

In SIP dialogs, only the initial requests known as "dialog-initiating requests" "discover" a path to the final recipient. Every proxy that receives such a request makes its routing decision on where to forward it to. Transporting of responses and subsequent requests cannot be altered at the server's discretion, since it is governed by definition of the protocol. In particular, responses always follow the reverse path of their requests. Subsequent requests normally follow the same path as the initial request, even though in special cases some "hops" may remove themselves from the path or new elements can be introduced to the path. This behavior is implemented using the *Via*, *Contact*, *Route* and *Record-Route* header fields, that have the following functionality:

- *Via*: the *Via* header field records the sender's transport address and indicates where a response should be sent. The UAC adds its own *Via* header field, and so does every SIP proxy server en route. This is important, particularly if a SIP server en route is completely stateless: information in the *Via* header field allows it to forward replies to the request originator.

  ```
  Via: SIP/2.0/UDP 192.168.1.101:2051;branch=z9hG4bK-zdpsywsq30d5;rport
  ```

- *Contact*: the *Contact* header describes the whereabouts of a client. This is particularly important for REGISTER requests in which the *Contact* field contains the current address of a SIP telephone that is to be bound to the permanent SIP address. The *Contact* information is gathered in a URI that includes IP address, transport protocol type and port number. The *Contact* header field may include additional parameters that describe the priority of this contact if multiple contacts are registered (q parameter) and the contact's validity interval (expires parameter). If the latter is set to 0, it indicates the client's wish to "unregister". The *Contact* header field may also have a slightly different meaning if used in a response: then it lists contacts of all currently active clients. In the case of other requests, the *Contact* header indicates the current location of the user agent and the location at which the user agent expects to receive requests from the callee.

  ```
  Contact: <sip:bedas@192.168.1.101:3238;transport=tcp>;q=1;expires=600
  ```

- *Record-Route* and *Route*: the *Record-Route* header is used to remember the path a dialog-initiating request has discovered on its way from the User Agent Client to the User Agent Server. All proxy servers en route add a *Record-Route* header field with their address in it, unless they wish to be removed from the path for subsequent requests. All such server signatures are collected and returned in a final positive response. Both User Agents put the addresses in *Route* header-field of subsequent requests. This header-field dictates the path subsequent requests take and guarantees thus that the same path will be used again.

  ```
  Record-Route: <sip:10.0.0.1;lr>
  ```

3.6.4   Even More Header-fields

The core specification of the SIP protocol utilizes actually almost 50 header-fields serving additional purposes: troubleshooting, transaction and dialog management, and capability negotiation. About 50 other header fields are defined in various extensions to SIP, focusing primarily on PBX features, conferencing features, instant messaging and presence and features for support of SIP in mobile 3GPP networks. An up-to-date list is maintained by IANA at the following URI: http://www.iana.org/assignments/sip-parameters.

In the remaining part of this section, we review some frequently used header-fields.

- Loop detection. Every SIP request must include a counter that describes how many hops a request may still visit. This counter is named *Max-Forwards* and resembles IP's Time-to-Live. Its initial value is set by the client and decremented by every hop the request visits. The hop that receives the request with a *Max-Forwards* of zero must either process it without forwarding or reject it.

```
Max-Forwards: 70
```

- Troubleshooting hints. These include the *User-Agent* header field that identifies the type of sender or *Warning* header field that discloses more about the initiator of a response,

```
User-Agent: snom360/4.1
Warning: 392 213.192.59.75:5060 "request received from
    IP 88.103.74.161 with request URI sip:356332@iptel.org"
```

- Proprietary extensions. In real traffic you may also find a variety of proprietary extensions that some vendors consider useful but have not undergone the standardization process with them. The example below shows QoS characteristics as reported during call termination by some equipment:

```
X-RTP-Stat: CS=183;PS=963;OS=231120;SP=0/0;SO=0;PR=1440;
    OR=230400;CR=0;SR=0;PL=0;BL=0;RB=0/0;SB=-/-;EN=PCMA;DE=PCMA;
    JI=0;DL=0
```

Note that some of the header-field names also have a compact form: *Via* may appear as *v*, *Contact* as *m*, and so on.

3.6.5   SIP Message Body

While the SIP protocol manages an abstract notion of session, it is still important to exchange application data specific to such a session among participating User Agents. This is what the SIP message body is used for. In the simplest case, the message body includes a description of voice streams between two participants. In particular, voice sessions are described using the Session Description Protocol (Handley *et al.* 2006).

The Session Description Protocol (SDP) payload advertises the IP address and port number at which the party is prepared to receive RTP streams. Additionally it offers a list of codecs it supports. When a call is being set up or modified, both parties exchange their

```
v=0
o=root 2103355869 2103355869 IN IP4 192.168.1.102
s=call
c=IN IP4 192.168.1.102                                IP address for receiving RTP streams
t=0 0
m=audio 62730 RTP/AVP 0 18 101                        Port number and enumeration of
a=rtpmap:0 pcmu/8000                                  supported codecs with sampling rate:
a=rtpmap:18 g729/8000                                 0=PCM u-law, 18=G.711, 101=DTMF
a=rtpmap:101 telephone-event/8000
a=fmtp:101 0-15                                        Supported DTMF events (RFC2833)
```

Figure 3.20   SDP payload

SDP payloads. The client sends an "offer" of available capabilities, which is confirmed by server's "answer" according to the rules specified in (Rosenberg and Schulzrinne 2002a). The details of an example SDP payload are shown in Figure 3.20. This particular example lists three different codecs the User Agent is prepared to receive at UDP port 62370: PCM, G729 and DTMF tones.

Other payloads than SDP may be used as well or combined with each other using the Multipurpose Internet Mail Extensions (MIME; Freed and Borenstein 1996). Details of use are explained in RFC3261 and in specific detail in (Camarillo 2008). A prominent example is tunneling of PSTN signaling in trunking deployments: both SDP and the PSTN signaling bodies are attached to SIP messages (Zimmerer *et al.* 2001). SIP can also be used for instant messaging (Campbell *et al.* 2002) and presence (Rosenberg 2004a), in which case the payload will carry text messages and notifications of presence status changes.

3.6.6   SIP Methods

The SIP method specifies the purpose of a SIP message; so far we have learned the REGISTER method for registering a User Agent with a SIP server and INVITE for initiating a call. The SIP standard additionally specifies ACK, CANCEL, BYE and OPTIONS. Additional methods are specified in various SIP extensions. The complete list of SIP methods is given in Table 3.1.

3.7   SIP Dialogs and Transactions

So far we have been describing the SIP communication in terms of the actual SIP messages. For many reasonable purposes it is useful to abstract from the individual messages and refer to the abstracted notion of transactions. SIP is in fact a transaction protocol that delivers a final response or a failure notification for a request. The notion of transaction hides many underlying reliability details, such as retransmissions, time-outs, assignment of responses to proper requests and transport failure handling. The actual transport protocols for SIP, TCP, UDP and optionally SCTP (Rosenberg *et al.* 2005) are hidden underneath the transaction processing. Transactions are characterized by the request method they convey: a REGISTER transaction registers a User Agent with a registrar, an INVITE

Table 3.1 SIP methods

Method	Purpose	Normative reference
REGISTER	(Un)registers a phone's address with the user's address of record for a period of time	RFC3261
INVITE, ACK, CANCEL	INVITE attempts to establishes a call. INVITEs include SDP bodies that advertise where and what type of media the caller is prepared to receive. Optionally, it may include more bodies, typically in trunking scenarios to convey tunneled PSTN signaling. If an INVITE is accepted, the answer to it also includes one or more message bodies. Every INVITE is confirmed by sending an ACK for reliability purposes. If a call is in progress and has not been established yet, the caller may cancel it using the CANCEL method. The CANCEL and ACK requests are only used in association with INVITE requests. Multiple INVITEs can be sent within a session to change its status.	RFC3261
BYE	BYE requests tear down a previously established call	RFC3261
OPTIONS	OPTIONS allows clients to "probe" other SIP User Agents and servers to determine their capabilities and status	RFC3261
REFER	REFER allows a call participant to retarget her peer to another party. This is typically used for call transfer. However, REFER's generality allows for many other services, such as click-do-dial scenarios. In such a scenario a web-server uses a REFER request to instruct a SIP phone to dial a destination upon clicking a click-to-dial link	RFC3515
SUBSCRIBE, NOTIFY, PUBLISH	SUBSCRIBE-NOTIFY is a very general mechanism for asynchronous notification, used for example to notify SIP handsets of messages stored on voicemail server. A SIP UAC sends the SUBSCRIBE message to a server if it desires to learn whenever a specific piece of information (such as voicemail status) changes. The server then notifies the UAC of changes using the NOTIFY message. Another SIP UAC may also manipulate the status using the PUBLISH method, upon which obviously NOTIFYs are disseminated to make subscribed UACs aware of status change	RFC3265 RFC3903
INFO	INFO is an "escape code" in that it is designed to carry auxiliary information during an established call. In the field, INFO has been used to convey DTMF tones	RFC2976
MESSAGE	MESSAGE is used for delivering instant messages over SIP	RFC3428
PRACK	Auxiliary method used to achieve reliability for provisional responses, if such is needed	RFC3262
UPDATE	An INVITE-like method that allows a session to be modified "on the run"	RFC3311

transaction opens up a SIP session between User Agents, and a BYE transaction terminates a session. All these transactions but INVITE are served by the same type of state
machine.

The non-INVITE transaction machine is very simple from both client and server per-
spectives. On the client side, it sends a request, and keeps it retransmitting until either a
final response comes back or a timeout hits. On the server side it receives the original
requests, sends immediately a response back and filters any possible request retransmissions for a period of time. That's it.

The INVITE transaction is more complex. The client accounts for an unknown period of
time until a human callee answers a call. This leads to a more "patient" client state machine
waiting for final answer from the User Agent Server for a transaction in progress. The
final answer, once it comes, must be then confirmed by an acknowledgment for reliability
purposes. The acknowledgment is a very special request used only along with INVITE
requests.

An example of such a transaction is shown in 3.21: the Caller's User Agent Client
starts an INVITE transaction by sending an INVITE (#1). The callee's User Agent Server
responds with provisional progress replies indicating that the INVITE was delivered and
the phone began ringing (#2 and #3). When eventually the callee answers the phone call,
his User Agent Client sends a *200 OK* to the caller (#4), that is for reliability purposes
acknowledged with an ACK request (#5). The request for terminating this call is in fact

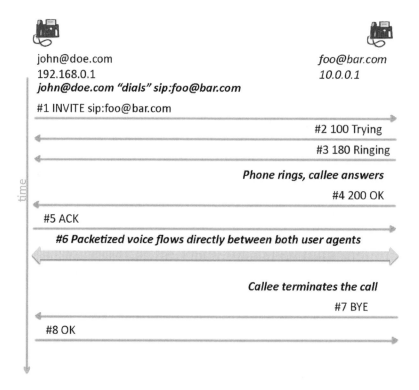

Figure 3.21   Example INVITE call flow

john@doe.com
192.168.0.1
john@doe.com "dials" sip:foo@bar.com

foo@bar.com
10.0.0.1

#1 INVITE sip:foo@bar.com

#2 100 Trying

#3 180 Ringing

Phone rings, callee does not answer

#4 CANCEL   *Caller gives up and hangs up*

#5 200/CANCEL (confirmation of CANCEL receipt)

#6 487 Request Canceled

#4 ACK

Figure 3.22   Cancelled INVITE transaction

much simpler—it only consists of a BYE request (#7) and an immediate final answer to it (#8).

There is yet another special feature in SIP transaction: the client can cancel a transaction in progress. This is done by sending a CANCEL request, which must only be used along with INVITE transactions to which it belongs. The CANCEL requests results in two answers: a *200 OK* confirms receipt of the CANCEL and a 487 termination of the pending INVITE transaction. See Figure 3.22 for call-flow details.

Another interesting aspect of the INVITE transaction is it is a dialog-initiating one. That means it creates a SIP session, or more accurately a SIP Dialog. SIP Dialog is a term for the relationship between two User Agents, that is created by an INVITE transaction and terminated by BYE transaction. The SIP Dialog context resides in User Agents, is identified by a unique dialog identifier, and comprises the SIP routing path between the two User Agents and sequencing information about related transactions.

There can be more dialog-iniating requests specified in extensions to RFC3261. In particular, (Roach 2002) defines a SIP method "SUBSCRIBE" that is used for subscription to asynchronous events, such as status updates in presence applications (Rosenberg 2004a). The most important aspect of dialog-initiating requests is that of SIP path discovery. As a dialog-iniating requests traverses the SIP trapezoid or a more complex SIP path, each hop along the path makes a routing decision on where to forward the request next. This is in contrast with the subsequent mid-dialog requests that follow the same path as the dialog-initiating request, or at least a subset of it.

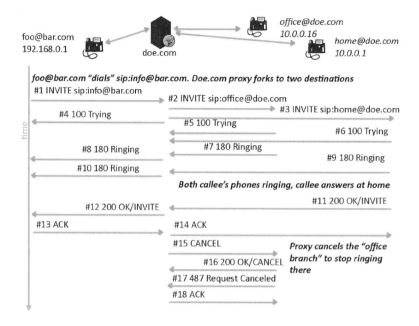

Figure 3.23   Forked INVITE call flow

Yet another interesting property of dialog-initiating request is that of forking: a request can be forwarded by a proxy server to more than one destination. This can be useful, for example, if a user wishes to be reachable under his address both at his office and at home. If a proxy server decides to fork a transaction, it must await answers from multiple destinations. The earliest positive answer wins then the "ringing race", and the other branches are cancelled by the proxy server. If only negative answers come back, the proxy is supposed to pick one for forwarding to client based on rules specified in (Rosenberg *et al.* 2002b). Figure 3.23 shows an example call flow in which an INVITE is forked to two SIP telephones, one at home and the other at office. The home phone answers first and creates a dialog with the caller. The proxy server then cancels the branch to the office phone to stop it ringing. Note that the actual order of messages can differ in reality due to forking's parallelism.

Forking adds a race condition: multiple destinations can positively answer the incoming INVITE in parallel. In such a case, all such positive responses propagate to the client, that ends up with multiple dialogs. It is then an implementation choice if the client terminates all but one using BYE requests, or keeps all open and switches between them.

The dialog-initiating transaction thus creates a path from the User Agent Client to the User Agent Server and a dialog between them. The dialog is represented by a unique dialog id. This comprises a client-contributed and server-contributed part, both of which are unique, sufficiently long random numbers. Including the server-contributed part helps to differentiate multiple dialogs if such are created during forking. It also adds additional robustness for poorly implemented or maliciously designed clients that do not provide sufficient randomness in their part. Without such robustness, INVITEs and BYEs could

bedas@doe.com   1000@bar.com

#1 INVITE sip:foo@bar.com

INVITE sip:1000@iptel.org SIP/2.0
Via: SIP/2.0/TCP 192.168.1.103:2049;branch=z9hG4bK-3qpbcso8guls;rport
From: "bedas" <sip:bedas@iptel.org;transport=tcp>;tag=**4204rexv1x**
To: <sip:1000@iptel.org;user=phone>
Call-ID: **3c271569e30d-17vqz5so2jdc@snom360**
CSeq: 2 INVITE
...

#2 200 OK

SIP/2.0 200 OK
Via: SIP/2.0/TCP 192.168.1.103:2049;received=88.103.74.161;branch=z9hG4bK-3qpbcso8guls;rport=2049
From: "bedas" <sip:bedas@iptel.org;transport=tcp>;tag=**4204rexv1x**
To: <sip:1000@iptel.org;user=phone>;tag=**3CFF1F42-48FD1B0C000344FB-B707ABB0**
Call-ID: **3c271569e30d-17vqz5so2jdc@snom360**
CSeq: 2 INVITE
...

**From tag and Call-Id supplied by UAC along with To-tag contributed by
UAS in its final answer form together a unique dialog identifier**

Figure 3.24   Creation of dialog ID

not be properly correlated and could result in bogus Call Detail Records. Syntactically, the dialog ID consists of *CallID* header field and "tag" parameter of the *From* header field on the client side, and "tag parameter" of the *To* header field contributed by the server side, as shown in Figure 3.24.

3.8   SIP Request Routing

The key functionality of SIP proxy servers is routing of SIP requests, i.e. determining where to forward incoming requests. Routing is used to locate the final destination(s), implement some basic services and also enforce some economic policies such as finding the least cost destination. A routing decision can have a variety of forms – which is chosen is a matter of local administrative policy. Dynamically registered SIP phone users are found using the user location database we have discussed in the section 3.4.2. This routing can be overridden by the callee's preference for call forwarding to an alternate destination. Static destinations that only rarely change, such as PSTN gateways or emergency response centers, can be routed to using preprovisioned routing tables. A final destination for an E.164 phone number can be looked up using the ENUM database, which translates telephone numbers into URIs.

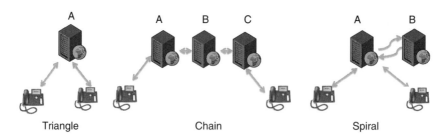

Figure 3.25 Variations of the SIP trapezoid

The actual routing decisions can lead to a different path than trapezoid. A simple and frequent case occurs when the caller's proxy server figures out that the recipient of a call is in the same domain. Then there is no other proxy server in the play and the geometry of the SIP paths simplifies itself to a triangle (see Figure 3.25). The path may, however, also become more complex. For example, a user can have requests for him forwarded to another domain. Then a call request from domain A continues through the domain B to the final domain C and the upper trapezoid's edge becomes a chain of proxy servers. It may also occur that such call-forwarding returns the request to a proxy server it has already visited before. If say *alice* in domain A calls *bob* in domain B, *bob* can still have forwarding set up to user *cyril* in the domain A. Such a path is called a "spiral".

In the following subsections we summarize routing techniques most widely used by proxy servers. Note that in actual deployments routing policies are typically composed out of several such techniques. For example a proxy may attempt to find a destination for an INVITE request to an E.164 number using its user location database, if not successful then using ENUM, and if ENUM fails too then using a Least-Cost-Routing table of PSTN gateways.

3.8.1   User Location Routing

Since SIP telephones' whereabouts frequently change, the SIP standard defines registration: a process in which telephones associate their current temporary whereabouts with a user's permanent Address of Record.This process, registration, is described in section 3.4.2. The bindings between the Address of Records and the whereabouts are stored in the registrar and available to the proxy server. When a SIP request comes to the proxy server, it looks up the final recipient and forwards the request there. If no telephone is registered, the proxy is supposed to respond to the request with the response code "480 Temporarily unavailable".

3.8.2   User-provisioned Routing

Users may also desire to influence the routing decision for their requests. This is typically the case with all sorts of call forwarding features: users have their calls forwarded to alternate destinations if they are busy, not answering, off-line, or permanently. Thus proxy implementations typically allow their users to specify such preferences in their profiles, either in a proprietary or in a standardized way. The standardized way is Call Processing Language (CPL), specified in Lennox *et al.* (2004).

3.8.3 ENUM: Public Phone Number Directory

Many SIP users resort to E.164 telephone numbers as their primary identities, so that they can be easily reached from the PSTN. That makes it harder, however, for the SIP callers: they do not know to which SIP address they should send a request for a particular telephone number. This is addressed by the DNS-based ENUM directory that allows callees to publish URIs associated with their telephone numbers.

The ENUM directory works as follows: "Owners" of E.164 numbers publish their URIs in the ENUM directory. The publication is subject to validation procedures that preclude unauthorized access to the directory. The ENUM validation architecture is described in detail in (Mayrhofer and Hoeneisen 2006). Once an entry is available in the public ENUM system, callers can query it using the following steps:

1. The SIP proxy translates a telephone number in a DNS name. It removes all non-numerical characters, puts dots between digits, reverses the order and appends the string ".e164.arpa" to the end. For example "+44-207-9460-148" is translated in this way into "8.4.1.0.6.4.9.7.0.2.4.4.e164.arpa".
2. The SIP proxy generates a DNS-NAPTR (Mealling and Daniel 2000) query for the DNS name gained in the previous step.
3. A DNS server returns a substitution expression that transforms the initial telephone number in the resulting URI. For example, the substitution expression "!^(.*)$!sip:john@johndoe.com!" will turn the initial telephone number into the following URI: sip:john@johndoe.com.
4. The SIP proxy performs the actual translation and sends the request to the new destination.

3.8.4 Interdomain Routing: DNS

Routing between administrative domains is governed by (Rosenberg and Schulzrinne 2002c). This specifications specifies DNS procedures to be used by SIP elements to find a destination for their requests, if these are targeted out of their domain. The procedure consists of the following steps:

1. The SIP proxy determines the transport protocol from the target domain name using a DNS NAPTR query. The response to the query returns a prioritized list of DNS names with transport protocols to be used. For example, "mydomain.com" will be translated to "_sip._udp.mydomain.com", which indicates that the target domain prefers to receive SIP traffic over UDP.
2. The SIP proxy initiates a DNS SRV query to find out port number and host name for the particular machine out of the result of the previous step. In our example "_sip._udp.mydomain.com" will be translated into "server01.mydomain.com:5060".

3. Eventually the SIP proxy determines the IP address of the destination by regular DNS-A procedures. In our example, "server01.mydomain.com" resolves to "213.195.44.12". Now we know the transport protocol (UDP), port number (5060) and IP address (213.195.44.12).

3.8.5 Routing Tables

Static routing tables are used to find the next-hop to destinations that, unlike SIP telephones, do not frequently change their address. Examples are PSTN gateways and media servers. The routing tables typically include some matching criteria and destination to be used for forwarding if the criteria match. The criteria can be based on SIP URI and other parts of the SIP requests, time of day, addresses the request is coming from, etc. A typical use-case is Least-Cost-Routing policy, which finds the most affordable destination for a particular call attempt. Another frequently used routing table is that with locations of the closest emergency response centers.

From some deployment size and number of separate administrative domains, it begins to be impractical to maintain manually routing tables. Thus protocols have been envisioned that allow multiple SIP components to exchange routing information dynamically. Particularly, that's what the "Telephony Routing over IP" (TRIP) protocol (Rosenberg *et al.* 2002a) has been designed for. We have not yet seen it deployed, probably because various SIP administrative domains frequently communicate with each other over the PSTN.

3.9 Authentication, Authorization, Accounting

Authentication, Authorization and Accounting (AAA) are three security procedures that are the responsibility of a SIP server. Determining the request originator's identity, Authentication, is important for several reasons: users like to know to whom they are talking, and service operators like to know who to charge and which users may obtain preferential treatment. Also spam-protection techniques based on black-listing would be rather worthless if there was no reliable notion of identity. Details of the authentication procedures are described in the subsection 3.9.1.

Once the identity of a request originator is known, the request is checked against site policies. This procedure is called Authorization. The authorization policies typically determine which user can use which identities to represent herself and which destinations she may call. The number of possible policies is virtually unlimited. In the subsection 3.9.2 we illustrate the authorization concept using a specific policy example for a public SIP service with PSTN termination.

The last AAA procedure is Accounting, which is as simple as producing usage records for the sake of producing Call Detail Records or just collecting statistics for monitoring and planning purposes.

All three AAA functions can be embedded in the SIP server or decomposed in an external AAA server. If an external AAA server is used, RADIUS (Fox and Gleeson 1999) or its successor DIAMETER (Calhoun *et al.* 2003) protocol is used for communicating to it. The RADIUS alternative has been deployed and standardized except for the accounting part (Nakhjiri *et al.* 2007); the DIAMETER alternative has been standardized in (Garcia-Martin *et al.* 2006), but rarely deployed. Additional mechanisms for sharing authorization policies in distributed environments using "traits" are being proposed in the IETF in (Tschofenig *et al.* 2008) and (Peterson *et al.* 2006).

3.9.1 User Authentication in SIP

The HTTP Digest authentication (Franks *et al.* 1999) is today the most common way to determine the originator's identity of a SIP request. It is based on cryptographic verification of a plaintext password shared between a client and a server. It is a challenge–response protocol that uses the MD5 cryptographic hash function (Rivest 1992) to enable the client to transmit the mangled password over an insecure network. The server can verify the hashed password but an eavesdropper cannot recover the original password. The SIP client calculates a 128-bit hash value using the password shared between itself and a SIP server using the MD5 algorithm. It, however, still takes some extra bits: if only the MD5 hash of a password was sent over a network, an eavesdropper could replay it and impersonate the message originator. Therefore the MD5 hash also covers the server challenge: a one-way unpredictable string previously issued by the server in order to make the hash variable and resilient against replay attacks. The MD5 hash also includes several bits reflecting the actual SIP message. The server verifies that the challenge is valid, calculates its own MD5 hash and compares it against the one received from client. If both are equal, the sender has proved the sender's identity.

The HTTP digest authentication mechanism comes in two types:

- User-to-user authentication–denotes the authentication procedure that takes place between a UAC and a UAS. The UAS may be, for example, the called party or the SIP registrar. In the former case the authentication occurs at the SIP call setup, whereas in the latter case it takes place during the SIP registration.
- User-to-proxy–runs between an UAC and a SIP proxy, which is usually its local outbound SIP proxy, and occurs during a SIP call setup.

3.9.1.1 User-to-user Authentication

The user-to-user authentication is illustrated in Figure 3.26 for the case of a SIP registration. Only the contents of headers that are relevant to the HTTP digest authentication have been illustrated.

1 Initial request   the UAC sends a request (e.g. a REGISTER) to the server (e.g. the SIP Registrar).

2 Challenge   the server (e.g. the SIP registrar) challenges the UAC by responding with a *401 (Unauthorized)* message containing a *WWW-Authenticate* header that includes the parameters:

- Realm–contains the host name or the domain name of the server performing the authentication and is used to hint at which username and password the user should use.
- Quality of protection (qop)–may be "auth" or "auth-int" or may be missing and influences the way the response is calculated [see equations (3.1), (3.2), (3.5) and (3.6)].

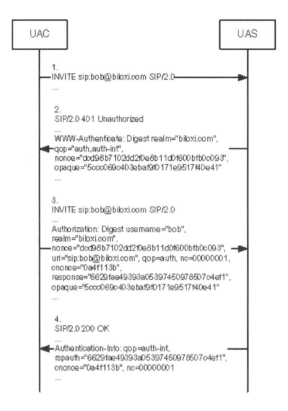

Figure 3.26 User-to-user MD5 digest authentication

- Nonce—represents the challenge generated by the server and is used by the client for the response calculation. We will also denote it as the server nonce, to distinguish it from the client nonce, which will be introduced later on. The server nonce consists of a timestamp concatenated with a hash calculated over the timestamp and a server secret key. This scheme offers replay protection by allowing the server to verify the freshness of the nonce, which is returned in the *Authorization* header of the subsequent requests.
- Opaque—contains opaque data that must be returned by the client unchanged in the *Authorization* header of the subsequent request. The server may use this field to store state information related to the authentication session;
- Algorithm—may be "MD5" or "MD5-sess" and determines the structure of the A1 term used in the response calculation [see equations (3.3) and (3.4)]. If the "MD5-sess" algorithm is used, following the sucessful authentication of the UAC, a "session key" is set up that includes in its calculation the initial server and client nonces. The "session key" allows the server to accept responses contained in subsequent UAC requests even if the server nonce is not fresh anymore. This reduces the number of round trips by allowing the UAC to proactively calculate and provide the responses without being challenged each time. A "session key" is valid throughout the lifetime of an authentication session, which ends when the server issues a new WWW-Authenticate or Authentication-Info response. If the *algorithm* parameter is missing, "MD5" is assumed.

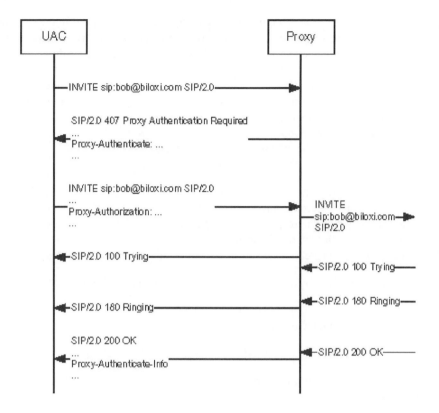

Figure 3.27 User-to-proxy MD5 digest authentication

- Stale–if "true", it indicates to the client that its previous request was rejected because the nonce was expired but the response was in fact calculated with the right password. In this case the client will use the same password but the new server nonce to re-calculate the response. If "false", then the user must be prompted to provide the password anew.

3 Response The UAC calculates the response and re-sends the request, this time including an *Authorization* header that contains the following parameters:

- Username–the username in the specified realm.
- Realm–the same as in the *WWW-Authenticate* header.
- Nonce–the same as in the *WWW-Authenticate* header.
- Digest-URI (uri)–the value of the Request-URI. The Request-URI itself may, however, change in transit.
- Quality of protection (qop)–indicates the quality of protection chosen by the UAC.
- Nonce count (nc)–indicates the number of distinct requests sent by the UAC using the nonce value in this request. The nonce count is used as input in the response calculation, enabling the server to detect replays by maintaining its own copy of the nonce count. A *cnonce* parameter must be present if a *qop* parameter has been specified.

- Client nonce (cnonce)–this is a nonce generated by the UAC and used in the calculation and validation of the response. Its purpose is to protect the password against chosen plaintext attacks, which can facilitate breaking cryptographic algorithms. A chosen plaintext attack can be mounted in this case by an attacker posing as a server and choosing the nonce values. The client nonce allows the UAC to introduce randomness in the output that is controlled only by himself. A *cnonce* parameter must be present if a *qop* parameter has been specified.
- Response–the UAC's response.
- Opaque–the same as in the *WWW-Authenticate* header.
- Algorithm–the same as in the *WWW-Authenticate* header.

We describe next how the A1, A2 and the response are calculated. The following notations are used: KD(secret, data) = H(secret $\|$ ":"$\|$ data), where H denotes applying the MD5 hash function to the respective input parameters, and unq(param), which denotes the unquoted value of the "param" parameter.

When the quality of protection is "auth" or "auth-int", the response is calculated as

$$\text{response} = \text{KD} \ (\text{H(A1)}, \text{unq(nonce)}\|":"\|\text{nc}\|":"\|\text{unq(cnonce)}\|":"\|\text{unq(qop)}\|":"\|\text{H(A2)}) \tag{3.1}$$

and when the *qop* parameter is missing,

$$\text{response} = \text{KD} \ (\text{H(A1)}, \text{unq(nonce)}\|":"\|\text{H(A2)}) \tag{3.2}$$

When the "MD5" algorithm is used or no algorithm is explicitly specified, the A1 term is calculated as

$$\text{A1} = \text{unq(username)}\|":"\|\text{unq(realm)}\|":"\|\text{passwd} \tag{3.3}$$

where "passwd" denotes the user's password. When the "MD5-sess" algorithm is used, the formula becomes

$$\text{A1} = \text{H(unq(username)}\|":"\|\text{unq(realm)}\|":"\|\text{passwd})\|":"\|\text{unq(nonce)}":"\|\text{unq(cnonce)} \tag{3.4}$$

It may be observed that the response calculation includes a hash value calculated over the user's password (among other parameters) but not the explicit password. This enables scenarios where the authenticator is a third-party server that obtains the H(A1) from the authenticating server.

When the quality of protection is "auth" or unspecified, the A2 term is calculated as:

$$\text{A2} = \text{method}\|":"\|\text{URI} \tag{3.5}$$

where "method" denotes the SIP method name. If "auth-int" is selected, the formula becomes:

$$\text{A2} = \text{method}\|":"\|\text{URI}\|":"\|\text{H(body)} \tag{3.6}$$

where "body" denotes the body of the SIP request, which is, in this way, integrity protected.

4 Mutual authentication   assuming the user is successfully authenticated, the server may include an *Authentication-Info* header, the main role of which, in the context of SIP, is to provide mutual authentication. The *Authentication-Info* header contains the following parameters.

- Nextnonce–provides the nonce value that the UAC shall use to authenticate the next request;
- Quality of protection (qop)–indicates the quality of protection provided by the server to the response message;
- Client nonce (cnonce)–the same as the one provided by the client in the *Authorization* header;
- Nonce count (nc)–the server's copy of the nonce count;
- Response authentication (rspauth)–a response calculated by the server in order to prove that it knows the user's password. The same equations are used as for the client's response, except for the A2 term, which has a different structure.

When the quality of protection is "auth" or unspecified, the A2 term for the server response authentication is calculated as:

$$A2 = \text{":"} \| URI$$

If "auth-int" is selected, the formula becomes:

$$A2 = \text{":"} \| URI \| \text{":"} \| H(body)$$

where "body" denotes the body of the SIP response, which is, in this way, integrity protected.

3.9.1.2   User-to-proxy Authentication

Figure 3.27 illustrates the message sequence that takes place during the user-to-proxy authentication between the UAC and a SIP proxy, which is usually the UAC's outbound SIP proxy. The semantics of the *Proxy-Authentication*, *Proxy-Authorization* and *Proxy-Authentication-Info* headers are similar to the *WWW-Authentication*, *Authorization* and *Authentication-Info* headers.

The procedure consists of the following steps:

- If a SIP request does not contain a *Proxy-Authorization* header, the proxy challenges the UAC with a *407 Proxy Authentication Required* error response. Upon receiving a *407* response, the UAC extracts the challenge from the *Proxy-Authentication* header and calculates the response, using the procedure described in section 3.9.1.1, and includes in the *Proxy-Authorization* of a new SIP request.
- The SIP proxy verifies the UAC's response and may also send a *Proxy-Authentication-Info* header in the final *200 OK* response in order to authenticate itself towards the UAC. The usage of the *Proxy-Authentication-Info* for SIP is, however, not yet standardized (Dotson and Hoggan 2008).

- If the next nonce is present in the *Proxy-Authentication-Info* header, the UAC stores it and uses it in conjunction with an appropriately incremented "nonce count" to authenticate subsequent SIP requests.

3.9.1.3 Security Considerations Related to the Use of the HTTP Digest Authentication

The HTTP digest authentication features a couple of weaknesses but remains the most popular authentication technique used along with SIP. Cryptographic weaknesses of MD5 have been studied extensively. In particular collision attacks aimed at finding two values producing the same MD5 hash received attention from cryptanalysts (Boer and Bosselaers 1994; Dobbertin 1996; Liang and Lai 2007; Wang and Yu 2005a,b). In 2006 Vlastimil Klima published in a revised version of his paper an algorithm and software for finding MD5 collisions on an off-the-shelf PC within 31 seconds (Klima 2006). While such research suggests that stronger hash functions may be needed in the future, the resulting risk for SIP applications remains acceptable: the ability to produce two values with the same hash is not known to be exploitable to facilitate breaking the user's password (Hawkes *et al.* 2004).

In other applications it is, however, strongly recommended to consider a stronger hash function. For example using MD5 hash as a checksum for nonrepudiation is too weak a choice. Stevens, Lenstra, and Weger published in a jokingly named webpage, "Nostradamus"[2] 12 different PDF-formatted predictions for the winner of US presidential campaign. They are all signed with same MD5 checksum to "prove" they were genuinely produced before the election.

A more practical problem is use of the plaintext passwords in a server's authentication database. The MD5 hash verification process actually needs cleartext passwords as input. The passwords are thus stored entirely in the SIP server's database and whoever has access to it can use the passwords to impersonate the users. The danger is further aggravated by users who unwisely use a single password with multiple services. To perform at least some damage control, the secret is actually stored in server database as an MD5 hash of password, username and domain name (realm). While this value can still be used to impersonate a user within a given administrative domain, the attacker no longer has a cleartext password he could use elsewhere. The user's credentials are then calculated as MD5 hash, which takes the hashed password and nonce on input. To keep security risks manageable, the password database must be administratively well protected.

Also, it may be observed that, besides user authentication, the HTTP digest mechanism does not offer any protection to the SIP messages, except for integrity protection for the Digest-URI (which may differ from the Request-URI) method and possibly the SIP message body (if the "auth-int" quality of protection is selected). Therefore a man in the middle can for instance easily alter the content of the *Contact* header of a REGISTER message. For the purpose of protecting the integrity of the SIP messages, two mechanisms have been proposed (Rosenberg *et al.* 2002b): S/MIME and TLS. S/MIME (see section 6.2) offers end-to-end data authentication, nonrepudiation and partial confidentiality for the SIP signaling.

---

[2] http://www.win.tue.nl/hashclash/Nostradamus/

TLS offers hop-by-hop confidentiality and integrity protection, which corresponds to a trust model that can easily map the interoperator relationships that govern most of the networking infrastructure. In order to indicate that secure transport of the SIP messages is desired, the SIPS URI scheme is used (Rosenberg *et al.* 2002b). Delivering a SIP request to a SIPS URI indicates that each hop over which the request is forwarded, until the request reaches the SIP proxy responsible for the domain portion of the Request-URI, must be secured with TLS. Once established, TLS connections are reused for subsequent SIP requests and responses traveling in both directions (Mahy *et al.* 2008).

3.9.2 Authorization Policies

A key responsibility of a SIP server is to make sure that only eligible requests are processed and forwarded to their destinations and that forwarded requests bear credible identity. This procedure is called authorization and is executed by servers after the authentication procedure has determined the request originator's identity. In this section we are trying to give an example of a typical policy for a public SIP service. In the real world, actual policies may widely vary and depend on a variety of factors, such as the user's privileges and prepaid account status, blacklists and whitelists, and peering arrangements with other service providers.

Obviously the biggest security concern is how well is such an authorization policy is formulated. Leaks in it can allow attackers to impersonate other users, get access to services they have no authorization for and/or have their call charges debited to someone else's account. In particular, an attacker may subvert an authorization policy by route poisoning or by bypassing.

Route poisoning is a technique exploiting weak policies that do not detect requests requiring additional privileges as such. The requests may appear as being for another domain, an in-dialog request to be forwarded by record-routing rules or a request to be routed based on user location. At the same time, the routing is "poisoned" to have the request forwarded to a protected destination, such as PSTN gateway. The "other domain's" DNS name can actually point to the gateway, and so can REGISTER-uploaded "user-location contacts", or preloaded *Route* header fields. A weak authorization policy will then forward the requests to the PSTN gateway without additional checks.

An attacker can also exploit inconsistencies in distributed environments. If for example in an installation a protected PSTN gateway relies on its proxy to perform access control, the trust relationship must be reliably enforced. Otherwise an attacker could pretend to be a proxy server, bypass it and gain direct unauthorized access to the gateway. Such attacks are typically prohibited by appropriate firewall rules.

Both route poisoning and bypassing attacks on authorization policies show that formulating a bullet-proof authorization policy is a delicate matter requiring a lot of care and attention. In the future, we believe that administrators will be aided by automated policy auditing tools. Also, the use of systems such as traits (Peterson *et al.* 2006) can help in creating a consistent authorization policy in a distributed system.

3.9.2.1 An Authorization Policy Example

The following example shows an instance of a policy. The example verifies the caller's identity, legitimate use of URIs she owns and permissions for calls to PSTN. Eventually it checks if an unprivileged attacker has tried to subvert the policy and have her request relayed to a PSTN gateway. It is formulated in SER's routing language, which resembles C-shell. Before authorization can begin, we must know the identity of the request originator. This is determined using digest authentication. Note that there is a policy choice of which requests to be authenticate. In many cases, it is sufficient to authenticate only REGISTERs and dialog-initiating requests, such as INVITE. CANCEL and ACK cannot by challenged by protocol definition.

```
# inspect request method; never authenticate CANCEL and ACK
# requests;
if (method==CANCEL || method==ACK) { break; }
# if the client is sending a request from another domain (domain in
# From ID  does not exist in list of served domains), we cannot
# challenge  him
if (!$f.did) { break;}
# otherwise let's proceed with authentication; check if caller's
# credentials are valid...
if (!proxy_authenticate ("$fd.digest_realm", "credentials")) {
        # ... and decline the request if not
        sl_reply("407", "Proxy Authentication Required");
        drop;
}
```

Once we know that a request is sent from a legitimate user, we are going to verify that she is representing herself using a correct URI in the *From* header field.

```
# lookup ID of the user owning the requests From URI;   if
# there is  no such a URI in our domain, decline the request
if (!lookup_user("$fr.uid", "@from.uri")) {
        sl_reply("403", "User Unknown");
        drop;
}
#ok, the URI in From exist but does it really belong to the
# authenticated user? check if the user owning the URI in
# From is the same as the authenticated user...
if ($fu.uid != $fr.uid) {
        # ... if not, decline the request
        sl_reply("403", "Fake Identity");
        drop;
}

# when we got here, the URI in From is legitimate; go on with
# request processing
```

Now we know that the identity as determined using digest authentication and presented in the From URI is legitimate, we can proceed with checking if the request originator is authorized to place a call to the final destination. If so, we continue processing and introduce the *P-Asserted-Identity* header field to the request.

```
# load callers profile, that includes user privileges and telephone
# numbers
load_attrs("$fu", "$f.uid");
# if this is a dialog-. initiating INVITE.(that's the request
# causing costs!), check permissions of the caller for initial
# INVITEs.
if (method == "INVITE" && !@to.tag) {
        # allow forwarding to numerical destinations only
        # if the variable  gw_acl is set to 1 in callers
        # profile
        if (uri= "^sip:[0-9]+@" && !$f.gw_acl == "1") {
                sl_reply("403", "PSTN Not Permitted");
                drop;
    }
}
# access control checks passed too, lets print callers phone number
# as stored in his user profile using the Asserted Identity
# specification ...
xlset_attr("$rpidheader","<sip:%$asserted_id@%@ruri.host>;
screen=yes");
# .. and append it to the forwarded request
replace_attr_hf("Remote-Party-ID", "$rpidheader");
# remember that we have passed the PSTN Access Control
setflag(FLAG_PSTN_ALLOWED);
```

Finally, we may need to check that an attacker has not managed to poison our routing entries. This could be done, for example, by uploading the PSTN gateway's IP address to the user location database using REGISTER, or by using a DNS name pointing to such a gateway. The proxy server could then fail to recognize the request as being for PSTN, bypass PSTN access control and allow a malicious user to gain access to the gateway.

```
# Final checks before a request is sent out.
onsend_route
{
# Doublecheck that calls to our PSTN gateway have passed the
# Access Control; if the request is not internally marked as
        # if the request has been already recognized and marked
        # as for PSTN we can break this extra checks and proceed
        # to sending
        if (isflagset(FLAG_PSTN_ALLOWED)) break;
        # however if not, and the message is about to be
        # sent to our
        # PSTN gateway, there's a problem
        if (to_ip==$g.gw_ip) {
                # we ignore it for ACK, CANCEL, and
                # in-dialog requests...
                if (isflagset(FLAG_TOTAG) || method=="ACK" ||
                   method=="CANCEL") break;
                # but otherwise we drop the request
                # and issue an alarm
```

```
        log(1, "ALERT: non authorized packet for PSTN,
                 dropping...\n%mb\n");
            drop;
    }
}
```

3.9.3 Accounting

Accounting is the last step of the three As in the lifetime of a transaction. The output of the accounting process is usage reports also known as Call Detail Records (CDRs). The reports are used for charging, trending and troubleshooting. SIP proxy servers produce such reports on a per-transaction basis, B2BUAs can also produce more aggregated reports on a per-dialog basis. The reports must uniquely identify a call (i.e. they must include the complete SIP dialog ID consisting of the triple *CallId*, *From* tag and *To* tag), when it started, when it stopped, who originated it, who received it and how the call was terminated.

Examples of reports for call initiation and termination as collected by a RADIUS accounting server and displayed in human-readable Attribute-Value-Pair form are shown in Figure 3.28. The reports belong to a call initiated by user with SIP address *sip:gh@192.168.2.16*, calling user *sip:jiri@192.168.2.16*. The call was successfully established (SIP-Response-Code *200*) and took just one second: it started at 00:20:55 and terminated at 00:20:56.

It is also important to note that the accounting functionality can be placed in other components than the SIP proxy. Particularly, if only Call Detail Records for calls to/from the PSTN are of interest, PSTN gateways can be used to produce them. That has the advantage that these can include more detailed information that a proxy server does not by definition know: media status and PSTN status.

From a security point of view, robustness of the software for processing the CDRs is very important. That stems from the fact that data in the CDRs is coming directly from SIP messages, which can be easily handcrafted by malicious users. For example, attackers can attempt to mount injection attacks by putting fragments of SQL statements in their SIP messages. If the software is not robust enough, the SQL queries will be eventually executed. Another exploit is attacks on uniqueness of dialog ids–an attacker may send additional bogus requests for call setup and termination that confuse CDR correlation and result in favorable call length appearing in the final reports. Also, call termination requests can be handcrafted in such a way that they are processed and accounted by a proxy server but dropped by the User Agent server–if the CDR processing software is not robust enough, it will declare the call as finished (it has seen a BYE) but the call will continue (the BYE was broken and dropped). Again, the key for dealing with this type of attack is a robust CDR processing software that does not get confused when abnormal SIP requests appear.

3.10 SIP and Middleboxes

Network Address Translators (NATs) and firewalls have been one of the biggest obstacles to deployment of SIP. Both of them prohibitively impair communication between User Agents, the former as an unfortunate side-effect, the latter also to some extent as a matter

```
Tue Jun 24 00:20:55 2003
        Acct-Status-Type = Start
        Service-Type = 15
        Sip-Response-Code = 200
        Sip-Method = 1
        User-Name = "gh@192.168.2.16"
        Calling-Station-Id = "sip:gh@192.168.2.16"
        Called-Station-Id = "sip:jiri@192.168.2.16"
        Sip-Translated-Request-URI = "sip:jiri@192.168.2.36"
        Acct-Session-Id = "b9a2ffaa-
            0458-42e1-b5fd-59656b795d29@192.168.2.32"
        Sip-To-Tag = "cb2cfe2e-3659-28c7-a8cc-ab0b8cbd3012"
        Sip-From-Tag = "a783bd2f-bb8d-46fd-84a9-00a9833f189e"
        Sip-CSeq = "1"
        NAS-IP-Address = 192.168.2.16
        NAS-Port = 5060
        Acct-Delay-Time = 0
        Client-IP-Address = 127.0.0.1
        Acct-Unique-Session-Id = "9b323e6b2f5b0f33"
        Timestamp = 1056406855

Tue Jun 24 00:20:56 2003
        Acct-Status-Type = Stop
        Service-Type = 15
        Sip-Response-Code = 200
        Sip-Method = 8
        User-Name = "jiri@192.168.2.16"
        Calling-Station-Id = "sip:jiri@192.168.2.16"
        Called-Station-Id = "sip:gh@192.168.2.16"
        Sip-Translated-Request-URI = "sip:192.168.2.32:9576"
        Acct-Session-Id = "b9a2ffaa-
            0458-42e1-b5fd-59656b795d29@192.168.2.32"
        Sip-To-Tag = "a783bd2f-bb8d-46fd-84a9-00a9833f189e"
        Sip-From-Tag = "cb2cfe2e-3659-28c7-a8cc-ab0b8cbd3012"
        Sip-CSeq = "4580"
        NAS-IP-Address = 192.168.2.16
        NAS-Port = 5060
        Acct-Delay-Time = 0
        Client-IP-Address = 127.0.0.1
        Acct-Unique-Session-Id = "b2c2479a07b17c95"
        Timestamp = 1056406856
```

Figure 3.28 Examples of Start and Stop Usage Reports for a SIP-Call in human-readable AVP form

of policy. In both cases the problem is quite large: homes without NATs and enterprises without firewalls are rather an exception.

NATs and firewalls cause difficulties by being built to convey static symmetric client-server traffic, such as Web, email and FTP. Such applications send their requests to fixed port numbers and receive answers from the same place. NATs and firewalls have been able to deal with such traffic for decades. Unlike these applications, SIP negotiates port numbers dynamically in SDP payload and frequently receives traffic at a different port number than it is sending from.

The firewall problem is, at least in theory, easier to deal with: firewalls are installed in corporate networks that are typically highly controlled environments. SIP equipment and the firewalls are in the hands of the IT department. While IT departments are frequently for some reasons difficult to talk to, they should be at least in theory able to harmonize SIP equipment with their firewall policy. This can be achieved by configuring the SIP equipment to use well-known port ranges, placing it in separate subnets and enabling a less restrictive policy for the "VoIP subnet". Some firewalls even have built-in Application Level Gateways (ALGs) that can maintain by default more restrictive firewall policy and open up "pinholes" that are dynamic on demand. We do not think, however, that this approach brings substantial security benefits. On the contrary, these frequently cause interoperability issues as the extent of their SIP support varies from that of the application.

NATs are harder: they are beyond control. The SIP server may be administered by an ASP, NAT by an ISP or consumer, and the SIP telephone by the consumer. There is no single point of control and thus more robust traversal techniques are needed. In theory, the communication impaired by NATs could be facilitated by built-in ALGs, like with a firewall. Unfortunately this type of solution suffers from the same problem as firewall ALGs: application support may vary, which is even harder to manage with variety and number of consumer equipment. SIP itself can traverse NATs relatively easily, if deployed with the extensions that make it work symmetrically: (Rosenberg and Schulzrinne 2003), (Mahy et al. 2008) and (Jennings and Mahy 2008). The much bigger problem is transporting media over NATs. While outbound media leaves NATs to the public Internet without any problems, NATs typically drop media coming back unless they use the same, reversed path. Today's solutions typically use the concept of media relay, which transforms two asymmetric RTP streams between two User Agents into a pair of symmetric streams, each of them between a public Internet. This works as follows: a SIP server controls a "media proxy", a server that receives and forwards UDP packets. When an INVITE and later *200 OK* with SDP comes in, the SIP server forcibly replaces User Agents' IP addresses and port numbers with the media server's. Both User Agents send the media to the server, which forwards them reversely to the other party.

Obviously, the use of a media relay makes the voice path "longer" in that the voice must now visit the relay before being forwarded to the final destination. Also the media relay is subject to bandwidth scalability concerns. Another drawback of such a solution is that forcible rewrite of SDP breaks message integrity.

Therefore more sophisticated methods are being developed that minimize use of media relays. Thus an "Internet way" has been devised in which the end-devices take the most responsibility for communication. The methodology is known as "Interactive Connection Establishment". The key idea is allowing a client to learn all its addresses, which it can use to communicate, test the connectivity pro-actively and use the best-connected IP address for eventual communication. The IP addresses may be equipment's own, or an address learned using STUN (Rosenberg et al. 2008b) or if direct communication fails, a media server's address learned by a client using TURN protocol (Jennings and Mahy 2008). The detailed call-flows and usage guidelines for SIP NAT traversal can be found in (Rosenberg et al. 2008a).

3.11 Other Parts of the SIP Eco-system

Making SIP work alone is not sufficient to have a fully functional network. In this section we present a list of additional aspects that need to be considered carefully when building a SIP network.

Interoperability is a very practical concern. While many SIP products claim full standard compliance, various incompatibilities still do occur that need to be handled. Typical concerns include codec compatibility, compatibility of advanced PBX features and interworking with PSTN. There is never enough testing that can be done in advance. Interested readers may follow the work of the IETFs BLISS working group, which is concerned with improving interoperability.

QoS is an important aspect even though frequently is overemphasized. In most cases, overprovisioning is the simplest and most-effective answer to QoS concerns. Choice of an Internet-ready codec such as iLBC (Andersen *et al.* 2004) is recommendable. On the contrary, deployment of complex resource reservation protocols has been considered largely ineffective. Also, the use of popular Session Border Controllers is known to frequently increase voice latency. This is caused by use of media relay in such equipment. The media relay causes triangular voice routing that normally increases the end-to-end latency.

Scalability and availability needs to be addressed if large deployments are anticipated. Typically a combination of DNS-based load-balancing and special-purpose load-balancers is used.

Sound IT infrastructure for operating the SIP servers is a prerequisite for successful service operation. This requires, among other things, an adequate firewall policy, maintenance procedures, operation of supporting services such as DNS and NTP, monitoring facilities and regular security audits against administrative errors and software vulnerabilities.

Provisioning will be a challenge in massive deployments unless configuration of end devices is automated and requires interaction neither with users nor with administrators. Since there are today no widely accepted SIP provisioning standards, deployments typically resort to a proprietary provisioning system supplied with a particular SIP phone brand.

Compliance to regulatory requirements can have massive impact on the actual deployments. In many countries, public SIP services are becoming subject to regulatory requirements. The requirements may include support for emergency services and legal interception.

3.12 SIP Protocol Design and Lessons Learned

The SIP architecture has been devised with several design principles and goals in mind (Rosenberg and Schulzrinne 2005). Here we summarize them to establish better understanding of why SIP is doing things one way and not another.

Importantly, the SIP architecture follows the puzzle principle in which each specific task is carried out by a specific protocol. As illustrated in Figure 3.29, signaling is done using the SIP protocol, transport of the media using RTP protocol, domain name resolution using DNS, NAT traversal using STUN, TURN and ICE, and so on. A complex application, like a VoIP device, then supports many such protocols in parallel. In this way it is relatively

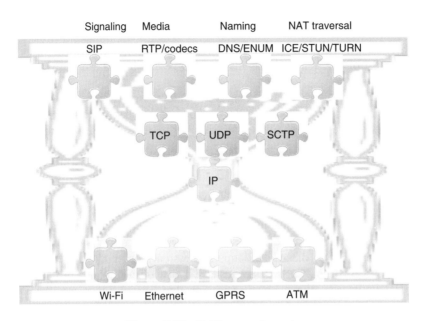

Figure 3.29   VoIP protocol puzzle

easy to update the complex system piecewise: a change to a task has only a partial impact on the whole application.

The puzzle pieces are organized in a so-called hourglass model. In this model, applications are isolated from physical networks by the IP layer. By virtue of the hourglass model the Internet applications living on top of different link technologies, such as Wi-Fi, Ethernet and GPRS, can easily talk to each other.

Another fundamental principle on which both the Internet and SIP rely is that of end-to-end design. This principle was formulated in the early 1980s (Saltzer *et al.* 1984) and addresses the dilemma of whether to implement features inside a network or inside terminals attached to it. The end-to-end design suggests putting service intelligence in terminals wherever applicable. The authors argue that leaving unnecessary intelligence in the network would introduce hard scalability problems, impair Internet's robustness against failures and make introduction of new applications difficult. SIP design follows the end-to-end principal as well. SIP proxy servers connecting SIP equipment are rather thin compared with terminals. They only temporarily store SIP transaction status in memory and remain unaware of dialog status and media processing. Thus they can scale better and their failure has minimum impact on existing sessions. In particular, a failure of a SIP proxy server has no impact on ongoing calls. To a large extent the servers dispatch SIP messages transparently, i.e. the servers can dispatch even message types for new applications that did not exist when the server software was written.

Extensibility has been one of the key design goals behind SIP. As the Internet applications develop rapidly, it is important for the protocol specification to keep pace with them. SIP design includes many features that allow introduction of applications that did not exist at the protocol design time. New header-fields can be introduced to convey information for new applications: proxy servers must still be able to pass them on transparently and end-devices unfamiliar with new elements silently ignore them. Similarly new message types and message bodies can be introduced and equipment unfamiliar with them will process them transparently or just ignore them. However, if a new message element has semantic that must be understood by the recipient, it can mandate it by using a *Require* header field. If this header field describes a feature unavailable in the recipient's User Agent, it must decline the request.

The possibility to integrate SIP applications with other Internet applications has been carefully kept in mind as well. This is seen in SIP's addressing scheme, which resembles Web and email addresses and the textual nature of the protocol, and identifies administrative domains by DNS names.

PSTN compatibility also largely formed the SIP architecture. The compatibility was key to connecting SIP users with hundreds of millions of PSTN subscribers. On the down-side it also allowed rather bizarre artifacts to appear in the protocol design. So-called "early media", sometimes humorously nick-named "late charging", is an example of such artifacts. This is a capability to play an announcement (e.g. "please wait until an attendant is available") before a call begins to be charged. Including support for it in SIP has caused some unusual behaviors such as intermixed announcements when a SIP call was forked, or premature termination of calls when a call in the "early media" stage was taking too long for a proxy server (see (Camarillo and Schulzrinne 2004) for additional details). Another artifact is DTMF tones: one may argue that in the age of webpages users do not bother to speak to automated voice menus but the installed base speaks for itself.

There are of course some SIP protocol design decisions that could have been done better during the IETF standardization. It may still be interesting to study such decisions even though today it hardly makes sense to try to revisit them.

We think that the biggest single shortcoming was the initially lack of support for NATs. In the early protocol version SIP just did not work over NATs, which was a large problem given NAT penetration. While NATs help to conserve scarce IP address space, they also impair transparency of the network (see Carpenter 2000). With NATs, IP addresses of packets change en route, which makes the system more brittle. The IETF has thus considered NATs harmful and focused on engineering and encouraging IPv6 deployment to eliminate the need for NATs. This strategy did not work very well. IPv6 has not quite happened yet and early SIP adopters were compelled to seek tedious NAT workarounds. These have been typically based on the notion of a "media relay", a foundation of today's aftermarket Session Border Controller products. The media server helps to transform asymmetric peer-to-peer RTP streams into a pair of two NAT-compatible symmetric client-server streams relayed via a media server on the public Internet. In parallel, long-term solutions have evolved that minimize the need for a media server using the ICE and STUN protocols. Fortunately most of the other SIP shortcomings are rather transient or marginal.

Many believe that the SIP protocol has been too talkative: one thing can be formulated in too many ways. This design has followed famous Postel's "be liberal receiver" principle (Postel 1981b) and paid somewhat less attention to "be conservative sender". The design in favor of liberalism has caused the SIP receivers' parsers to be fairly complex, impacting performance, code stability and interoperability. With a variety of mature implementations today, we consider the problem of transient and historical nature.

SIP's transaction model could have been designed more simply too. SIP maintains two types of transactions: short-term transactions between machines, such as BYEs, and long-term transactions between humans, which are represented by a special-case INVITE transaction machine. The INVITE transaction is designed to wait for a called party for some amount of time to answer a call invitation. SIP proxy servers in the network have to hold the transaction context for a human-driven period of time, possibly indefinitely. Such indeterminism can easily amplify denial-of-service attacks. Additionally, INVITE special-casing makes the code more complex and thus, at least for some transient period of time, less stable and interoperable. One could have designed call setup using the SUBSCRIBE-NOTIFY elements, in which a caller rings the callee by a SUBSCRIBE request. The SUBSCRIBE transaction completes instantly, leaving the network free of transaction context. When later the callee answers, a NOTIFY is sent to the caller. Nevertheless, while such a simplification is nice to have, it is too late for it now.

Many also consider support for early-media unnecessary complexity, inspired by the PSTN. With early media, the callee may emit announcements before its INVITE transaction is finished. This causes a variety of interoperability oddities, such as interference with forking that can result in multiple voice streams switching among each other and annoying the caller. If we had the comfort to travel back in time, staying with simple SUBSCRIBE-NOTIFY transactional model and not allowing early media would appear simpler. Now, some of the early media aspects are being clarified in RFC3960.

The secured transport protocol has just not been thought through: the behavior of the network for requests whose confidentiality is requested using the sips discriminator is underspecified. This could have as a result undesirable disclosure of SIP traffic. Work is underway in the IETF to provide a normative addendum to the SIP specification (Audet 2008). A complete list of all known specification insufficiencies is maintained at the following Web address: http://bugs.sipit.net.

In summary, the SIP protocol has leveraged past Internet experience and has been designed to scale and be future-proof through built-in extensibility.

Thus it has actually shown a surprisingly low number of shortcomings, most of which have been fixed in the field and/or the standards process. The biggest shortcoming, insufficient support for dealing with NATs, is being addressed by the aftermarket and at the same time in the IETF.

4

Introduction to IMS

The Internet and mobile telephony services have been the two major driving forces in the communication area in the last 20 years. On the one hand, the Internet has evolved from a tool for research and academia to a commodity that is the basis for entertainment, communication and data exchange in various aspects of our daily life today. Mobile phones, on the other hand, have moved from an expensive gadget available only to a few people to an affordable appliance that is used by people of different ages, nationalities and wealth.

The IP Multimedia Subsystem has resulted from the work of the Third Generation Partnership Project (3GPP) that aimed at specifying an all-IP communication service infrastructure. The 3GPP is a collaboration agreement that was established in December 1998 between a number of telecommunication standards bodies; namely ARIB, CCSA, ETSI, ATIS, TTA and TTC. Mainly looking at the needs and requirements of mobile operators, the 3GPP first specified IMS as a service architecture, combining the Internet's IP technology and the wireless and mobility services of current mobile telephony networks. Through the work of the TISPAN, the IMS architecture was extended to include fixed networks as well. The Telecoms and Internet converged Services and Protocols for Advanced Networks (TISPAN) is a standardization body of ETSI, specializing in fixed networks and Internet convergence and was formed in 2003.

4.1 SIP in IMS

By deciding to use SIP as the signaling protocol for session establishment and control in IMS instead of developing its own set of protocols, 3GPP has opened the door toward a tight integration of the mobile, fixed and Internet worlds. However, for a SIP-based solution to replace the current mobile and fixed telecommunication infrastructure it needs to offer the same capabilities; namely secure and efficient access to high-quality multimedia services regardless of the user's location. To achieve this goal, IMS introduces a number of additional SIP headers (Garcia-Martin *et al.* 2003) and requires a specific deployment architecture. While these additions can be generally used in an ISP and VoIP provider environment as well, they are tailored to meet the requirements of traditional providers of telecommunication services.

SIP Security   Dorgham Sisalem, John Floroiu, Jiri Kuthan, Ulrich Abend and Henning Schulzrinne
© 2009 John Wiley & Sons, Ltd

The main difference between SIP as specified in IETF and described in Chapter 3 and SIP as used in IMS is listed in the following subsections.

4.1.1 Quality of Service Control

One of the major differences between VoIP and traditional telephony services is the decoupling of the media and signaling paths. That is, unlike traditional telephony in which media information follows the same path as the signaling information, the path traversed by the VoIP signaling packets will in general be completely independent of the path traversed by the media packets. On the one hand, this decoupling allows for the establishment of new business models in which a service provider can offer VoIP services without having to own the physical network itself. On the other hand, this implies that the provider will not be able to support any kind of traffic prioritization or resource reservation that would be needed to offer VoIP services with predictable quality of service (QoS) levels.

IMS services are expected to be based on a network that can offer different QoS levels (23.107 TS 2007) with the desired QoS level being signaled by the user equipment. To be able to provide certain QoS levels to a multimedia session, the IMS specifications extend the session establishment procedures as described by (Rosenberg *et al.* 2002b). In the IMS the session establishment process is coupled tightly with the reservation of resources required for achieving the desired QoS level (29.208 TS 2007). Further, certain IMS SIP components have an additional interface that allows them to control and communicate with the underlying physical infrastructure.

4.1.2 Support for Roaming

For the user to utilize IMS services from any location, she will require at a minimum access to an IP service. While such access is becoming increasingly ubiquitous, using the service of a different provider than the VoIP provider the user is subscribed to bears a number of limitations:

- Multiple bills – similar to current providers of telephony services, IMS service providers will most likely be providing their services in geographically limited locations. Therefore, when out of these locations, a user wishing to use IMS services will have to use the IP service of some other provider. This usually entails having to pay a provider for the right to use the IP access at that location. Hence, the user will be paying two bills: one for the IMS service and one for the IP access. This not only diverges from our current experience with mobile telephony services, but makes access to IMS services in roaming situations much more complex and possibly more costly.
- QoS – For an IP service provider to support high quality IMS services, the provider will need to (23.107 TS 2007):
  - provide resources and mechanisms allowing for QoS control;
  - identify the requirements and characteristics of the IMS services and enable the correlation between the session establishment and reservation of QoS resources;
  - unless the provider can route the traffic between the caller and callee entirely in her own network, the provider will have to cooperate with other IP service providers so as to achieve end-to-end QoS control. Regardless of the approach used for providing

QoS and identifying the service requirements (Sisalem and Kuthan 2004), the need to support QoS control in an end-to-end fashion will require some form of cooperation between providers of IP services and providers of IMS services. These agreements will deal with issues like pricing, service level agreements and security.

To overcome these limitations, the IMS introduces the concept of home and foreign service providers in a similar manner to the current mobile telephony system. A home service provider maintains a contractual relation with the user as well as various user-related information required for authenticating the user and offering her certain services. A foreign provider is the provider offering access to the IMS services in geographical locations not covered by the home provider. In order to enable a user to roam to geographical locations not covered by his own provider and still get access to IMS services in a simple and transparent way, roaming agreements between the home and foreign providers are established. These agreements govern whether a user is allowed to access IMS services in a foreign location and the costs of such access. The IMS specifications introduce the proxy call session control function (P-CSCF) that is mainly responsible for correlating the session establishment requests with QoS reservations as well as securing the user's traffic. This component might belong to either the home or foreign provider, depending on the user's location and service architecture.

To avoid the multiple bills issue, the providers also define aspects of billing in their roaming agreements and include dedicated billing headers in the exchanged signaling messages.

4.1.3 Security

The authentication mechanisms (33.102 TS 2008) used in current mobile telephony networks (23.101 TS 2007) on the one hand enable the network provider to authenticate the subscribers. On the other hand, these mechanisms enable the subscriber to authenticate the network and give her the assurance that she is not communicating with a malicious entity. With the IETF SIP specifications (Rosenberg *et al.* 2002b), this level of authentication can only be achieved indirectly. Users are authenticated toward the service provider using the HTTP digest mechanism (Franks *et al.* 1999). The Transport Layer Security (TLS) (Dierks and Rescorla 2008) protocol provides an approach for authenticating the service provider toward the user and securing the traffic exchanged between the user and the service provider. In order to support roaming, the security model in IMS requires also the establishment of a trust relation between the user and the foreign service provider as well as a trust relation between the foreign provider and the home provider. IMS supports similar authentication mechanisms to those used in current mobile networks as well as digest-based authentication. Further, with the extension of IMS to support fixed networks, additional security mechanisms were specified for IMS that reflect the specific needs and characteristics of these networks. Details on the different security mechanisms used with IMS are described in Chapter 5.

4.1.4 Efficient Resource Usage

IMS was initially designed to be used mainly in mobile networks such as UMTS (23.101 TS 2007). In such networks the maximum available access bandwidth is limited by the available frequencies, the distance between the user and the access point, the energy and CPU available for the user's devices and the number of users sharing the resources of the access point. One of the major design guidelines for IMS was the efficient usage of available bandwidth. This led to some discrepancy between IMS and SIP as specified in (Rosenberg *et al.* 2002b) in some areas.

4.1.4.1 Registration Before Invitation

To prevent fraud and bill VoIP calls, the user should be authenticated before forwarding the request to its final destination. The authentication procedure would, however, increase the call setup delay and result in the sending of additional signaling packets for the purpose of authentication over the network. In order to avoid having to authenticate each request, the IMS requires the user to register before initiating any IMS sessions. During the registration step a security association is established and is subsequently used to authenticate the user's future requests. In the case of SIP as defined in (Rosenberg *et al.* 2002b), a user can initiate a call without having to register her IP address and SIP URI first.

4.1.4.2 Signaling Compression

SIP and SDP protocols are text-based protocols that use the UTF-8 character set (Yergeau 1998). While this encoding has some advantages for the implementation and debugging processes, it can be inefficient in terms of message sizes compared to binary encoding methods such as ASN.1 (ITU-T Rec. X.680 2002). To reduce the size of SIP messages and the bandwidth used by SIP, IMS supports the usage of signaling compression (SigComp) (Price *et al.* 2003). SigComp enables a SIP entity to compress a message before sending it and decompress a received message (Bormann *et al.* 2007). Measurements conducted in (Bin *et al.* 2006) suggest that, by using signaling compressing, the size of a SIP message can be reduced by 80% depending on the compression algorithm used. This is especially the case for repetitive messages such as reINVITE or REGISTER requests. As compressed, i.e. smaller, packets introduce a smaller transmission delay the overall session establishment delay is reduced as well.

SigComp introduces a new packet format that includes the compressed content and some additional information such as:

- The decompression algorithm–with SigComp the compressor sends the binary code of the algorithm used for decompressing the messages with the first message sent to the decompressor. This algorithm is then used by the Universal Decompression Virtual Machine (UDVM) for decompressing messages belonging to the same SIP dialog. With this approach the compressor can use any compression algorithm it wishes without having to negotiate this with the decompressor in advance or standardize which algorithms should be supported by the decompressor.
- State information–the compressor sends to the decompressor various information such as dictionary entries that can be used for the decompression. Once this state information is installed at the decompressor the compressor can send pointers to this information in future messages instead of retransmitting the information. This can greatly improve the compression ratio.

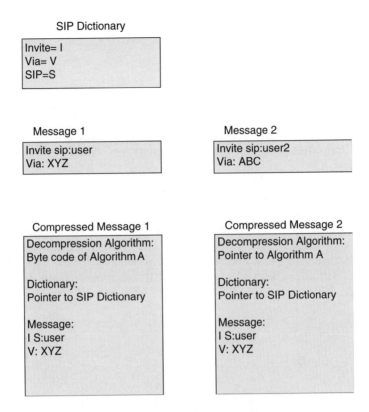

SIP Dictionary

Invite= I
Via= V
SIP=S

Message 1

Invite sip:user
Via: XYZ

Message 2

Invite sip:user2
Via: ABC

Compressed Message 1

Decompression Algorithm:
Byte code of Algorithm A

Dictionary:
Pointer to SIP Dictionary

Message:
I S:user
V: XYZ

Compressed Message 2

Decompression Algorithm:
Pointer to Algorithm A

Dictionary:
Pointer to SIP Dictionary

Message:
I S:user
V: XYZ

Figure 4.1   Example of compressed messages

Figure 4.1 presents an example of compressed messages. The first message carries the byte code of the algorithm used for decompression as well as the parts of the dictionary used for the compression and decompression. The second message which uses the same dictionary and decompression algorithm does not need to carry these parts any more and can just use references to them. If later messages use dictionary entries that have not already been sent to the decompressor, then the compressor will have to send these parts as well.

4.1.4.3   Network-centric Call Control

Current mobile telecommunication networks provide different capabilities that enable the operators to terminate a user's active communication session when the prepaid account of a user becomes empty or terminate her subscription if she does not pay her bill for some time. To offer similar capabilities, the SIP components used in an IMS network maintain sufficient dialog and registration information so as to be able to terminate a running session by sending a BYE request to the caller and callee.

Further, the user of a mobile phone has an indication of her registration status to the network. A similar feature is specified also for IMS user equipment that enables the user to detect for example whether she is still registered to the network and capable of receiving calls or has been unregistered due to loss of wireless coverage, for example.

4.2   General Architecture

As illustrated in Figure 4.2, the IMS architecture consists of a fairly large number of components. These components should be seen as logical functions and not as separate physical components, as more than one of those components could be provided by a single physical device. These components can be roughly divided into the following functional areas:

- User equipment (UE)–the UE is a SIP user agent with IMS-specific features such as support for QoS reservations and the IMS authentication methods as well as the capability of accessing 3GPP network technology.
- Signaling components–the IMS supports different call session control functions that are responsible for routing the SIP messages between the caller and callee.
- Interworking components–the interworking components enable the IMS subscribers to communicate with subscribers of other IMS or SIP-based networks as well with legacy circuit-switched (CS) networks.

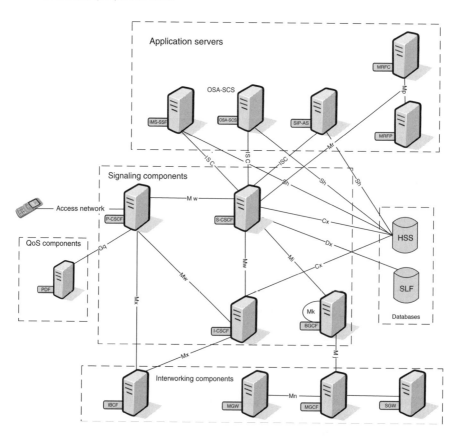

Figure 4.2   Functional architecture of IMS networks

- QoS – the QoS-related components enable the signaling components to interact with the underlying transport network.
- Application and service provisioning servers.
- Database.

The communication between different components in IMS is described in terms of reference points. A reference point indicates which protocols, message flows and contents are used when two components communicate with each other. Unless stated otherwise, all of the reference points mentioned in the following sections are based on SIP.

4.2.1 Subscriber and User Equipment

The user equipment represents the devices and applications used by the subscriber to initiate and receive multimedia sessions in IMS. In general, the UE resembles a user agent as defined in (Rosenberg *et al.* 2002b) with a number of extensions required for IMS networks. These extensions include a number of SIP extensions that are mandatory for IMS as well as support for IMS-related security and networking technologies.

4.2.1.1 User and Service Identities in IMS

In order to make sure that a call is routed to the correct user, the user must have a unique identity that cannot be shared by any other user. In the PSTN, these identities are phone numbers. Hence, when a certain phone number is dialed in PSTN, a phone that belongs to a certain user will ring. While in PSTN user devices, i.e. phones, are assigned unique identities, in IMS, the users themselves are assigned unique identities. These identities are then mapped to one or more devices through an explicit registration process. IMS supports two kinds of user identities: public and private identities. The public identity is used to route a call to the right user. Private identities are used by the operator for internal purposes such as generating billing records and user authentication.

IMS Public User Identity (IMPU) Similar to a phone number or email address, the IMS public user identity is a unique address that can be used by a caller to contact the subscriber. Each IMS subscriber will have one or more public identities. These identities can take the form of a SIP or SIPS URI or a Tel URI (see Chapter 3). The public identities are used by the IMS components for routing SIP requests from a caller to a callee as well as for identifying certain services used by the callee. Hence, in some sense the public identity resembles the MSISDN (Mobile Subscriber ISDN Number) in GSM (23.003 TS 2008). Before the user is able to receive or initiate a call at least one public identity must be registered with the IMS network. As the subscriber might have several public identities, IMS defines implicit registration sets. An implicit registration set consists of a number of public identities. Once any one of the public identities in the set is registered or deregistered, all other identities of the set are registered or deregistered as well.

IMS Private User Identity (IMPI) The private user identity is used by the network operator for various internal operations mainly concerning user authentication, authorization, accounting and administration purposes. Each subscriber will have at least one unique identity and possible more. The private identity has the format of a network address identifier (NAI) (Aboba and Beadles 1999). That is, a private identity can be presented as user@operator.com, for example. Unlike the public identity, a user's private identity is not used for routing purposes and is probably not even known to the user. In some sense it resembles the usage of the IMSI (International Mobile Subscriber Identity) in GSM (23.003 TS 2008).

Public Service Identities (PSI) Public service identities are defined by 3GPP so as to enable users to directly access public services such as a conferencing bridge or voicebox. These identities can be a SIP URI or a Tel URI and indicate a certain service at an application server. Unlike public user identities, a PSI is not related to a certain user.

4.2.1.2 Signaling Extensions at the UE

For a mobile device to function correctly in an IMS environment it has not only to act as a SIP UA according to (Rosenberg *et al.* 2002b) but also to support various extensions. These extensions include support for the IMS authentication procedures, support for signaling compression and support for a number of SIP extensions, namely:

- Preconditions–a precondition (Camarillo *et al.* 2002) is a set of constraints about the session which are indicated in the caller's offer. Such constraints can, for example, indicate that the callee should not send a final response or alert the user about the call until another condition is fulfilled. In IMS the establishment of a session is coupled with the reservation of QoS resources in the network. In this case the precondition is that the user should not be alerted about the call until reservation of network resources was successful. This avoids the case that the user answers a call but then the call itself fails due to lack of bandwidth in the network.
- Update–the UPDATE method (Rosenberg 2002) allows a client to update parameters of a session (such as the set of media streams and their codecs) but has no impact on the state of a dialog. In that sense, it is like a re-INVITE, but unlike re-INVITE, it can be sent before the initial INVITE has been completed.
- 100rel–by specifying the 100rel (Rosenberg *et al.* 2002b) in the list of its supported SIP extensions in a *Supported* or *Required* header, the UE indicates that it wishes all provisional responses except for *100* be sent in a reliable manner. In this case the server should retransmit the provisional response until it receives a Provisional Response ACKnowledgement (PRACK) method from the client.

These extensions are required for supporting session establishment and QoS reservation in IMS. Further, the UE must support a number of additional headers namely:

- *Service-Route* and *Path*–the *Service-Route* header (Willis and Hoeneisen 2003) describes the route between the UE and its home provider and lists all the nodes that should be traversed by the requests sent by the UE. The *Path* header (Willis and Hoeneisen 2002) informs the home provider of the UE about all the nodes that must be traversed when forwarding a request to the UE.
- P-headers–P-headers (Garcia-Martin *et al.* 2003) are proprietary headers that are used by the IMS components for exchanging additional information such as the user's identity, charging information and identity of the access nodes, for example.

4.2.1.3 Subscriber Identity Modules (SIM)

The 3GPP specifications discuss not only IMS services but also IP-based services and 2G services. In 3GPP identity modules provide the needed information and mechanisms for authenticating the users and securing their access to the different services. Each of these services has specific requirements regarding the information needed for user authentication and controlling the access to the services as well as the features supported at the user equipment. Hence, 3GPP defines different identity modules for each type of service. These modules have rather similar requirements in terms of processing needs, memory and cryptographic schemes. The Universal Integrated Circuit Card (UICC) is a micro-controller-based access module that provides the platform for the different identity module (31.101 TS 2007). These identity modules are given below.

Subscriber Identity Module (SIM)   The SIM module (11.11 TS 2007) is required for accessing 2G services, i.e. circuit-switched-based GSM services. While in 2G devices there is no distinction between the SIM and UICC, in 3G equipment SIM is only an application running on top of the UICC and needs only to be present on the device if it is to also access GSM services. The SIM stores information required for accessing GSM services such as the phone number and passwords.

Universal Identity Module (USIM)   The USIM (31.102 TS 2008) can be used by 3G equipment to access both circuit-switched and packet-switched services, e.g. 2G and 3G services. It runs as an application on top of the UICC and stores among other the following information:

- International Mobile Subscriber Identity—the IMSI is a globally unique identity, i.e. each IMSI is only allocated to one subscriber. The IMSI is a private identity used by the operator for uniquely identifying the user and for internal administrative purposes and is hence generally not even known to the user. The first three digits in an IMSI are the Mobile Country Code (MCC), and the next digits (two in Europe and three in North America) for are the Mobile Network Code (MNC).
- Mobile Subscriber Integrated Services Digital Network Number—the MSISDN is the subscriber's phone number.
- Authentication functions and information—the user is authenticated toward the operator by proving that she knows a certain secret. The USIM stores a long-term secret that in conjunction with a number of cryptographic functions enables the user to authenticate herself toward the operator.
- Ciphering and integrity keys—the ciphering and integrity keys are used for securing the traffic sent by the user equipment. The USIM uses separate keys for accessing packet- and circuit-switched services
- Provider and access information—the provider and access information indicate the name of the user's provider, the preferred roaming partners and a list of forbidden access networks.
- Billing and charging information—the USIM might include information indicating the price of a calling unit and the number of consumed calling units.

- Special phone numbers–this includes numbers used for emergency calls, special service numbers or barred phone numbers.
- Short message service (SMS)–the USIM can include the list of received and sent short messages as well as various related information such as the address of the short message center to use or the supported protocols.
- Multimedia message service (MMS)–similar to the short message service, the USIM provides storage for sent and received multimedia messages as well as various configuration parameters for this service.
- Various application-related information such as email address, phonebook entries and configuration and location information.
- Administrative information such as the user's preferred language and parameters used for various access technologies.

IMS Subscriber Identity Module (ISIM)   The ISIM (31.103 TS 2008) is the identity module used for accessing IMS services. The ISIM contains the following information:

- Authentication functions and information–similar to the USIM, the ISIM includes the necessary secrets and cryptographic functions needed for authenticating the subscriber toward the operator.
- IMS Private User Identity (IMPI)–the ISIM stores one private user identity.
- IMS Public User Identity (IMPU)–the ISIM can store one or more public user identities.
- Home network domain name–the SIP URI of the user's home domain.
- P-CSCF–in general it is expected that the UE would discover a close-by P-CSCF to use. However, especially during the early deployment of IMS, users might roam to networks in which no P-CSCFs are deployed. In such a case the UE is expected to contact a P-CSCF in the home domain of the subscriber. As such P-CSCFs cannot be discovered dynamically, the ISIM can be configured with the names of one or more P-CSCFs to be used by the subscriber.

4.2.1.4   Generation and Storage of User Identities

Where the UE contains an ISIM application, then the user's public and private identities will be stored in the ISIM. However, especially during the early deployment stages of IMS, it is also possible to have a 3G UE with only a USIM application. To still be able to participate in the IMS authentication procedure, the IMPI and home network domain name will have to be generated using the IMSI.

If only a USIM application exists, the home network domain is generated as ims.mcc<mcc value>.mnc<mnc value>.3gppnetworks.org. So with an IMSI of 123456789123456 the home network provider identity would be set to ims.mcc123. mnc45.3gppnetwork.org. The private identity is generated as IMSI@home. Hence with the previous example the IMPI would be set to 123456789123456@ims.mcc123. mnc45.3gppnetwork.org. Similarly, a temporary public user identity would be set to sip:123456789123456@ims.mcc123.mnc45.3gppnetwork.org.

4.2.2   Signaling Components

The IMS components responsible for processing and forwarding SIP messages are called the "Call Session Control Function" (CSCF). Depending on their functionality, we can distinguish three types of CSCFs, namely proxy (P-CSCF), interrogating (I-CSCF) and serving (S-CSCF). The communication between control functions is conducted over the Mw reference point.

4.2.2.1 Proxy Call Session Control Function

The P-CSCF acts as the outbound proxy and the first point of contact for the subscriber's user equipment. Hence, all SIP traffic sent or received by the UE will traverse the P-CSCF. The communication between the UE and the P-CSCF is conducted over the Gm reference point.

Once a UE is started and has established its connection to the network, i.e. successfully obtained the right to send and receive IP packets, it needs to determine which P-CSCF to contact. Two approaches are suggested for discovering the P-CSCF (24.228 TS 2006):

- DHCP–Using the Dynamic Host Configuration Protocol (DHCP) (Droms 1997; Droms *et al.* 2003) a host can discover and contact a DHCP server. The DHCP server provides the host with an IP address as well as the addresses of a DNS server and a default router that can be used for routing traffic to the Internet. DHCP is extended for SIP (Schulzrinne 2002; Schulzrinne and Volz 2003) so that DHCP servers also inform the SIP user agents about the P-CSCF or an outbound SIP proxy to which they should address their SIP requests.
- Access specific–when deployed over a UMTS network the IMS terminal contacts the network in order to be authenticated and be allowed to send and receive IP packets over the network. During this exchange, GPRS PDP context activation (24.008 TS 2008), the terminal is informed about the address of the P-CSCF.

As the result of the discovery process the UE might receive multiple addresses out of which the UE must use only one. Once a P-CSCF has been discovered, all SIP traffic sent or received by the UE will traverse the discovered P-CSCF.

The P-CSCF has the following tasks:

- Receive SIP traffic from the UE, process it and forward it to another SIP entity, e.g. I-CSCF or S-CSCF.
- Receive traffic from other control functions and route it to the UE.
- If SigComp was deployed in the IMS domain, then the P-CSCF is responsible for compressing the SIP messages sent to the UE and decompressing the messages received from the UE.
- Establish zero or more IPSec security associations or TLS connections with the UE. These associations and connections are used to offer integrity protection and prevent an intruder from changing the content of the SIP messages sent by the user. As the security associations and connections are used by the P-CSCF to authenticate the user, it is not necessary to re-authenticate the user for each new call setup request.
- The P-CSCF asserts the identity of the user toward other nodes in the IMS domain. Hence, all other nodes in the IMS network that trust the P-CSCF do not have to authenticate the user, see section 6.3
- The P-CSCF can enforce some provider-related policies such as dropping packets that do not conform to certain rules or rejecting forwarding packets destined to domains with whom no roaming agreement exists.

- The P-CSCF can correlate the call establishment process with the reservation of the network resources necessary to achieve the QoS level required for the multimedia session. This is achieved by having the P-CSCF interact with the underlying physical network through the policy decision function (PDF), see section 4.2.4.
- Add billing information to the SIP messages and report billing information to the billing components.

The P-CSCF can be deployed at the user's home provider or some foreign provider. In case the P-CSCF is located in a foreign network then a roaming agreement must exist between the home and foreign provider. This agreement will govern among others issues of security and billing between the providers.

In order to support SigComp or IPSec with the UE the P-CSCF needs to maintain session and registration-related information. Once a P-CSCF is contacted by a user, this P-CSCF will be used by the user for the entire duration of her registration. If a P-CSCF fails, the user will become unreachable until she has chosen another P-CSCF and initiated another registration. To reduce the failure possibilities and avoid the need for discovering another P-CSF and starting another registration procedure, the P-CSCFs have to be deployed in a highly available manner. This can be achieved by deploying the P-CSCFs in an active-standby model with the active P-CSCF sharing all collected session state information with the stand-by server.

4.2.2.2 Interrogating Call Session Control Function

In release 5 and release 6 as well as the first drafts of release 7, the I-CSCF was the entry point of an IMS domain toward other IMS networks. That is, if a request was destined to user@example.com, then resolving example.com using DNS leads to an I-CSCF belonging to the operator of example.com. Since Version 7.2.0 of (24.229 TS 2005), the I-CSCF is only the first entry point if no IBCF, see section 4.2.3.1, has been deployed.

The I-CSCF has the following responsibilities:

- When receiving a registration request the I-CSCF checks with the Home Subscriber Server (HSS), see section 4.2.6, whether the registering user is allowed to register and is roaming to a network with which a roaming agreement exists. The communication between the I-CSCF and the HSS is conducted over the Cx reference point.
- It forwards received requests to an S-CSCF.
- Up to IMS version 7.2.0 (24.229 TS 2005) the I-CSCF could also act a Topology Hiding Inter-network Gateway (THIG). In this role, the I-CSCF was responsible for encrypting network topology related information such as the information in *Record-Route* or *Via* headers included by the IMS elements in the I-CSCF's domain.
- The I-CSCF can add billing-related information to the SIP messages and reports billing events to the billing components. Further, when receiving SIP messages from or sending messages to untrusted IMS domains the I-CSCF is responsible for removing all billing-related information from the messages.

The I-CSCF can be located in either the foreign or home networks or both depending on the provider's topology. Further, an IMS network will usually include a number of I-CSCFs for reliability and scalability reasons. As the I-CSCFs are not required to maintain session information, DNS-based load balancing and fallover, e.g. by using DNS round robin or priorities and weights, can be used for distributing the traffic between different I-CSCF instances.

4.2.2.3 Serving Call Session Control Function

The S-CSCF acts as the central component in an IMS network. Similar to a SIP proxy, the major task of an S-CSCF is to receive requests, check the user location and forward the requests to the callee or application servers.

When a user registers a SIP identity, an S-CSCF is allocated to the user and this S-CSCF acts as the home S-CSCF to this user for the duration of the user's registration. The procedure for assigning a certain S-CSCF to a certain user is not defined and is left to the provider. When a registration request is received at the S-CSCF, the S-CSCF downloads authentication information from the HSS and uses this information to authenticate the user. Once a user is authenticated, all requests generated by this user or destined to this user will traverse the user's home S-CSCF. Besides downloading authentication information from the HSS, the S-CSCF downloads a service profile from the HSS for the registered SIP identity of the user. The service profile indicates which application servers are to be contacted when a message destined to the user or originated by the user is received at the S-CSCF. The service profile includes the Initial Filter Criteria (IFC) which include the triggers that determine which SIP requests are forwarded to which application servers (29.228 TS 2008). The IFC rules could, for example, indicate that all requests destined to the user and arriving from a certain caller should be forwarded to an application server that acts as a voice box. Further, the service profile can include certain provider policies that might indicate which kinds of media streams the user is allowed to use, e.g. the user is allowed to use audio but not video telephony. The communication between the S-CSCF and the HSS is conducted over the Cx reference point.

The major task of the S-CSCF is to route messages to and from the user so that when a request arrives at the S-CSCF then:

- If the request is destined to a user registered with the S-CSCF, then the S-CSCF uses the registration information it has collected during the registration phase to determine the location of the user and the next node to send the request to.
- If the request was originated by a user that is registered with the S-CSCF, then the information included in the request is used to determine the next hop to send the request to. If the callee's address has been specified in the form of an MSISDN or a phone number, i.e. a Tel URI (Schulzrinne 2004) then the S-CSCF will have to first map this address to a SIP URI. This is usually done using a service such as ENUM (Peterson *et al.* 2004).
- If the callee's address has been specified as a SIP URI, the request is routed toward an entry point in the callee's home (terminating) domain.

Besides authentication, authorization, service interaction and routing, the S-CSCF supports the following functionalities:

- Inform the user and the P-CSCF to which the user is attached regularly about the registration status of the user. This way the user can be alerted when his registration has been terminated for whatever reason and can restart the registration process again and reduce the time period during which he is not reachable.

- Terminate a user's session based on the operator's request. That is, if the administrator of the IMS network decides to terminate a user's session, for example after the expiration of the user's prepaid account, the S-CSCF can terminate the user's active dialogs by sending BYE requests to the parties involved in these dialogs. This requires the S-CSCF to maintain a sufficient amount of dialog-related state information such as dialog identity and route information to the involved parties for the duration of the dialogs.
- Terminate a user's registration based on the operator's request. That is, if the administrator of the IMS network decides to terminate the user's registration, for example due to the termination of the user's contract with the carrier, the S-CSCF notifies the user about the termination of the registration.
- Enforce operator policies such as preventing users from using certain media types.

Once an S-CSCF is allocated to a user, this S-CSCF will be used by the user for the entire duration of her registration. Hence, if this S-CSCF fails the user will become unreachable until she has initiated another registration and chosen another S-CSCF. To reduce the failure possibilities and avoid the need for starting another registration procedure, the S-CSCFs have to be deployed in a highly available manner with each S-CSCF having a hot stand-by server with whom the S-CSCF shares all collected registration-related state information. Further, the active and stand-by S-CSCF servers must share dialog-related state information. This is needed so as to allow the stand-by server to fully assume the role of the failed server and be able to control the active dialogs, e.g., terminate a dialog.

4.2.3 Interworking Components

As IMS networks will only be gradually deployed, they will have to interact not only with other IMS networks but also with the current dominant technologies, namely circuit-switched networks, e.g. SS7-based PSTN networks, and SIP services conforming to (Rosenberg *et al.* 2002b).

As illustrated in Figure 4.3, two types of interworking components exist; namely components connecting the IMS network to circuit-switched networks which include the BGCF, MGCF, SGW and MGW and components connecting the IMS network to other IMS networks or networks conforming to the SIP specifications (Rosenberg *et al.* 2002b), which include the IBCF.

4.2.3.1 Interconnection Border Control Function

The Interconnection Border Control Function (IBCF) acts as the interconnection component between IMS networks and other IMS or SIP networks (29.162 TS 2008). To be able to accommodate different kinds of interaction scenarios, the IBCF provides the following functionalities:

- TrGW—early designs of IMS specified IPv6 as the network protocol for IMS-based communication. This was mainly driven by the assumption that, due to the shortage of IPv4 addresses, IPv6 will be needed in order to support large numbers of IMS terminals. With the advance of NAT traversal solutions for SIP, see section 3.10, and

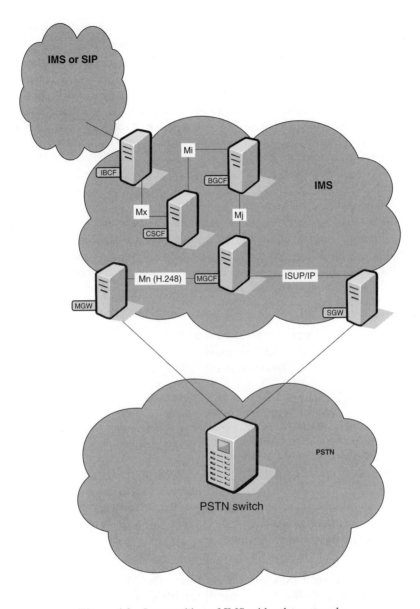

Figure 4.3   Interworking of IMS with other networks

the slow introduction of IPv6, 3GPP decided to also allow the usage of IPv4 in IMS. The transmission gateway (TrGW) is a NAT-PT/NAPT-PT (Network Address Port Translator-Protocol Translator) (Tsirtsis and Srisuresh 2000). The TrGW has a range of IPv4 and IPv6 addresses that it dynamically allocates to IMS sessions. The TrGW also translates the IP headers at the media level (RTP/RTCP).

- IMS-ALG–with the application level gateway functionality the IBCF acts as a back-to-back user agent. The IMS-ALG maintains two independent signaling legs:

one toward the internal IMS network, i.e. the network to which the IBCF belongs, and the other toward another IMS or SIP network. This is needed in the following cases:

- The two networks use different versions of IP, e.g. IPv4 and IPv6. In this case each session leg would run with a different IP version. The IMS-ALG manipulates the SDP part of the messages. The IMS-ALG interacts with the TrGW to request the binding of an IPv4 address and port to an IPv6 address and port and vice versa.
- The external network is a non-IMS network, i.e. a SIP network conforming to (Rosenberg *et al.* 2002b). In this case the user agents in that network will most likely not support some of the IMS mandatory features such as UPDATE (Rosenberg 2002), *100rel* (Rosenberg and Schulzrinne 2002b) and preconditions (Camarillo *et al.* 2002). So in this case, the IMS ALG acts as an IMS compliant user agent on the one side and as a non-IMS client on the other.

- Security-related features–the IBCF is also tasked with screening SIP messages and rejecting messages which do not fulfill certain operator defined rules, e.g. include headers that are not allowed in the operator's network. Further, the IBCF can remove certain IMS-related headers such as charging and authorization headers when sending messages to a non-trusted network. Finally, the IBCF can provide the topology hiding functionality (THIG).

As illustrated in Figure 4.4 the IBCF interacts with the session control functions over the Mx reference point. The IMS-ALG communicates with the TrGW over the Ix reference point. This reference point is, however, not specified in 3GPP currently.

4.2.3.2   Border Gateway Control Function

The Border Gateway Control Function (BGCF) is a SIP proxy that routes SIP requests destined to a user in the circuit-switched domain using a dial plan. This dial plan indicates which PSTN/CS gateways or other networks serve a certain phone number. When a phone number is served by another network, then the BGCF forwards the request to a BGCF in that network over the Mk reference point. Otherwise, the request gets forwarded to an MGCF in the same network as the BGCF over the Mj reference point.

4.2.3.3   Media Gateway Control Function

The Media Gateway Control Function (MGCF) is responsible for mapping session establishment protocols between IMS and CS networks. The MGCF converts SIP signaling into ISUP (ITU-T Rec. Q.761 1999) over IP or BICC over IP (ITU-T Rec. Q.1902.3 2003). On the IMS side, the MGCF acts as an IMS UE that initiates and terminates SIP sessions. Besides converting the protocols, the MGCF also triggers the establishment of a media path at the Media Gateway. The protocol between the MGCF and the MGW is H.248 (ITU-T Rec. H.248.1 2005).

4.2.3.4   Signaling Gateway

While the MGCF is responsible for converting SIP messages into ISUP/BICC and vice versa, the Signalling Gateway (SGW) performs lower layer protocol conversion. The SGW translates the SCTP/IP transport (Stewart 2007) used in IMS into MTP (ITU-T

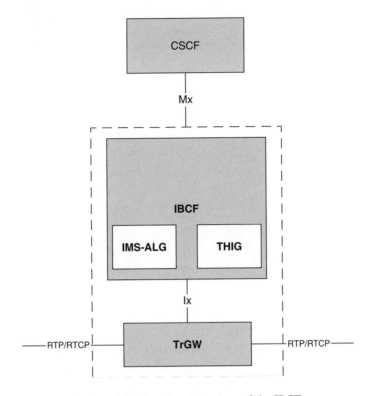

Figure 4.4   Functional structure of the IBCF

Rec. Q.701 1993) used in SS7 networks. The SGW does not interpret the application
layer messages but may have to interpret the underlying layer to ensure proper routing of
the signaling.

4.2.3.5   Media Gateway

While the MGCF converts the signaling part of a call between an IMS user and a CS
user, the Media Gateway (MGW) converts the media, e.g. audio or video, exchanged
between the two. The MGW terminates the CS bearer and generates Real Time Transport
Protocol (Schulzrinne *et al.* 2003) packets based on the received content and vice versa.
Besides unpacking the media content from the one technology and packing it into the
other, the MGW might perform transcoding in case the codec used by the IMS terminal
is not supported by the CS terminal.

4.2.4   QoS-related Components

As already mentioned previously, one of the major differentiators of IMS compared to
plain VoIP services is the ability of the IMS providers to not only provide session establishment but also offer the network resources needed to guarantee the QoS of the established
sessions. To achieve this, the IMS signaling components, i.e. control plane, will have to

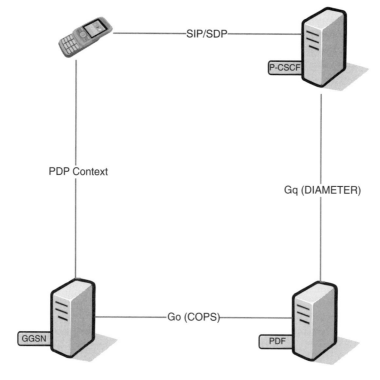

Figure 4.5 Control and user plane correlation in IMS

be able to communicate with the components responsible for providing the user with the IP access, i.e. user plane (29.208 TS 2007), see Figure 4.5. Access to IP resources in a UMTS network (23.101 TS 2007) is controlled by the Gateway GPRS Support Node (GGSN). A user who wants to send and receive IP packets over a UMTS network must establish a PDP (Packet Data Protocol) context with the GGSN. During the establishment of the PDP context the GGSN authenticates the UE and authorizes the QoS resources requested by the UE. As these resources should mirror the media description signaled in the SDP part of the SIP messages, a close correlation between the session establishment process and the PDP context establishment must exist. This correlation is achieved with the policy decision function (PDF), which mediates between the P-CSCF and the GGSN. When the P-CSCF receives a request for session establishment, it informs the PDF about the SDP content in the SIP messages over the Gq reference point (29.209 TS 2007), which is based on the DIAMETER protocol (Calhoun *et al.* 2003). When a GGSN receives a PDP context activation or modification request from the UE, it checks with the PDF whether the requested resources mirror those that were signaled in the SDP body during the session establishment. The communication between the PDF and GGSN is achieved over the Go reference point which is based on the Common Open Policy Service (COPS) protocol (Durham *et al.* 2000). The PDF authorizes the request based on local policies and provides its decision to the signaling and transport components (29.207 TS 2005).

4.2.5 Application and Service Provisioning-related Components

Application servers (AS) are SIP servers that host and execute services. Application servers interface the S-CSCF over the SIP-based Service Control Interface (ISC). In general one can distinguish between native SIP application servers that host and execute services based on SIP and application servers that provide an interface between the IMS and legacy service infrastructures such as the OSA-SCS and the IM-SSF. The Open Service Access-Service Capability Server (OSA-SCS) provides an interface to the OSA framework application server (29.198 TS 2001), which offers the capability to access the IMS securely from external networks. The IP Multimedia Switching Function (IM-SFF) (29.278 TS 2005) allows the IMS to reuse CAMEL (Customised Applications for Mobile networks Enhanced Logic) (Noldus 2006) developed for GSM. Both the OSA-SCS and IM-SSF act toward the IMS as SIP application servers and as native OSA or CAMEL entities toward the OSA or CAMEL infrastructures.

A special kind of an application server is the Multimedia Resource Function (MRF). The MRF provides media services such as multi-party conferencing sessions, announcements and transcoding services and consists of two components:

- Multimedia Resource Function Processor (MRFP)–this component provides the resources needed for mixing streams for conferencing, generating media streams for announcements or processing media streams for transcending.
- Media Resource Function Controller (MRFC)–this component deals with the signaling required for enabling media services such as the conference establishment. Based on the signaling requests, the MRFC triggers the MRFP to actually provide the media services. The MRFC controls the MRFP via the Mp reference point, which is based on the H.248 protocol.

The MRF communicates with the S-CSCF or an AS over the Mr reference point, which uses SIP.

4.2.6 Database-related Components

In IMS there are two main databases: the Home Subscriber Server (HSS) and the Subscription Locator Function (SLF). The HSS is the main database for storing all subscriber information and service-related data. This includes the subscriber's identity, the access information needed for authenticating the subscriber and the subscriber's service profile, which describes which services the user is subscribed to and which application servers should be used when the subscriber is called or is establishing a call. Further, the HSS maintains the subscriber's registration status and the address of the S-CSCF the subscriber is registered with. The subscriber's access, registration and profile information are used by the S-CSCF for authenticating and authorizing the subscriber's requests for services.

The I-CSCF, S-CSCF and AS contact the HSS when they need to authenticate or authorize a request or require some subscriber-related information. Where more than one HSS is deployed in an IMS network, the SLF acts as a resolution mechanism that enables the I-CSCF, S-CSCF and AS to find the HSS that holds the subscriber data for a certain user identity.

The I-CSCF and S-CSCF communicate with the HSS via the Cx reference point and with the SLF via the Dx reference point (29.228 TS 2008). The AS communicates with the HSS via the Sh reference point and with the SLF via the Dh reference point (29.328 TS 2008). The Cx, Dx, Sh and Dh reference points are based on the DIAMETER protocol (Calhoun *et al*. 2003).

4.3   Session Control and Establishment in IMS

Before a subscriber can actually use an IMS service, the subscriber must first authenticate and register herself at her home network. In section 4.3.1 the registration process is described. The sequence of messages exchanged during the establishment of a call in IMS is described in section 4.3.2. The description of the messages exchanged in both the registration and session establishment phases is based on the 3GPP specifications in (24.228 TS 2007).

4.3.1   UE Registration in IMS

With the IETF specification of SIP (Rosenberg *et al*. 2002b) a user agent can establish SIP calls without having to register first as the registration is only required if the user agent wants to also receive calls. In IMS the registration process is a precondition for both making and receiving calls. The registration process in IMS consists of two steps: the exchange of the REGISTER request and the subscription to the registration.

4.3.1.1   Exchange of the REGISTER Request

Figure 4.6 illustrates a successful registration process. Note, that the registration process is tightly coupled with the authentication process. So if the UE is not registered yet then the first attempt to register is rejected and the UE is requested to authenticate itself first. Only after repeating the registration process and including the right credentials in the REGISTER request is the registration accepted. If the UE was already registered then the REGISTER message includes the appropriate credentials and the registration process is conducted in one round only.

To ease the description of the registration process, we separate the description of the routing aspects from those of the authentication process. Figure 4.6 and the following description include only message routing-related details. Aspects of authentication are described in Chapter 5. All security-related headers are left empty in our description here.

1 UE to P-CSCF   To start the registration process the UE sends a REGISTER request to its home domain and includes its public identity in the *From* and *To* headers. The address of the home domain as well as the public identity are retrieved from the ISIM application or constructed using the IMSI where only a USIM is available to the UE. The REGISTER request is sent to a P-CSCF that was previously discovered during the PDP context activation or through DHCP. In the *P-Access-Network-Info* the UE provides information related to the access network.

2 P-CSCF to I-CSCF   The P-CSCF receives the REGISTER request, strips out the *Security-Client* header and *sec-agree* option tag and forwards the request to an I-CSCF in

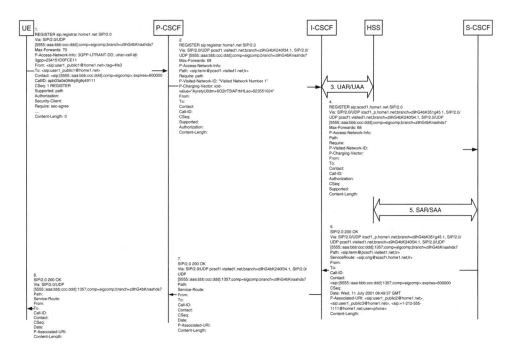

Figure 4.6 Registration process in IMS

the subscriber's home network. The I-CSCF is discovered by resolving the *Request-URI*, i.e. registrar.home1.net, using DNS. To inform the subscriber's home S-CSCF that all requests sent to the UE should be first forwarded to the P-CSCF, the P-CSCF adds a *Path* header including its own address. The P-CSCF also adds the *P-Visited-Network-ID* header including an identity referring to the network it is located in as well as a *P-Charging-Vector* with a charging identity.

3 I-CSCF to HSS When receiving a REGISTER request the I-CSCF contacts the HSS via the Cx reference point using the DIAMETER User-Authorization Request (UAR) to query the status of the subscriber's registration. The HSS indicates with the User-Authorization Answer (UAA) whether the subscriber is already registered and which S-CSCF is to handle the subscriber's registration (29.228 TS 2008).

4 I-CSCF to S-CSCF The HSS replies to the I-CSCF with the address of one or more S-CSCFs that can act as the home S-CSCF of the subscriber. The I-CSCF chooses an S-CSCF, includes the address of the S-CSCF in the *Request-URI* and forwards the REGISTER request to the chosen S-CSCF.

5 S-CSCF to HSS The S-CSCF informs the HSS using the DIAMETER Server-Assignment Request (SAR) that the user is now registered and receives with the Server-Assignment Answer (SAA) the user's profile. The user profile includes the user's public identities that are associated with the registered private identity and the Initial Filter Criteria.

6 S-CSCF to I-CSCF In case authentication is required the S-CSCF replies with a *401* response. The details of the authentication process are described in Chapter 5. Here, it is assumed that no authentication is required and that the S-CSCF sends back a final positive *200 OK* response. The response includes among others the received *Path* header and the S-CSCF adds a *Service-Route* and a *P-Associated-URI* headers. The *Service-Route* header includes the S-CSCF's URI and is added so that non REGISTER requests are routed from the UE to the UE's home S-CSCF without having to consult the HSS each time. It also includes a character string in the user part to differentiate mobile originating requests from mobile terminating requests. The *P-Associated-URI* indicates one or more public identities of the UE that are associated with the registered address. The content of the *Path* header is saved at the S-CSCF, so that the S-CSCF knows how to forward incoming requests to the UE.

7 I-CSCF to P-CSCF The I-CSCF routes the *200* response to the P-CSCF. As the I-CSCF in our example is not supposed to stay on the path of future requests it does not add itself to the *Service-Route* header.

8 P-CSCF to UE The P-CSCF saves the value of the *Service-Route* header and associates it with the UE. The *200* response is then forwarded to the UE.

4.3.1.2 Subscription to the Registration State

GSM mobile terminals inform the user whether the terminal is under radio coverage and whether the user is registered to the network. To support a similar capability in IMS, the UE subscribes to its registration state at its home S-CSCF, see Figure 4.7. This is achieved by sending a SUBSCRIBE request for the `reg` event package (Rosenberg 2004b). The SUBSCRIBE request is sent by the UE to the P-CSCF with a *Route* header indicating the P-CSCF which was discovered before the registration process and the S-CSCF indicated in the *Service-Route* header received in the response to the REGISTER request. In return, the S-CSCF sends a NOTIFY request to the UE through the P-CSCF.

The P-CSCF keeps also some UE related information needed for the signaling compression and the establishment of security associations. This information should be deleted when the registration of the UE is terminated. In order to be informed in real time when a registration is terminated by the operator, the P-CSCF also subscribes to the status of the registered public identities of the UE.

4.3.2 Session Establishment in IMS

Figure 4.8 illustrates a basic session establishment in the IMS with the caller and callee roaming to foreign networks. The example is based on the scenario provided in (24.228 TS 2006).

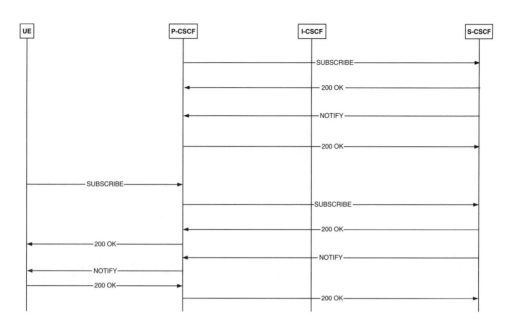

Figure 4.7 Subscription to the registration state

1–2 INVITE processing at the visited originating P-CSCF The calling UE sends an INVITE request to the P-CSCF that has been previously discovered and already used for the registration process. In the example illustrated in Figure 4.9 the UE includes in the *Request-URI* an E.164 number, includes itself in the *Via* header and indicates that it wishes to use signaling compression. The *Route* header includes the address of the P-CSCF as well as the address of the home S-CSCF which was indicated in the *Service-Route* header received during the registration process. The *Privacy* header indicates that no privacy is needed; see Chapter 6 for more details. As the user might have multiple identities, the *P-Preferred-Identity* header (Jennings *et al.* 2002) includes the identity preferred by the user for this session. The SDP part describes the audio and video compression algorithms supported by the UE and at which IP address the UE wishes to receive media data. After receiving an INVITE request from the UE the P-CSCF checks the following headers and parts:

- *Route*–does the *Route* header include the address of the home S-CSCF of the UE as indicated by the *Service-Route* header received by the P-CSCF during the registration process of the UE. If that is not the case then the P-CSCF might either reject the request or modify the *Route* header.
- *P-Preferred-Identity*–the P-CSCF maintains a security association with each UE and knows that any request arriving over that association is from the UE itself. By subscribing to the registration state of the users it also knows the public identities registered for each UE. Before sending the INVITE to the S-CSCF, the P-CSCF adds a *P-Asserted-Identity* header which asserts to other CSCFs that this request has been received over a secure association from a UE that has successfully registered this identity; see section 6.3. The user can indicate in the *P-Preferred-Identity* header which

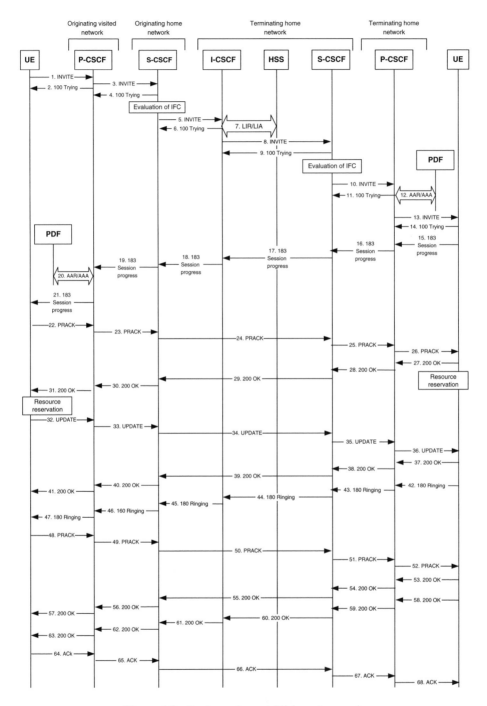

Figure 4.8 Basic session establishment scenario

```
INVITE tel:+1-212-555-2222 SIP/2.0
Via: SIP/2.0/UDP [5555::aaa:bbb:ccc:ddd]:1357;comp=sigcomp;
   branch=z9hG4bKnashds7
Max-Forwards: 70
Route: <sip:pcscf1.visited1.net:7531;lr;comp=sigcomp>,
   <sip:scscf1.home1.net;lr>
P-Preferred-Identity: "John Doe" <sip:user1_public1@home1.net>
P-Access-Network-Info: 3GPP-UTRAN-TDD; utran-cell-id-3gpp=
   234151D0FCE11
Privacy: none
From: <sip:user1_public1@home1.net>;tag=171828
To: <tel:+1-212-555-2222>
Call-ID: cb03a0s09a2sdfglkj490333
Cseq: 127 INVITE
Require: precondition, sec-agree
Proxy-Require: sec-agree
Supported: 100rel
Security-Verify: ipsec-3gpp; q=0.1; alg=hmac-sha-1-96;
   spi-c=98765432; spi-s=87654321; port-c=8642; port-s=7531
Contact: <sip:[5555::aaa:bbb:ccc:ddd]:1357;comp=sigcomp>
Allow: INVITE, ACK, CANCEL, BYE, PRACK, UPDATE, REFER, MESSAGE
Content-Type: application/sdp
Content-Length: (...)
v=0
o=- 2987933615 2987933615 IN IP6 5555::aaa:bbb:ccc:ddd
s=-
c=IN IP6 5555::aaa:bbb:ccc:ddd
t=0 0
m=video 3400 RTP/AVP 98 99
b=AS:75
a=curr:qos local none
a=curr:qos remote none
a=des:qos mandatory local sendrecv
a=des:qos none remote sendrecv
a=rtpmap:98 H263
a=fmtp:98 profile-level-id=0
a=rtpmap:99 MP4V-ES
m=audio 3456 RTP/AVP 97 96
b=AS:25.4
a=curr:qos local none
a=curr:qos remote none
a=des:qos mandatory local sendrecv
a=des:qos none remote sendrecv
a=rtpmap:97 AMR
a=fmtp:97 mode-set=0,2,5,7; mode-change-period=2
a=rtpmap:96 telephone-event
a=maxptime:20
```

Figure 4.9   INVITE from UE to P-CSCF

```
Record-Route: <sip:pcscf1.visited1.net;lr>
P-Asserted-Identity: "John Doe" <sip:user1_public1@home1.net>
P-Charging-Vector: icid-value="AyretyU0dm+6O2IrT5tAFrbHLso
   =023551024"
```

Figure 4.10   INVITE processing at the visited originating P-CSCF

identity should be used with the *P-Asserted-Identity*. If the identity included with the
P-Preferred-Identity header has not been registered by the UE or the INVITE request
did not include the *P-Preferred-Identity* header at all, then the P-CSCF chooses one of
the registered public identities. The *P-Preferred-Identity* header is removed from the
INVITE request before forwarding.
• SDP–the SDP part indicates the audio and video codes supported by the UE. The
  P-CSCF might have certain policies about which codes should be used and by whom. In
  case the UE wishes to use a code that is prohibited then the P-CSCF rejects the request.

As illustrated in Figure 4.10, the P-CSCF also adds a *P-Charging-Vector* header and
a *Record-Route* header with its own address and forwards the INVITE request to the
S-CSCF. It further replies to the received INVITE request with a *100 Trying* response.

3–4 INVITE processing at the home originating S-CSCF   The subscriber's home
S-CSCF evaluates the initial filter criteria so as to decide to which application servers the
INVITE request should be forwarded. Similar to the P-CSCF, the S-CSCF also checks
the SDP part and rejects the request if it includes media parameters which are not allowed
for the subscriber. That is, if the user has subscribed to the operator's audio service but
not the video service, the SDP part should not include video parameters.

The request is forwarded using the *Request-URI*. If the *Request-URI* includes a Tel
URI then the phone number must be mapped to a SIP URI first. This is done using
ENUM for example. If the phone number cannot be mapped to a SIP URI and can not
be routed using some pre-defined rules, then the request is terminated in the PSTN and
is thus forwarded to the BGCF.

After mapping the Tel URI to a SIP URI, the S-CSCF resolves the domain part of the
URI using DNS. The DNS query results in the address of the I-CSCF acting as the entry
point for the callee's home network. Before forwarding the request, the S-CSCF adds
a *Record-Route* header with its own address. The S-CSCF further sends a *100 Trying*
response to the P-CSCF.

5–7 INVITE processing at the home terminating I-CSCF   If the I-CSCF at the callee's
home network does not know the home S-CSCF of the callee then it queries the HSS with
a DIAMETER Location-Info Request (LIR) and receives back a Location-Info Answer
(LIA). The S-CSCF indicated in the LIA message is then added to a *Route* header and
the request is forwarded to the S-CSCF. The I-CSCF further sends a *100 Trying* response.

8–9 INVITE processing at the home terminating S-CSCF   The callee's home S-CSCF
evaluates the callee's initial filter criteria in order to decide to which application servers

```
P-Called-Party-ID: <sip:user2_public1@home2.net>
```

Figure 4.11   INVITE processing at the home terminating S-CSCF

the INVITE request should be forwarded. The S-CSCF includes a *Route* header with the address of the P-CSCF received in the *Path* header during the callee's registration and includes a *Record-Route* header with its own address. As illustrated in Figure 4.11, the S-CSCF further adds a *P-Called-Party-ID* with the URI included in the *Request-URI* and replaces the *Request-URI* with the *Contact* information registered by the callee. Adding the *P-Called-Party-ID* header enables the callee to know to which public user identity the INVITE was originally intended. The S-CSCF further sends a *100 Trying* response to the I-CSCF.

10–12 INVITE processing at the visited terminating P-CSCF   The P-CSCF forwards the request using the address in the *Request-URI* and includes a *Record-Route* header with its own address. The P-CSCF also asks the PDF using a DIAMETER AAR request to authorize the media resources to be used for this session as described by the SDP part. The PDF replies with an AAA reply including a media authorization token (Calhoun *et al.* 2005). The P-CSCF includes the token in a *P-Media-Authorization* header field (Marshall 2003) before forwarding the request to the UE, see Figure 4.12. This token is then added by the UE to the messages used for reserving resources. The P-CSCF sends back a *100 Trying* response to the S-CSCF.

13–15 INVITE processing at the callee's UE   The callee's UE receives the INVITE request and replies back with a *100 Trying* response. As the INVITE included a *Require* header set to *precondition* the UE must send back a *183 Session Progress* response including the SDP answer of the callee. As illustrated in Figure 4.13, the *183 Session Progress* response includes also a *Require* header set to *100rel* so as to indicate that the reception of provisional responses must be acknowledged.

16–20 Processing of the *183* response at the CSCFs   The *183 Session Progress* response traverses backwards the same CSCFs that were traversed by the INVITE request. The P-CSCF at the visited terminating network adds a *P-Asserted-Identity* header, which includes the identity it received in the INVITE request in the *P-Called-Party-ID*. The P-CSCF at the visited originating network adds a *Media-Authorization-Token* header with a token that was generated by the PDF in an AAR/AAA exchange. The token is used by the calling UE as part of the resource reservation process.

```
P-Media-Authorization:
   0020000100100101706466322e76697369746564322e6e6574000
     c020139425633303732
```

Figure 4.12   INVITE processing at the home terminating P-CSCF

```
SIP/2.0 183 Session Progress
Via: SIP/2.0/UDP pcscf2.visited2.net:5088;comp=sigcomp;branch
   =z9hG4bK361k21.1, SIP/2.0/UDP scscf2.home2.net;branch
   =z9hG4bK764z87.1, SIP/2.0/UDP icscf2_s.home2.net;branch
   =z9hG4bK871y12.1, SIP/2.0/UDP scscf1.home1.net;branch
   =z9hG4bK332b23.1, SIP/2.0/UDP pcscf1.visited1.net;branch
   =z9hG4bK240f34.1, SIP/2.0/UDP [5555::aaa:bbb:ccc:ddd]:1357;
   comp=sigcomp;branch=z9hG4bKnashds7
Record-Route: <sip:pcscf2.visited2.net:5088;lr;comp=sigcomp>,
   <sip:scscf2.home2.net;lr>, <sip:scscf1.home1.net;lr>,
   <sip:pcscf1.visited1.net;lr>
P-Access-Network-Info: 3GPP-UTRAN-TDD; utran-cell-id-3gpp
   =234151D0FCE11
Privacy: none
From:
To: <tel:+1-212-555-2222>;tag=314159
Call-ID:
CSeq:
Require: 100rel
Contact: <sip:[5555::eee:fff:aaa:bbb]:8805;comp=sigcomp>
Allow: INVITE, ACK, CANCEL, BYE, PRACK, UPDATE, REFER, MESSAGE
RSeq: 9021
Content-Type: application/sdp
Content-Length: ()

v=0
o=- 2987933623 2987933623 IN IP6 5555::eee:fff:aaa:bbb
s=-
c=IN IP6 5555::eee:fff:aaa:bbb
t=0 0
m=video 10001 RTP/AVP 98 99
b=AS:75
a=curr:qos local none
a=curr:qos remote none
a=des:qos mandatory local sendrecv
a=des:qos mandatory remote sendrecv
a=conf:qos remote sendrecv
a=rtpmap:98 H263
a=fmtp:98 profile-level-id=0
a=rtpmap:99 MP4V-ES
m=audio 6544 RTP/AVP 97 96
b=AS:25.4
a=curr:qos local none
a=curr:qos remote none
a=des:qos mandatory local sendrecv
a=des:qos mandatory remote sendrecv
a=conf:qos remote sendrecv
a=rtpmap:97 AMR
a=fmtp:97 mode-set=0,2,5,7; mode-change-period=2
a=rtpmap:96 telephone-event
a=maxptime:20
```

Figure 4.13 *183 Session Progress* response from the callee

```
PRACK sip:[5555::eee:fff:aaa:bbb]:8805;comp=sigcomp SIP/2.0
Via: SIP/2.0/UDP [5555::aaa:bbb:ccc:ddd]:1357;
  comp=sigcomp;branch=z9hG4bKnashds7
Max-Forwards: 70
P-Access-Network-Info: 3GPP-UTRAN-TDD; utran-cell-id-3gpp
  =234151D0FCE11
Route: <sip:pcscf1.visited1.net:7531;lr;comp=sigcomp>,
 <sip:scscf1.home1.net;lr>, <sip:scscf2.home2.net;lr>,
 <sip:pcscf2.visited2.net;lr>
From: <sip:user1_public1@home1.net>;tag=171828
To: <tel:+1-212-555-2222>;tag=314159
Call-ID: cb03a0s09a2sdfglkj490333
Cseq: 128 PRACK
Require: precondition, sec-agree
Proxy-Require: sec-agree
Security-Verify: ipsec-3gpp; q=0.1; alg=hmac-sha-1-96;
  spi-c=98765432; spi-s=87654321; port-c=8642; port-s=7531
RAck: 9021 127 INVITE
Content-Type: application/sdp
Content-Length: ()

v=0
o=- 2987933615 2987933616 IN IP6 5555::aaa:bbb:ccc:ddd
s=-
c=IN IP6 5555::aaa:bbb:ccc:ddd
t=0 0
m=video 3400 RTP/AVP 98
b=AS:75
a=curr:qos local none
a=curr:qos remote none
a=des:qos mandatory local sendrecv
a=des:qos mandatory remote sendrecv
a=rtpmap:98 H263
a=fmtp:98 profile-level-id=0
m=audio 3456 RTP/AVP 97 96
b=AS:25.4
a=curr:qos local none
a=curr:qos remote none
a=des:qos mandatory local sendrecv
a=des:qos mandatory remote sendrecv
a=rtpmap:97 AMR
a=fmtp:97 mode-set=0,2,5,7; mode-change-period=2
a=rtpmap:96 telephone-event
a=maxptime:20
```

Figure 4.14 PRACK request sent by UE

21–22 Processing of the *183* response at the caller's UE   The *183 Session Progress* response includes the callee's SDP answer with the description of the media capabilities of the callee. The caller can now determine which codes to use and creates her own modified SDP offer. As the callee requested that the *183* response be sent reliably, the caller acknowledges the reception of the *183* response by sending a PRACK request. As illustrated in Figure 4.14 the modified SDP offer is included in the PRACK request. Further, the caller's UE starts the resource reservation process. The exact mechanism for the resource reservation depends on the used access technology.

23–25 Processing of the PRACK at the CSCFs   The PRACK is routed through all the CSCFs that have included a *Record-Route* header with their addresses in the INVITE request.

26–31 Processing of the PRACK request at the callee's UE   The PRACK includes the final SDP offer and the callee's UE can start reserving the needed resources for the session. The successful reception of the PRACK request is acknowledged by sending a *200 OK* response back to the caller.

32–41 Processing of the UPDATE request   Once the caller's UE has successfully reserved the needed resources for the session it generates an UPDATE request. The UPDATE request indicates that the reservation process was successful and includes the SDP offer on basis which the resource reservation was done. The UPDATE request traverses all the CSCFs that have included a *Record-Route* header in the INVITE request. That is the P-CSCFs and the S-CSCFs in our example. The callee acknowledges the successful reception of the UPDATE request by sending a *200 OK* response back to the caller. In conformance with the offer/answer model used with SIP for the negotiation of the used media capabilities (Rosenberg and Schulzrinne 2002a) the *200 OK* response will include the SDP answer of the callee.

42–57 Processing of the 180 Ringing   The callee should only be alerted about the incoming call request after having successfully reserved the needed resources for ensuring the needed QoS level for the call. Establishing the call first and then starting the resource reservation process might lead to the situation where after alerting the callee the call might need to be terminated as the needed resources are not available. The callee determines that the reservation process was successful when its own reservation process was successful and an UPDATE request was received from the caller, indicating the success of the reservation process at the caller. At this stage the callee's UE can alert the subscriber and notify the caller that the callee is being alerted by sending back a *180 Ringing* response. The caller acknowledges the reception of the *180 Ringing* response by sending a PRACK request which is then acknowledged with a *200 OK* response.

58–68 *200 OK* and ACK exchange   When the callee finally accepts the call, the UE sends a *200 OK* response to the caller, which is then acknowledged by the caller with an ACK request. At this stage the session establishment was successful and the caller and callee can start exchanging the media traffic.

5

Secure Access and Interworking in IMS

The 3GPP specifications considering security aspects in IMS can be roughly divided into two parts: user access control and network security. The user access control discussion (33.203 TS 2008) which is presented in section 5.1 addresses aspects of user authentication and ensuring the security of the user's signaling traffic. Network security specifications (33.210 TS 2008), discussed in section 5.2, look at aspects of securely transporting the IMS signaling traffic between the different components of an IMS network as well as between different IMS networks. Additionally, there is currently work in progress regarding the specification of a data plane security architecture in (33.828 TR 2008). This is still, however, at an early stage without final specifications.

5.1 Access Security in IMS

Before being granted access to the IMS services, the subscriber must authenticate herself to the operator. The access security mechanism originally defined in IMS uses AKA to achieve the mutual authentication of the users and access networks. However, as IMS is to be deployed in fixed networks and non-UMTS-based wireless networks it must also support terminals that do not include an ISIM or USIM application. This has introduced the need for the IMS to interoperate with "non-AKA" access security mechanisms, which fall into two main categories:

- access-bundled security mechanisms, which are described in section 5.1.2;
- security mechanism based on the HTTP Digest, which is described in section 5.1.3.

5.1.1 IMS AKA Access Security

Using AKA the IMS service provider can authenticate the subscriber. Additionally, in order for the subscriber to be assured that she is communicating with her operator and not with an attacker, the operator authenticates himself toward the subscriber as well. Further, to ensure that the subsequent signaling between the operator and the subscriber is secured and protected without having to repeat the authentication process, the AKA authentication

SIP Security Dorgham Sisalem, John Floroiu, Jiri Kuthan, Ulrich Abend and Henning Schulzrinne
© 2009 John Wiley & Sons, Ltd

process results in the establishment of a secure tunnel between the subscriber and the P-CSCF over which all messages sent to or from the subscriber are transported. The mutual authentication process is described in section 5.1.1.1, whereas the establishment of the secure tunnel is described in section 5.1.1.3.

5.1.1.1 Mutual Authentication

Mutual authentication in IMS is the process by which the network authenticates the user and the user authenticates the network. The authentication mechanism is based on the AKA-based HTTP digest authentication mechanism (Niemi *et al.* 2002), which extends the HTTP digest authentication scheme (Franks *et al.* 1999) by using AKA for one-time password generation.

The mutual authentication in IMS takes place during the UE registration and consists of two REGISTER exchanges. In the initial exchange the IMS client provides the user identity and receives a challenge, and the second exchange carries the response to the challenge and the result of the authentication.

Figures 5.1 and 5.2 illustrate the IMS registration procedure. The messages depicted provide samples of those SIP headers that are relevant to the mutual authentication (*Authorization*, *WWW-Authenticate*) and secure tunnels setup (*Security-Client*, *Security-Server*, *Security-Verify*) procedures.

1–4 Sending the initial REGISTER message   The first REGISTER message sent by the UE (message 1 in Figure 5.1) conveys the following information that identifies the user and her home domain:

- The *Request-URI* –contains the UE's home domain name. The P-CSCF that receives the REGISTER uses the DNS mechanisms (Rosenberg and Schulzrinne 2002c) to locate an inbound I-CSCF in the UE's home domain. The I-CSCF then initiates a Diameter User-Authorization-Request/User-Authorization-Answer exchange with the HSS (exchange 3 in Figure 5.1) in order to determine the S-CSCF where the REGISTER must be forwarded to. Note that the routing path may also include an outbound I-CSCF in the visited domain, which has not been represented in Figure 5.1 because it does not play any role in the authentication procedure.
- The *Authorization* header includes the following information:
  - the registration URI–which is also carried in the *uri* parameter of the *Authorization* header;
  - the realm of authentication–a Fully Qualified Domain Name (FQDN) (Mockapetris 1987) carried in the *realm* parameter of the *Authorization* header and containing the name of the realm where the username is authenticated;
  - the public user identity–a SIP Address-of-Record carried in the *From* and *To* headers; as the result of a successful SIP registration, the public user identity is bound to the IP address or host name contained in the *Contact* header;
  - the private user identity–a Network Access Identifier (Aboba *et al.* 2005) carried in the *username* parameter of the *Authorization* header; the private user identity identifies the user for authentication purposes.

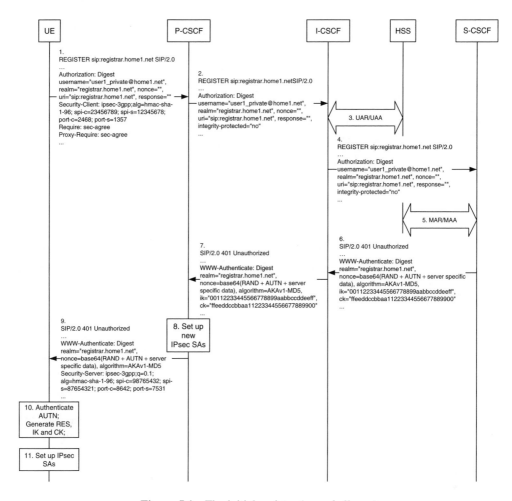

Figure 5.1   The initial register (user challenge)

The *Authorization* header in the first REGISTER message also contains an empty *nonce* and *response* parameters as well as an *integrity-protected* parameter, whose value is set to "no" if the message is not protected by IPsec tunnels, or "yes" if an IPsec tunnel set up during a previous authentication session is available.

5 Retrieving an AKA AV   Upon receiving the initial REGISTER (message 4 in Figure 5.1), the S-CSCF requests from the HSS one or more AKA AVs corresponding to the private user identity being authenticated. This is achieved by means of the Diameter Multimedia-Authentication-Request/Multimedia-Authentication-Answer commands (MAR/MAA, exchange 5 in Figure 5.1). Note, however, that the MAR/MAA exchange may be absent if the S-CSCF still has AKA AVs available from a previous authentication session.

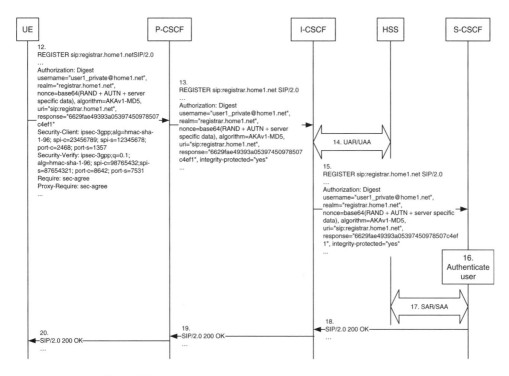

Figure 5.2 Second REGISTER exchange (user authentication)

The RAND and AUTN components of the AKA AV are used to challenge the user, the IK and CK keys are delivered to the P-CSCF to serve as keying material for the secure tunnels setup and XRES is retained by the S-CSCF for the response validation.

6, 7 Challenging the user The S-CSCF challenges the UE by replying to the initial REGISTER with a *401 Unauthorized* response (message 6 in Figure 5.1), which contains a WWW-*Authenticate* header carrying the following information:

- the realm of authentication–has the same value as in the initial REGISTER;
- the challenge–is contained in the *nonce* parameter and consists of a concatenation of the RAND, AUTN and server specific data encoded in base 64 (Josefsson 2006);
- the session keys–carried in the *ik* and *ck* parameters; they are removed by the P-CSCF, which retains the values and uses them as keying material for the IPsec tunnels that the P-CSCF establishes between itself and the UE;
- the algorithm name–specified in the *algorithm* parameter and its value set to "AKAv1-MD5".

The S-CSCF creates then an authentication session, awaiting the UE's response.

8 Setup of the IPsec SAs on the P-CSCF IPsec security associations are set up on the P-CSCF. They will be used to protect the second REGISTER exchange carrying the UE's response, as well as subsequent re-registration, de-registration, user re-authentication and call setup signaling sequences.

9 Challange forwarding   The P-CSCF forwards the *401 Unauthorized* response to the UE. This response is not protected because the UE has not yet learned the protected port numbers the P-CSCF is going to use.

10 Network authentication   Upon receiving the *401 Unauthorized* response (message 9 in Figure 5.1), the UE extracts the challenge, authenticates the network by verifying the MAC (see section 2.3.2) and, assuming the MAC validation is successful and the SQN is in the right sequence, calculates the response RES. The UE calculates then A1 [see equation (3.3)] using the private user identity as username and the RES as password, A2 [see equation (3.5)] and finally the response [see equation (3.2)].

11 Setup of the IPsec SAs on the UE   IPsec security associations are set up on the UE. The establishment of the IPsec tunnels is discussed in more detail in section 5.1.1.3.

12–15 Sending the response   The *Authorization* header in the second REGISTER message (message 12 in Figure 5.2) carries the UE's response in the *response* parameter. Its content is similar to that of the first REGISTER, with the following differences:

- the challenge received in the *401 Unauthorized* message is returned in the *nonce* parameter;
- the *response* parameter contains now the UE's response;
- an *algorithm* parameter is included, whose value is "AKAv1-MD5";
- The *integrity-protected* parameter, whose value is set by the P-CSCF to "yes", indicates that it has been received through an IPsec tunnel.

16–20 Succcesful user authentication and registration completion   Upon receiving the second REGISTER (message 15 in Figure 5.2), the S-CSCF uses the XRES parameter of the current AKA AV as one-time password to calculate the expected response and compares it with the received response (step 16 in Figure 5.2). If they match, then the user is successfully authenticated. As a result, the user's registration state on the S-CSCF is updated, the user's profile is downloaded using the Server-Assignment-Request/Server-Assignment-Answer commands (exchange 17 in Figure 5.2) and the REGISTER request is answered with a *200 OK* message (messages 18 in Figure 5.2).

It may be observed that there are a couple of differences between the HTTP digest authentication based on AKA specified at the IETF (Niemi *et al.* 2002):

- The IMS client already uses an *Authorization* header in the initial REGISTER message in order to convey authentication-relevant information to the S-CSCF (most importantly, the private user identity). This helps the HSS to decide on the S-CSCF responsible for processing the request and the S-CSCF to retrieve one or more AKA Authentication Vectors (AVs) for the respective private user identity.

- The final response from the S-CSCF does not include an *Authentication-Info* header. The *Authentication-Info* header enables the client to (i) authenticate the server, (ii) to integrity protect the server's response and (iii) to allow the server to provide the next nonce that the client is supposed to use for the next authentication attempt (see Section 3.9.1). These functions are instead implemented in IMS in the following way: (i) AKA enables the IMS client to authenticate the server by validating the MAC parameter of the AUTN, (ii) in IMS the second REGISTER exchange is protected between the UE and the P-CSCF by the IPsec tunnels established as result of the initial REGISTER exchange, (iii) an user re-authentication is initiated by the UE by sending an initial REGISTER request and triggering the S-CSCF to send a fresh challenge, while all other REGISTER messages (re-registrations) are protected by the IPsec tunnels and challenging them is therefore not necessary.

5.1.1.2 Resynchronization

If during the AUTN validation (step 10 in Figure 5.1) the UE determines that the SEQ_{HE} is stalled, a resynchronization sequence is initiated (see Figure 5.3). The UE will not set up new IPsec SAs in this case (step 11 in Figure 5.1 will be missing). Instead, the UE will start a re-synchronization sequence.

a–d Resynchronization   The re-synchronization sequence starts with the UE generating an AUTS using its own SEQ_{MS} value (see Section 2.3.2) and sending it to the S-CSCF in a new REGISTER message, in the form of an *auts* parameter carried in the *Authorization* header. For the response calculation the UE uses the same algorithm as in step 10 but with an empty string as password.

e Retrieval of a fresh AKA AV   The S-CSCF passes the AUTS to the HSS. The HSS performs SQN re-synchronization (see Section 2.3.3) and responds with a fresh AKA AV set.

f, g User re-challenge   The S-CSCF re-challenges the UE.

h Setup of IPsec SAs on the P-CSCF   The P-CSCF replaces its previous set of IPsec SAs with new ones (at least the IK and CK parameters will differ from the previous REGISTER exchange).

i Re-challenge forwarding   The P-CSCF forwards the *401 Unauthorized* response to the UE, unprotected.

j, k Network authentication   The UE authenticates the AKA AV received, checks that it is a fresh one and, assuming both are successful, calculates the response and sets up the IPsec SAs. The signaling sequence continues then as illustrated in Figure 5.3, protected by the freshly set IPsec SAs.

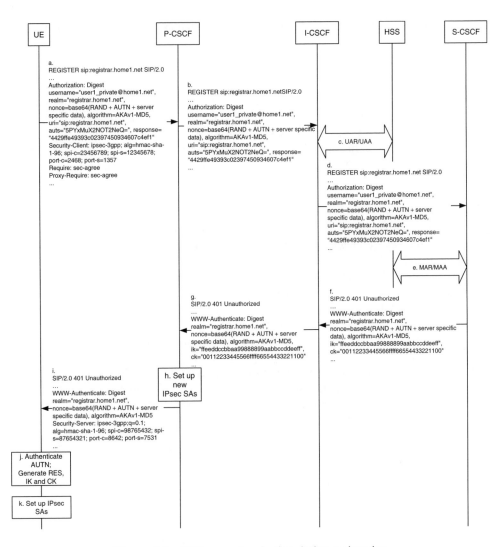

Figure 5.3   SQN resynchronization during registration

   User re-authentication is triggered by the S-CSCF by sending *reg* event package notifications to the UE. When this happens, the UE sends a new initial REGISTER, which may be protected by the IPsec tunnels that are already in place; in this case the *integrity-protected* parameter must be set to *"yes"*

   Re-registrations and de-registrations are performed in order to extend the registration lifetime of public user identities and to deregister public user identities. The content of their *Authorization* header must be identical to that of the last successfully authenticated REGISTER request. Also, re-registrations and de-registrations should be protected (*integrity-protected="yes"*), otherwise they will be challenged by the S-CSCF and a new user authentication sequence will unfold.

5.1.1.3 Secure Tunnels Setup

The IPsec tunnels that protect the SIP signaling in the access network are set up between UE and a P-CSCF during the SIP registration. RFC3329 (Arkko *et al.* 2003) defines the *Security-Client*, *Security-Server* and *Security-Verify* SIP headers that enable the negotiation of a security mechanism between an UA and its first hop SIP entity, which in the case of IMS is the P-CSCF. The security mechanism may be one of the following:

- "ipsec-ike", indicating that IKE will be used to negotiate the keying material and the security parameters of the IPsec SAs;
- "ipsec-man" (IPsec with manual keying) which uses an out-of-bound mechanism to negotiate the keying material and the security parameters of the IPsec SAs;
- "tls" for TLS;
- "digest" for HTTP-Digest;
- "ipsec-3gpp", where the keying material is provided by AKA (in the form of the IK and CK session keys) and the IPsec SA parameters negotiation (SPI, cryptographic algorithms, etc.) is done through the SIP signaling. This mechanism is mandatory to be implemented in IMS and will be discussed in detail in this section.

The negotiation procedure is illustrated in Figure 5.4. The client announces the security mechanisms that it supports, selects the most preferred security mechanism that it implements from the list provided by the server and initiates that mechanism (e.g. the

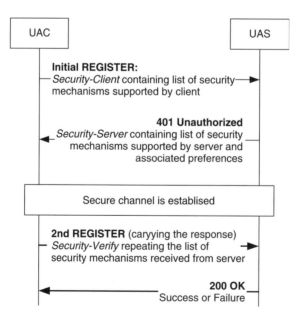

Figure 5.4 SIP Security mechanism negotiation

establishment of a TLS connection or IKE session requires specific signaling to take place).

The *Security-Client* and *Security-Server* headers are used to carry the client's and server's proposals, whereas the *Security-Verify*, which repeats the content of the *Security-Server* header, is transported over the secure channel and has the role of preventing bid-down attacks. A bid-down attack occurs when a man in the middle removes the "stronger" security mechanisms in an attempt to force the UAC and UAS to use a "weaker" and hence easier to attack security mechanism. In IMS the *Security-Client* header is also repeated by the server in the protected response during the second REGISTER exchange.

The actual security mechanism negotiation requires one round trip and is followed by a verification exchange that takes another round trip and occurs over the just-established secure channel. This message sequence is consistent with the IMS registration-with-authentication procedure consisting of two REGISTER exchanges (see Figures 5.1 and 5.2).

The IPsec tunnels established between the UE and the P-CSCF use the transport mode ESP encapsulation (see section 2.2.1.1). Also, the lifetime of the IPsec SAs has a fixed duration of 2^{32-1} seconds and refreshing of the shared keys occurs as the user re-authenticates. Finally, a successful mutual authentication allows both the UE and the P-CSCF to end up with the same set of shared secrets. Therefore, the "ipsec-3gpp" security mechanism only negotiates the authentication algorithm, the encryption algorithm and the Security Parameter Indexes.

In order to force a SIP message to be sent through an IPsec tunnel and respectively to make sure that a SIP message has been received through an IPsec tunnel, the protected SIP signaling is exchanged between the UE and the P-CSCF through "protected port numbers". The protected port numbers are different from the SIP standard 5060 and 5061 port numbers and are announced in the *Security-Client* and *Security-Server* headers in the *port-c* and *port-s* parameters. The *Security-Client* and *Security-Server* headers also contain in the *spi-c* and *spi-s* parameters the SPIs for the IPsec SAs that have to be established in both directions, as well as the cryptographic algorithms to be used by the IPsec SAs.

We will denote by port-uc, port-us, spi-uc and spi-us the *port-c*, *port-s*, *spi-c* and respectively *spi-s* parameters specified by the UE in the *Security-Client* header, and by port-pc, port-ps, spi-pc and spi-ps the *port-c*, *port-s*, *spi-c* and *spi-s* parameters specified by the P-CSCF in the *Security-Server* header.

Figure 5.5 describes the mapping between the IPsec SAs and the traffic selectors when sending and receiving SIP requests and responses over UDP and TCP. There are four IPsec SAs bound to the traffic exchanged between the ⟨UE, port-uc⟩ and ⟨P-CSCF, port-ps⟩ and between the ⟨UE, port-us⟩ and ⟨P-CSCF, port-pc⟩ transport endpoints, two of them in each direction.

The SPI values provided by the UE and P-CSCF represent the values the UE and P-CSCF expect to see in the incoming IPsec datagrams. It may be observed that the behavior differs between UDP and TCP in that SIP responses sent over TCP use the same TCP connection (hence client–server port numbers pair) as the SIP requests that triggered them. REGISTER messages received on a protected port number are marked by the P-CSCF with the *integrity-protected* = *"yes"* attribute in the *Authorization* header, indicating to the S-CSCF that the user need not be challenged.

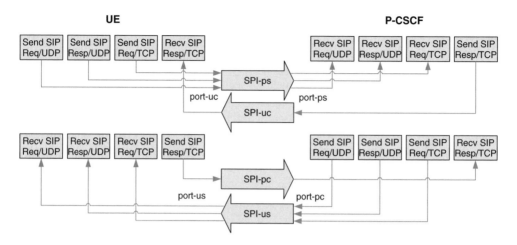

Figure 5.5 Overview of the IPsec SAs and traffic selectors

The IPsec SAs established by a previous successful authentication session are used for all subsequent signaling, including fresh user re-authentication attempts, which, if successful, will result in the set up of "new" IPsec SAs. In such a situation, the "old" IPsec SAs are kept until all pending SIP transactions that were using them have been completed or have expired.

From the UE's perspective, the new IPsec SAs have been successfully established after receiving the final *200 OK* (message number 20 in Figure 5.2), while from the P-CSCF's perspective the new IPsec SAs have been successfully established as soon as a SIP message arrives from the UE through the new incoming IPsec SAs (message number 12 in Figure 5.2).

It is also possible to set up "incomplete" IPsec SAs on the P-CSCF, as for instance in the case where the challenge (message 9 in Figure 5.1) gets lost and the UE fails to set up the matching IPsec SAs on its side. In such a situation the UE would typically repeat the initial REGISTER (possibly using the "old" IPsec SAs) and the P-CSCF would discard the "incomplete" IPsec SAs and set up fresh IPsec SAs as soon as a new challenge (message 8 of the next REGISTER attempt, see Figure 5.1) is received.

When the P-CSCF receives the challenge (message 7 in Figure 5.1), it uses the SPI values exchanged in the *Security-Client* and *Security-Server* headers to set up the IPsec SAs (SAD entries) illustrated in Table 5.1. The P-CSCF also uses the port numbers exchanged in the *Security-Client* and *Security-Server* headers and the IP addresses seen in the IP header of the REGISTER to set up the IPsec policy rules (SPD entries) illustrated in Table 5.2.

When the challenge is finally received by the UE, and assuming that the UE has successfully authenticated the network, the UE sets up IPsec policy rules and IPsec SAs mirroring those on the P-CSCF (the writing of which we leave as an exercise to the reader).

The second REGISTER request containing the response, as well as all the subsequent SIP messages, is sent using ESP transport mode encapsulation. When receiving the second REGISTER request the P-CSCF will match it against SA 1, based on the destination IP address, protocol and SPI value. After the integrity protection check and decryption, the

Table 5.1 IPsec SAs on the P-CSCF

SA no.	Source (IP)	Destination (IP)	Protocol /mode	SPI
1	UE	P-CSCF	ESP/transp	SPI-ps
2	UE	P-CSCF	ESP/transp	SPI-pc
3	P-CSCF	UE	ESP/transp	SPI-uc
4	P-CSCF	UE	ESP/transp	SPI-us

Table 5.2 IPsec policy rules on P-CSCF

Rule no.	Direction	Source (IP, port no.)	Destination (IP, port no.)	Transport protocol	Verdict	Bound to SA	Protocol
1	in	UE, port-uc	P-CSCF, port-ps	udp	ipsec		ESP/ transp
2	in	UE, port-uc	P-CSCF, port-ps	tcp	ipsec		ESP/ transp
3	in	UE, port-us	P-CSCF, port-pc	tcp	ipsec		ESP/ transp
4	out	P-CSCF, port-pc	UE, port-us	udp	ipsec	SA4	ESP/ transp
5	out	P-CSCF, ; port-pc	UE, port-us	tcp	ipsec	SA4	ESP/ transp
6	out	P-CSCF, port-ps	UE, port-uc	tcp	ipsec	SA3	ESP/ transp

IP datagram will be checked against the policy rules. Rule 1 will match and the IPsec stack on the P-CSCF will check that the correct protection (ESP transport mode) has indeed been applied to the REGISTER request.

When the final *200 OK* is received from the downstream CSCF entity, the P-CSCF forwards it to the UE using the protected source and destination port numbers. This results in the outgoing *200 OK* matching Rule 4 and consequently being processed according to SA 4 (as resulting from Tables 5.1 and 5.2).

5.1.2 Access-bundled Authentication

Access-bundled authentication mechanisms are used when no direct authentication can be performed at the IMS layer between the IMS client and the S-CSCF. In such cases the authentication at the IMS layer must be bundled to the authentication mechanisms available at the transport layer, like for instance in the early IMS systems and the Next Generation Networks (NGNs).

5.1.2.1 Early IMS Authentication

Early IMS systems denote IMS implementations that do not fully support the requirements specified in the 3GPP technical specifications. The definition of what "early" means is rather fuzzy. Originally, one major criterion was the IPv6 support, which was mandatory in IMS; this is however no longer valid starting with Release 5, when support for IPv4 has been introduced. In terms of security, early IMS refers to deployments that do not implement the ISIM applications and IPsec and hence cannot perform key agreement to secure the signaling between the UE and the P-CSCF.

The security functions in the early IMS deployments (33.978 TR 2007) are restricted to defending against user identity impersonation, which means not allowing a user who acquires IP connectivity to register at the SIP layer his contact IP address under the identity of another legitimate user, so that the victim will be falsely charged for the service. This is achieved by creating a binding between the private and public user identities and the IP address assigned to the terminal using the following procedure:

- When a PDP context is created as result of a GPRS attach procedure (23.060 TS 2008), a Radius client operating in the GGSN sends the HSS the mapping between the UE's IMSI and the allocated IP address.
- The GGSN checks that the source IP addresses in the datagrams sent by the UEs match those allocated to the respective PDP context. This protects against IP address spoofing.
- When an UE initiates an IMS registration:
  - The public user identity that the UE uses in the REGISTER message is derived from the IMSI using the procedures described in (23.003 TS 2008). An *Authorization* header is, however, not present and therefore the private user identity cannot be explicitly conveyed in the REGISTER message. As a result, when invoking the HSS, the I-CSCF and S-CSCF must derive the UE's IMSI-based private user identity from the IMSI-based public user identity contained in the REGISTER mrequest. An early IMS subscriber in therefore identified in the HSS by its IMSI-based private and public user identities.
  - The P-CSCF receiving a registration request checks that the datagram IP source address is the same as the IP address in the topmost *Via* header. This ensures that the REGISTER message originates from the legitimate owner of that IP address.
  - When the REGISTER message finally reaches the S-CSCF, the S-CSCF invokes the HSS using the IMSI-based public user identity retrieved from the *From* header and the private user identity derived from it. The HSS returns the corresponding IP address, which is checked against the topmost *Via* header. If the match is successful, then a binding between the sender of the REGISTER message, the IMSI and the IP address has been proved.

The user authentication mechanism used in the early IMS deployments has a number of limitations:

- Because the GGSN needs to convey to the HSS PDP context information, it has to be located in the home domain.
- An IMSI-based public user identity cannot be simultaneously registered by multiple terminals using early IMS authentication methods and an UE cannot use more than one IP address (which must instead be shared by all active PDP contexts).
- There must not be any NAT devices anywhere between the UE and the S-CSCF since the IP address assigned during the PDP Context Activation must remain consistent with the IP address contained in the *Via* header.

5.1.2.2 NASS-IMS-bundled Authentication

The Next Generation Network architecture (282001 ES 2008) enables the access to IMS services from fixed broadband access networks, such as those using Digital Subscriber Line technologies (generically denoted as xDSL). NGNs represent so far the most important example of non-UMTS IMS-convergent networks. This convergence between fixed and mobile networks at the IMS layer is possible because the IMS architecture was conceived to be largely independent of the underlying transport mechanisms.

NGN terminals that support ISIM will make use of the AKA-based authentication procedure for the purpose of mutual authentication. For NGN terminals not supporting ISIM, the authentication at the IMS layer needs to be bundled with authentication that takes place at the NGN transport layer. The resulting authentication mechanism is known as NASS Bundled Authentication (NBA) (187003 TS 2008). The NASS, which stands for Network Attachment Sub-System, is the NGN transport-layer architectural entity responsible for the authentication of the network access and has the following functions:

- dynamic allocation of IP address and other terminal configuration parameters (e.g. using DHCP);
- user authentication, prior or during the IP address allocation procedure;
- authorization of network access, based on the user profile;
- access network configuration, based on the user profile;
- maintenance of the binding between the allocated IP address and the location information.

Authentication at the NASS layer may be performed using for instance EAP-based (Aboba *et al.* 2004) authentication mechanisms. Assuming a NGN terminal has already been authenticated at the NASS layer, the SIP registration procedure is extended in the following way:

- When the P-CSCF receives a REGISTER message, it contacts the Connectivity Session Location and Repository Function (CLF) (282004 ES 2008). The CLF is the NASS entity that provides to the service control subsystems and applications an interface to the NASS. The CLF binds together the various pieces of information available from other functional entities operating within the NASS, including the allocated IP address, the physical access ID, the logical access ID, the NASS user identity, QoS profile, etc. In case of xDSL, the physical access ID represents an identifier of the subscriber's physical line, while the logical access ID identifies the port, ATM Virtual Path and/or the ATM Virtual Circuit that carry the traffic. The P-CSCF supplies to the CLF the terminal's IP address and retrieves the terminal's location information, which it then adds into a *P-Access-Network-Info* header. The *P-Access-Network-Info* (*PANI*) header must also contain the *network-provided* parameter in order to distinguish it from the *PANI* header that may be supplied by the terminals.
- Upon receiving the REGISTER, the S-CSCF contacts the User Profile Server Function (UPSF), which is the correspondent of the HSS in the NGN architecture. The UPSF returns the location information that corresponds to the user indicated by the private user identity and the public user identity contained in the REGISTER message. This

location information is compared with the one contained in the *PANI* header and if they match then the user is considered successfully authenticated.

It must be ensured that access from an NGN domain can only be performed through a "*PANI*-aware P-CSCF". A *PANI*-aware P-CSCF must check the content of existing *PANI* headers and remove those containing the *network-provided* parameter. This condition is necessary because *PANI* headers are forwarded transparently by the *PANI*-unaware P-CSCFs. A malicious user may take advantage of this and impersonate its victim by using the victim's private user identity and location information but registering its own IP address. A *PANI*-unaware P-CSCF will leave the forged *PANI* header rather than retrieving itself the correct location information from the NASS, based on the subscriber's IP address. As a result the S-CSCF will successfully validate the registration using the victim's user profile data and will unknowingly register the attacker's IP address.

5.1.3 HTTP Digest-based Access Security

The HTTP digest (see section 3.9.1) is used by UEs that lack an ISIM application but can be authenticated using an username and password. When the HTTP digest is used, the SIP signaling is in fact unprotected. The only security measure the P-CSCF can implement in this case is to maintain an "IP address check table" that maps the IP address of an UE to its authenticated private user identity and its corresponding registered public user identities. The "IP address check table" provides a sufficient proof of origin authentication for the SIP signaling only in those deployments where the IP address spoofing, IP address re-assignment without the notice of the SIP application and the presence of man-in-the-middle entities can be ruled out.

If this is not the case, then two alternatives may be used to enhance the level of protection provided to the SIP signaling:

- The S-CSCF may perform user re-authentication with each new REGISTER request and use the HTTP digest proxy authentication mechanism to authenticate the non-REGISTER requests (see section 3.9.1.2);
- The HTTP digest may be used in conjunction with TLS (see section 5.1.3.1).

The user authentication procedure using HTTP Digest is illustrated in Figure 5.6. It differs from the basic HTTP digest authentication scheme described in section 3.9.1.1 in that the initial REGISTER message carries an *Authorization* header. This is necessary because the I-CSCF requires the user private identity in order to locate the S-CSCF and the S-CSCF requires it as well in order to retrieve a "SIP Digest Authentication Vector" (SD-AV) (instead of an AKA AV) from the HSS.

The SD-AV contains the following elements: the realm, the authentication algorithm ("MD5"), the quality of protection ("auth") and the $H(A1)$ parameter. It may be observed that, by storing the $H(A1)$ and in fact not possessing the actual user password, the S-CSCF acts as a third-party authenticator.

The interactions between the involved entities are similar to the message sequence described the section 5.1.1.1, with the observations that in this case:

- no security mechanism is negotiated between the UE and the P-CSCF;

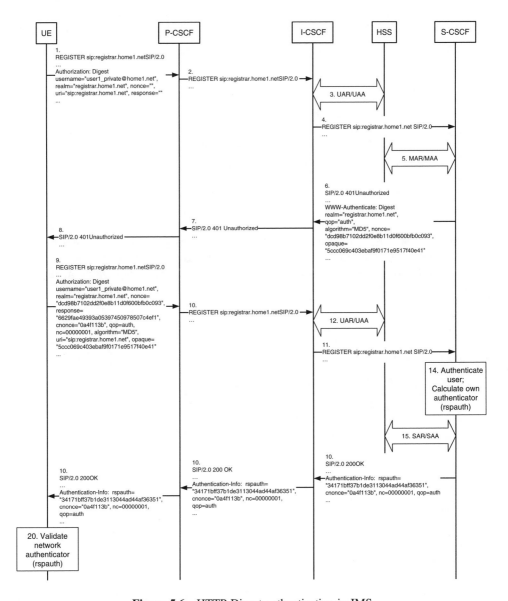

Figure 5.6 HTTP Digest authentication in IMS

- the mutual authentication is achieved by means of the *Authentication-Info* header, which is sent in the final *200 OK* response and enables the UE to authenticate the server.

For the authentication of the non-REGISTER requests, the HTTP digest proxy authentication procedure (see section 3.9.1.2) is used between the UE and the S-CSCF. After being challenged the first time, the UE should store the nonce received in the *Proxy-Authentication* and use it to authenticate subsequent non-REGISTER requests, along with an accordingly incremented nonce counter.

5.1.3.1 TLS with HTTP Digest Authentication

When the HTTP Digest authentication is used, the signaling may be integrity protected and encrypted between the UE and the P-CSCF using TLS. Setting up a TLS connection in an IMS access domain is not triggered by the use of a SIPS URI, as specified in (Rosenberg *et al.* 2002b), but rather as result of the UE and the P-CSCF negotiating the use of the "tls" security mechanism (see section 5.1.1.3).

The TLS connection may not use encryption but must use integrity protection and must not use compression. The type of the protection and the algorithms used are negotiated at the TLS layer, making it unnecessary to exchange them by means of the *Security-Client* and *Security-Server* headers.

A registration procedure that uses HTTP Digest user authentication and negotiates the setup of a TLS connection is illustrated in Figures 5.7 and 5.8; only the headers that are relevant to the authentication and security setup mechanisms have been illustrated.

In our example the UE indicates support for both "ipsec-3gpp" and "tls" security mechanisms and the P-CSCF selects "tls" by indicating a higher preference (higher q value).

In this scenario, the P-CSCF authenticates itself to the UE by presenting a valid certificate that chains up a trusted root certificate and may possibly include intermediate CA

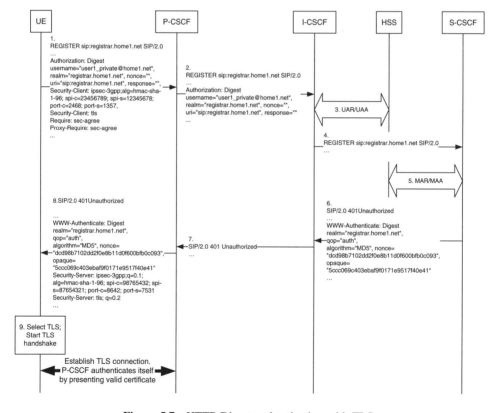

Figure 5.7 HTTP Digest authentication with TLS

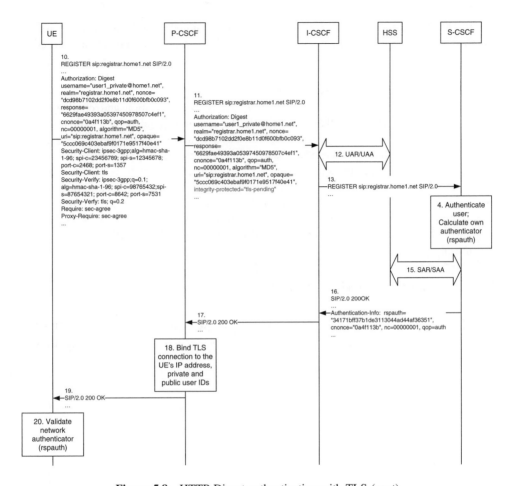

Figure 5.8   HTTP Digest authentication with TLS (cont)

certificates (see section 2.1.2.4). The certificate's subject name must contain the P-CSCF's FQDN. The REGISTER message containing the UE's response is sent over the TLS connection, however, the UE is not yet authenticated at this time and therefore the P-CSCF will mark the REGISTER with *integrity-protected="tls-pending"*.

If the HTTP digest validation is successful, the S-CSCF will respond with *200 OK*. As a result of receiving the *200 OK* response, the P-CSCF will consider the UE authenticated and it will associate the UE's IP address, user private identity and all successfully registered public user identities with the TLS connection. All subsequent non-REGISTER messages must be sent and received over the TLS connection and they must match one of the associated public user identities. The same applies to re-registrations, with the exception that the matching is performed against the private user identity. In addition, the P-CSCF marks the re-REGISTER messages as *integrity-protected="tls-yes"* to indicate to the S-CSCF that the UE has been authenticated. Only initial REGISTER messages and emergency calls are accepted unprotected.

The HTTP Digest authentication over TLS is a "tunneled authentication protocol", which is vulnerable to user impersonation attacks if the same user authentication credentials (username and password) are used outside the secure channel (Asokan *et al.* 2002).

The attack works in the following way: the attacker impersonates his victim, initiates an authentication session with a server and receives a challenge. In the next step, the attacker poses to his victim as a legitimate server requesting authentication and challenges the victim with the very challenge that he received. The victim responds with a valid response, which the attacker forwards then to the server and gets access to the desired service.

The major difference from the AKA-based authentication is that AKA provides a cryptographic binding between the authentication data (challenge and response) and the session keys (IK and CK) used to secure the communication channel. If such a cryptographic binding exists, then even if an attacker manages to obtain a valid response, he will not be able to communicate over the secure channel (set up between himself and the server), because he will not be able to derive the correct session keys by just knowing the challenge and the response.

5.1.4 Authentication Mechanism Selection

We have described so far a number of authentication mechanisms that may be used in different deployments depending on the UE capabilities (ISIM application availability, support for IPsec or TLS) and the type of the access network (NGN, early IMS or IMS-compliant):

- IMA AKA;
- Early IMS authentication;
- NASS-bundled authentication (NBA);
- HTTP Digest;
- HTTP Digest over TLS.

In this section we describe the criteria that the P-CSCF and S-CSCF must use to determine the appropriate authentication mechanism. If an *Authorization* header is present in the REGISTER message and if it contains a *Security-Client* parameter, the IPsec or TLS will be used, depending on the negotiated security mechanism, "ipsec-3gpp" or "tls" (see section 5.1.1.3).

Early IMS (see section 5.1.2.1) or NBA (see section 5.1.2.2) authentication procedures are executed if the REGISTER originates from a 3GPP or NGN network. The P-CSCF can determine the type of the access network based on the IP address range of the UE originating the REGISTER or by having the different access networks connected to different network interfaces.

Upon receiving a REGISTER request, the S-CSCF first checks the presence of the *Authorization* header. If it is present, and if it contains an *integrity-protected* parameter, then the AKA authentication is performed (see section 5.1.1).

If the *Authorization* header is present but it does not contain an *integrity-protected* parameter, the HTTP digest authentication mechanism is used (see section 5.1.3). If the second REGISTER contains an *integrity-protected="tls-pending"* parameter, then the

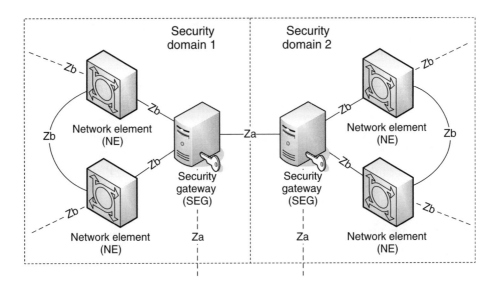

Figure 5.9   The network domain security architecture

S-CSCF will conclude that TLS is being used between the UE and the P-CSCF (see section 5.1.3.1) and the challenging of non-REGISTER requests is not necessary.

If a *P-Access-Network-Info* containing a *network-provided* parameter is present, then the NBA authentication procedure is also employed (see section 5.1.2.2). If the *Authorization* header is missing, then the UE is authenticated using the early IMS mechanism (see section 5.1.2.1).

5.2   Network Security in IMS

The IMS network domain security is concerned with protecting the signaling between the various IMS control plane entities that are located in the core network, denoted as Network Elements (NEs).

3GPP defines for this purpose the Za and Zb interfaces (33.210 TS 2008), which are IPsec tunnels negotiated with IKE. The network domain security is built around the concept of security domains. A security domain defines the set of NEs that are under the control of the same administrative entity, like for instance a certain operator. The communication between two different security domains is done through Security Gateways (SEGs) over the Za interface, whereas NEs and SEGs located within the same security domain communicate over the Zb interface (see Figure 5.9). The architecture does not exclude the possibility that an NE may be physically collocated with an SEG.

The traffic between two NEs located in two different domains will be protected in a hop-by-hop fashion along an NE–SEG Zb interface in the originating domain, one inter-domain SEG–SEG Za interface and one SEG–NE Zb interface in the terminating domain.

The Za and Zb interfaces use tunnel mode ESP. For Za, support for confidentiality is recommended and integrity protection is mandatory. Implementation of Zb is optional; if implemented, integrity protection is mandatory and encryption is optional.

The mutual authentication method between Za/Zb endpoints specified in (33.210 TS 2008) is based on preshared keys. There are, however, well-known limitations associated with the use of preshared keys, related to:

- Scalability–when a different preshared key is used for the mutual authentication of each pair of Za/Zb endpoints, the number of preshared keys in the system is growing in a quadratic fashion with the growth of the number of Za/Zb endpoints.
- Key management–a secure out-of-bound mechanism is necessary in order to distribute the preshared keys to the peers.

An alternative to the preshared key mutual authentication is the set-up of a PKI infrastructure based on cross-certification (33.310 TS 2008).

Cross-certification denotes the procedure by which two local CAs that decide to trust each other sign each other's public key (see section 2.1.2.3) and make available the resulting cross-signed certificates to the entities in their own domain. Cross-certification enables two SEG entities located in two different trusted domains to establish an IKE session using the signature-based authentication scheme (see section 2.2.1.2).

Setting up a PKI infrastructure requires each IMS domain to be configured with a number of logically different CAs (whose functions may be combined):

- A SEG CA that issues certificates to the local SEG entities.
- A TLS client CA and a TLS server CA that issue certificates to the TLS clients and TLS servers. While not mandated by the IMS network domain security architecture, TLS may be used to interconnect IMS with non-IMS networks.
- An Interconnection CA (ICA), which issues cross-certificates to the SEG CAs, TLS client CAs and TLS server CAs located in those domains with whom the local domain establishes a trust relationship.

The trust relationship between IMS domains using a PKI infrastructure may be established in two possible ways, using:

- direct cross-certification–a direct trust relationship is established through cross certification between every two domains that decide to trust each other;
- cross-certification with a bridge CA–multiple domains establish a direct trust relationship with a bridge CA, so that every domain can then establish a transitive trust relationship with any other domain that has cross-certified with the same bridge CA.

Figure 5.10 illustrates how a trust relationship is established between domains A and B through direct cross-certification and how two Za endpoints SEG_A and SEG_B get mutually authenticated as a result.

First, the SEG CAs receive certificate requests from the local SEGs (steps 1a, 1b) and issue certificates to them (steps 2a, 2b). These certificates are denoted $\{SEG_A\}SEGCA_A$ and $\{SEG_B\}SEGCA_B$. Each SEG is also configured with the root certificate of the

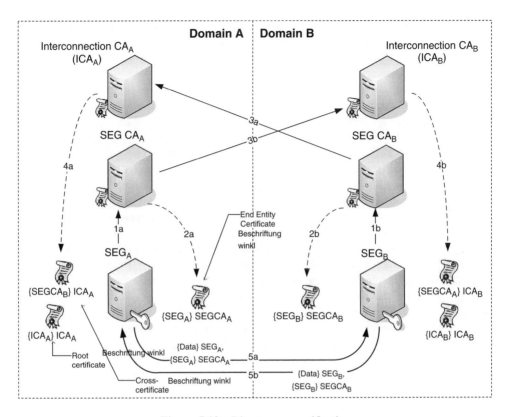

Figure 5.10   Direct cross-certification

local ICA, which is the self-signed certificate of the ICA, denoted as {ICA$_A$}ICA$_A$ and {ICA$_B$}ICA$_B$.

As part of establishing a roaming agreement, each of the two Interconnection CAs receives a certificate request from the SEG CA in the peer domain (steps 3a, 3b) and generates a cross-certificate, {SEGCA$_B$}ICA$_A$ and {SEGCA$_A$}ICA$_B$. The cross-certificates are stored by each ICA in a certificate repository from where the entities in the local domain can retrieve them (steps 4a, 4b).

During IKE Main mode using signature-based authentication, the SEGs exchange their certificates (steps 5a, 5b) and use the cross-certificates retrieved from the local repository to build a chain of certificates that trace back to the local root certificate. In this way, SEG$_B$ builds {SEG$_A$}SEGCA$_A$, {SEGCA$_A$}ICA$_B$ and {ICA$_B$}ICA$_B$, and SEG$_A$ does similarly.

Cross-certification with a Bridge CA (BCA) leads to a more scalable architecture because each security domain only has to cross-certify with the BCA regardless of the total number of trusted domains. There are, however, also a number of shortcomings associated with the use of a BCA.

First, the certificate chain gets longer. Considering an example similar to that in Figure 5.10 but where domain A and domain B cross-certify with a BCA (rather than directly), the certificate chain that SEG$_B$ will have to verify in order to authenticate SEG$_A$ will be {SEG$_A$}SEGCA$_A$, {SEGCA$_A$}BCA, {BCA}ICA$_B$, {ICA$_B$}ICA$_B$.

It may be observed that the certificate chain is one link longer than in the case of direct cross-certification, increasing the complexity of the certificate verification process. Also, the IKE messages will now have to carry two certificates instead of one, leading to undesirable IKE message fragmentation.

The second and more delicate aspect is that a domain that cross-certifies with a BCA will automatically trust any other domain that subsequently cross-certifies with the same BCA. This requires an additional access control function to be implemented on the SEG entities in order to ensure that IKE sessions are established only with the legitimate roaming partners. This decoupling between the trust and the access control leads to a so-called "extended trust model".

For an IKE session to be authorized, the subject field of the certificate presented by the peer SEG must contain the identity of a roaming partner. The problem is, however, that a malicious domain that cross-certifies with the same BCA can issue to its SEG entities fake certificates containing a subject name that belongs to a legitimate roaming partner of the victim, rendering the access control scheme vulnerable to impersonation.

In order to protect against this type of theft of identity, the BCA must include a *name-Constraints* extension in the certificates that it cross-signs with each domain, for instance attaching the name constraint "A" to {SEGCA$_A$}BCA. The *nameConstraints* extension (see section 2.1.2.4) defines a name space within which all subject names in subsequent certificates in a certification chain must be located. In this way the domain A will not be able to provide in the {SEG$_A$}SEGCA$_A$ certificate subject names outside of the "A" name space.

6

User Identity in SIP

identity: noun (pl. identities) 1 the fact of being who or what a person or thing is. 2 the characteristics determining this. 3 a close similarity or affinity (Compact Oxford English Dictionary of Current English 2005).

In the context of VoIP and other SIP-based services, an identity can describe a user, a group of users, a service or a device. In some cases a user has only one identity that is used both for identifying the user toward other users as well as for all service-related aspects such as access control, charging and billing. In other cases such as in the IP Multimedia Subsystem, see Chapter 4, one distinguishes between a public identity that is used as the user's identification toward other users and a private identity that is used only by the service provider for charging, billing and access control issues.

When considering the security aspects of an identity one can distinguish three issues:

- Identity theft–by stealing the identity of another user, an attacker can pretend to be that user and benefit from any privileges this user might have. This might include getting access to certain services or misusing the trust that other users might have placed in the user.
- Subscription theft–subscription theft is a severe case of identity theft in which the attacker can not only pretend to be another user and charge the costs to that user but receive calls destined for that user as well.
- Privacy–while a callee would want to have some assurance that the identity of a caller is the true one, a caller might often want to hide her identity from the callee and untrusted service providers. To ensure the privacy of the caller, her identity information has to be concealed in such a manner as to hide the information from the callee or untrusted service providers but still enable the service provider of the user to correctly route and bill the user's calls.

6.1 Identity Theft

The ten-thousand foot view of a call establishment in both PSTN and SIP networks are rather similar. After dialing a number, the end devices send some signaling information to

SIP Security   Dorgham Sisalem, John Floroiu, Jiri Kuthan, Ulrich Abend and Henning Schulzrinne
© 2009 John Wiley & Sons, Ltd

the service provider of the caller. The operator of the caller then forwards some signaling information to the service provider of the callee which alerts the callee about the call.

In PSTN one distinguishes between user-to-network signaling, for example Q.931 (ITU-T Rec. Q.931 1998), and network-to-network signaling, SS7 (Russell 1995). The identity of the phone line, the phone number or calling line identification could in principle be manipulated by the caller, and the user-to-network signaling messages could contain a manipulated phone number. This identity information is, however, not used by the service provider. The service provider includes in the network to network signaling messages identity information that is retrieved from the subscription information of the owner of the calling line. This information is maintained by the service operator and cannot be manipulated by the caller. As service providers accept signaling messages only from other service providers, the caller cannot send manipulated signaling information to the service provider of the callee. This closed structure makes identity theft in PSTN networks difficult and requires the attacker to get access to the service provider's networking equipment in order to manipulate the network-to-network signaling messages.

Figure 6.1 illustrates a typical call signaling flow with SIP. The user agent sends an INVITE request to the proxy of the user's service provider with the user's identity included in the *From* header. The proxy authenticates the user using HTTP Digest, for example, and verifies that the identity included in the *From* header belongs to the user. If the user is using an identity that is not her own then the proxy can reject the request. If the user is successfully authenticated, the request is forwarded to the proxy at the callee's service provider, which finally forwards the call to the callee. In theory, the checking done by the caller's home proxy should prevent identity theft in SIP in a similar manner to PSTN. In reality, though, this is not the case.

While it can be considered as a good practice to authenticate all INVITE messages and check the identity included in the *From* header, there is no obligation to do so. Actually, some VoIP service providers only authenticate calls to the PSTN as such calls incur costs and require charging and billing. Calls to other VoIP users are often not authenticated as these calls are usually offered free of charge. Further, even though it would be grossly negligent not to do so, a service provider that authenticates all INVITE messages is not obliged to check the identity in the *From* header. Considering that the identity included in an INVITE request is usually configured by the user in the user agent, it becomes obvious that the level of trust that a callee can place in the identity included in the *From* header of an incoming INVITE depends on the trust the callee can place in the service provider of the caller.

To make things worse, a caller can actually even evade the authentication and checks at her service provider completely. The caller can locate the SIP component responsible for serving the callee, SIP proxy or SBC, using DNS and send the INVITE requests to that component directly, as illustrated in Figure 6.1. Unless this component has some restrictions on the sources of incoming SIP requests, it cannot distinguish whether a request has arrived from a SIP proxy or from a user agent directly. Even if such restrictions exist, that is calls from caller.com must arrive through the proxy serving caller.com, the attacker can easily overcome such restrictions. By spoofing the source IP address in the INVITE request to the IP address of the proxy of caller.com and adding an additional *Via* header pointing to that proxy in addition to its own, an attacker can convince the proxy at callee.com that the request went through the proxy of caller.com and hence is to

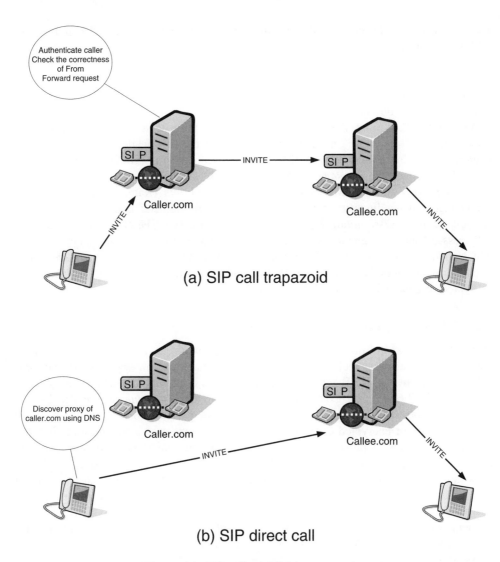

Figure 6.1   SIP call establishment scenarios

be trusted. The reply to the INVITE request will still reach the attacker as the proxy of callee.com would forward the reply based on the *Via* headers to the proxy of caller.com, which will forward it to the attacker's address listed in the second *Via* header.

6.2   Identity Authentication using S/MIME

The need for secure and authenticated identities was already identified in the early stages of the SIP standardization. The first version of the SIP RFC (Handley *et al.* 1999) specified the possibility of using Pretty Good Privacy (PGP) (Atkins *et al.* 1996) for securing the identity of the caller as well as signing and encrypting SIP messages. The usage of PGP was deprecated in the current SIP specification (Rosenberg *et al.* 2002b) in favor of S/MIME (Ramsdell 1999).

S/MIME provides the necessary mechanisms for ensuring the integrity and confidentiality of application-level information such as email or SIP messages. Using her own private key the user can sign her transmitted SIP messages. For encrypting a message the sender uses the public key of the receiver.

When S/MIME is used, the body of a SIP message must be structured as MIME bodies (Freed and Borenstein 1996). In order to protect the SIP headers, the SIP message is encapsulated in a *message/sip* MIME body (Rosenberg *et al.* 2002b). If confidentiality is desired, then the *message/sip* MIME body is carried in encrypted form inside a *application/pkcs7-mime* MIME body of type *enveloped-data* (Ramsdell 2004). The signature that provides integrity protection and data origin authentication is transported in an *application/pkcs7-signature* MIME body (Ramsdell 2004). The clear text or encrypted SIP message and the signature are encapsulated in the SIP message, the content type of which shall be *multipart/signed*. If confidentiality is provided to the SIP message, sensitive information may be removed from the headers of the SIP message that carries the encrypted original SIP message (see section 6.2.1).

6.2.1 Providing Encryption with S/MIME

Using the public key of the receiver, a sender of a SIP message can encrypt parts of or the whole SIP message and include the encrypted information as an *application/pkcs-mime* MIME part of the S/MIME type *enveloped-data*. As some of the SIP headers such as *From*, *To*, *Call-ID*, *CSeq*, *Contact*, *Via*, *Route* and *Record-Route* are needed by SIP proxies for correctly routing the messages, it is not possible to send a SIP message consisting only of encrypted content. To encrypt the content of a SIP message but still enable SIP proxies to process and forward the message, encrypted SIP messages are structured so as to contain an outer part and an inner one. The outer part includes all the headers needed for correctly processing and forwarding the SIP message. The inner part includes all headers of the outer part, additional headers that are not needed for the routing process and the SIP body parts such as the SDP part. To be able to treat the SIP headers as a MIME body that can be encrypted with S/MIME, these headers are included in the inner part as a MIME part of type *message/sip*.

As illustrated in Figure 6.2 the sender encrypts the SIP headers and SDP body and includes them as an *application/pkcs-mime* MIME part. If the caller wishes to keep his identity confidential, he provides in the *From* header in the outer part a SIP URI that contains no personal information, such as anonymous@anonymous.com. The actual user name that should be displayed to the receiver is then included in the encrypted part of the message.

When receiving a SIP request with an encrypted body the callee uses his private key to decrypt the message. Headers included in the encrypted part and that are not listed in the outer part or are not needed by the SIP proxies for routing should be used for any user agent related processing. These headers should not be manipulated by SIP proxies and include *Subject, Reply-To, Organization, Accept, Accept-Encoding, Accept-Language, Alert-Info, Error-Info, Authentication-Info, Expires, In-Reply-To, Require, Supported, Unsupported, Retry-After, User-Agent, Server* and *Warning*.

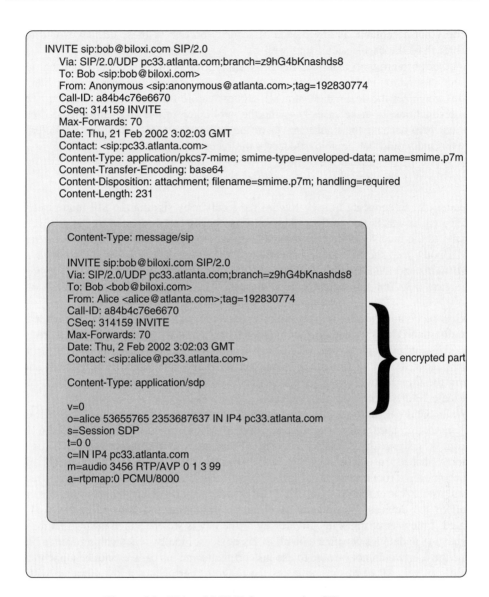

INVITE sip:bob@biloxi.com SIP/2.0
   Via: SIP/2.0/UDP pc33.atlanta.com;branch=z9hG4bKnashds8
   To: Bob <sip:bob@biloxi.com>
   From: Anonymous <sip:anonymous@atlanta.com>;tag=192830774
   Call-ID: a84b4c76e6670
   CSeq: 314159 INVITE
   Max-Forwards: 70
   Date: Thu, 21 Feb 2002 3:02:03 GMT
   Contact: <sip:pc33.atlanta.com>
   Content-Type: application/pkcs7-mime; smime-type=enveloped-data; name=smime.p7m
   Content-Transfer-Encoding: base64
   Content-Disposition: attachment; filename=smime.p7m; handling=required
   Content-Length: 231

Content-Type: message/sip

INVITE sip:bob@biloxi.com SIP/2.0
Via: SIP/2.0/UDP pc33.atlanta.com;branch=z9hG4bKnashds8
To: Bob <bob@biloxi.com>
From: Alice <alice@atlanta.com>;tag=192830774
Call-ID: a84b4c76e6670
CSeq: 314159 INVITE
Max-Forwards: 70
Date: Thu, 2 Feb 2002 3:02:03 GMT
Contact: <sip:alice@pc33.atlanta.com>

Content-Type: application/sdp

v=0
o=alice 53655765 2353687637 IN IP4 pc33.atlanta.com
s=Session SDP
t=0 0
c=IN IP4 pc33.atlanta.com
m=audio 3456 RTP/AVP 0 1 3 99
a=rtpmap:0 PCMU/8000

} encrypted part

Figure 6.2 Using S/MIME for encrypting SIP messages

Sending the SDP information encrypted improves the confidentiality of the communication as the communicating parties can hide the IP addresses used for exchanging the media streams. However, various components in a SIP network such as application layer gateways require this information for enabling the establishment of SIP sessions across NATs. Session border controllers might want to control the type and amount of media sent by the user. In such cases, encrypting the SDP bodies requires these components to have access to the private keys of the receivers or the session establishment would fail. Saving a user's private key on a session border controller, NAT or SIP proxy is, however, generally not acceptable as any breach of security on one of these devices would result in a breach of the user's security as well.

Another problem with using S/MIME for encryption is that the caller might not know the final destination in advance, because calls may be retargeted (that is, intermediate SIP entities may alter the destination of a SIP request before forwarding it) or forked to multiple destinations. In these cases knowing the public key of the originally intended callee does not help because the final receiver of the call may posses a different public/private key pair and would not be able to decrypt the received request.

6.2.2 Providing Integrity and Authentication with S/MIME

A sender can authenticate himself toward the receiver by signing the SIP request using a certificate that was issued to the sender by an authority that is trusted by the receiver. To indicate to the receiver which parts have been signed, the sender uses a *multipart/signed* MIME body with the SIP request including the SIP headers and SDP as a *message/sip* MIME parts and the signature itself as a MIME part of type *application/pkcs7-signature*, see Figure 6.3. The signature includes the user's certificate as well as the hash of the request.

When receiving a signed request the receiver retrieves the certificate of the signer from the request and checks whether it was signed by an authority trusted by him. The receiver checks also whether the owner of the certificate is the same as the identity listed in the *From* header. If any of these conditions are not valid then the SIP user agent should inform the user about these discrepancies and ask for the user's explicit approval of the certificate before processing the request.

While S/MIME provides a secure method for authenticating the caller, S/MIME is implemented in hardly any user agents. For S/MIME to function, the users need to have a public key signed by a trusted authority. There is, however, no one consolidated root authority that is trusted by all users, which means that a VoIP user agent will have to support multiple root certificates. Further, certificates are not usually available for free for private users. A user that has multiple VoIP devices would need to maintain his certificate on all of the devices. Certificates have a time limitation and have to be revoked and updated if the user changes his private key or the key is stolen, which means that the user will have to update his certificate on all of his devices. Finally, when calling from a public phone the user would not be able to use his certificate and his calls cannot be authenticated by the receiver. All of this complicates the usage of certificates and, combined with the financial aspects, makes using S/MIME less interesting for private users.

6.3 Identity Authentication in Trusted Environments

In order to support certain telephony services such calling line identity presentation (CLIP) and restriction (CLIR) (ITU-T Rec. I.250 1995) and fulfill certain regulatory requirements, such as the ability to trace back a call, operators require a trusted user identity. The identity information included in the *From* header can, however, be manipulated by malicious users or anonymized by users who want to keep their identity private and to hide it from the callee, see section 6.6.

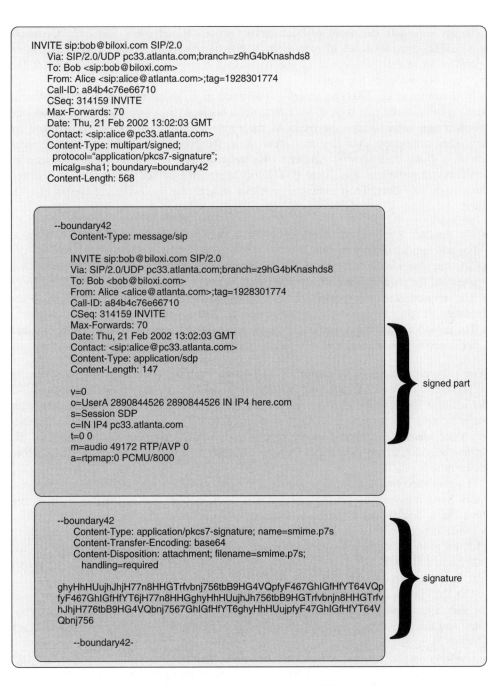

```
INVITE sip:bob@biloxi.com SIP/2.0
   Via: SIP/2.0/UDP pc33.atlanta.com;branch=z9hG4bKnashds8
   To: Bob <sip:bob@biloxi.com>
   From: Alice <sip:alice@atlanta.com>;tag=1928301774
   Call-ID: a84b4c76e66710
   CSeq: 314159 INVITE
   Max-Forwards: 70
   Date: Thu, 21 Feb 2002 13:02:03 GMT
   Contact: <sip:alice@pc33.atlanta.com>
   Content-Type: multipart/signed;
    protocol="application/pkcs7-signature";
    micalg=sha1; boundary=boundary42
   Content-Length: 568
```

```
   --boundary42
      Content-Type: message/sip

      INVITE sip:bob@biloxi.com SIP/2.0
      Via: SIP/2.0/UDP pc33.atlanta.com;branch=z9hG4bKnashds8
      To: Bob <bob@biloxi.com>
      From: Alice <alice@atlanta.com>;tag=1928301774
      Call-ID: a84b4c76e66710
      CSeq: 314159 INVITE
      Max-Forwards: 70
      Date: Thu, 21 Feb 2002 13:02:03 GMT
      Contact: <sip:alice@pc33.atlanta.com>
      Content-Type: application/sdp
      Content-Length: 147

      v=0
      o=UserA 2890844526 2890844526 IN IP4 here.com
      s=Session SDP
      c=IN IP4 pc33.atlanta.com
      t=0 0
      m=audio 49172 RTP/AVP 0
      a=rtpmap:0 PCMU/8000
```
} signed part

```
   --boundary42
      Content-Type: application/pkcs7-signature; name=smime.p7s
      Content-Transfer-Encoding: base64
      Content-Disposition: attachment; filename=smime.p7s;
       handling=required

   ghyHhHUujhJhjH77n8HHGTrfvbnj756tbB9HG4VQpfyF467GhIGfHfYT64VQp
   fyF467GhIGfHfYT6jH77n8HHGghyHhHUujhJh756tbB9HGTrfvbnjn8HHGTrfv
   hJhjH776tbB9HG4VQbnj7567GhIGfHfYT6ghyHhHUujpfyF47GhIGfHfYT64V
   Qbnj756

      --boundary42-
```
} signature

Figure 6.3   Using S/MIME for signing SIP messages

To accommodate the needs of both service providers and users, RFC3325 (Jennings *et al.* 2002) provides a set of extensions to the basic specifications of SIP that enable operators to assert the identity of the users in a trusted manner and still respect the privacy preferences of the users.

In (Jennings *et al.* 2002) an identity is asserted in a similar manner to PSTN. A user wishing to communicate with another must send his request to a proxy at his own service provider that authenticates the user. As the *From* header must not be manipulated by any other component than the user agent itself, the proxy adds the authenticated identity as a *P-Asserted-Identity* header to the request. The request is then forwarded to other service providers. Similar to PSTN, this approach assumes a trust relation between the service providers. Trust indicates here that at least the following conditions are fulfilled:

• The operator that is asserting an identity, i.e. adding the *P-Asserted-Identity* to a request, has authenticated the user first.
• The operator that is forwarding a request including a *P-Asserted-Identity* header has received the request from a trusted server.
• The request was transported in a manner that precludes manipulation by other parties.
• The provider receiving a request including a *P-Asserted-Identity* header will not misuse the asserted information and will respect the user's privacy preferences.

Where the user has multiple identities, then he can indicate in his request which identity he prefers to be used by including a *P-Preferred-Identity* header in the SIP requests. If the identity listed in the *P-Preferred-Identity* is a valid identity of the user, then after authenticating the user, this identity is asserted in the *P-Asserted-Identity*. Otherwise, any valid identity is included or the request is rejected depending on the policy of the service provider. In any case the *P-Preferred-Identity* is removed from the message before forwarding it.

A proxy receiving a request including a *P-Asserted-Identity* header from an untrusted entity must remove this header or replace it with another if it is capable of asserting the caller's identity, see Figure 6.4. To assert the identity all the way to the receiver the communication between the user agents and the proxies must be secured as well, using TLS, for example. Otherwise, an attacker that manages to send SIP requests directly to the callee could also add a *P-Asserted-Identity* with false content or manipulate the messages sent from the proxy to the user agent.

The IMS interworking structure is based on this trust model. Communication between the user agents and the network, represented by the P-CSCF in this case, is authenticated and the IMS domains exchange traffic between each other using IPsec tunnels (see Chapter 5).

Usability of the *P-Asserted-Identity* as well its predecessor, the *Remote-Party-ID* (Marshall *et al.* 2002), is limited to trusted environments. It does not offer a general identity model suitable for all kinds of inter-domain communication and is as strong as the weakest link in the chain. As the identities are not cryptographically certified they can be subject to forgery, replay and falsification if an attacker manages to get access to a proxy in the trusted environment or succeeds in intercepting the traffic exchanged between the trusted parties.

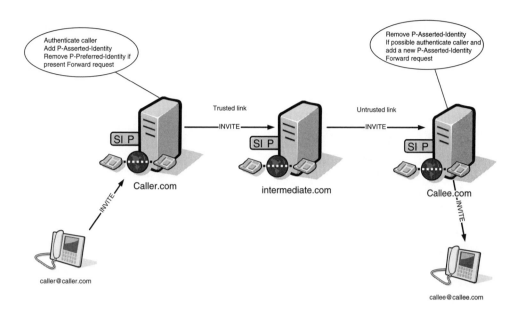

Figure 6.4   Hop-by-hop identity assertion

6.4   Strong Authenticated Identity

Using S/MIME for securing the identity requires the end users to own personal certificates. This has not found wide usage as it incurs costs for the users and requires the user devices to manage different root certificates. The *P-Asserted-Identity* is restricted to trusted environments and can be vulnerable to forgery or falsification. In (Peterson and Jennings 2006) a more general approach is described that provides a cryptographic assertion of the identity without requiring user certificates or a closed trust environment.

In (Peterson and Jennings 2006) a new entity called the authentication service is defined. The authentication service can be expected to be part of the SIP proxy but can be part of any component that has the capability of authenticating the users and possesses the private key of the service provider. If an end device possesses the private key of the service provider, which might be the case for application servers or PSTN gateways, then the end device can act as the authentication service.

As illustrated in Figure 6.5, a caller wishing to contact a callee sends the INVITE message to the authentication service. The authentication service authenticates the user using HTTP digest or by checking whether the request arrived over a secure connection that was established during the registration process. After successfully authenticating the user, the authentication server adds two new headers to the SIP message; namely the *Identity* and *Identity-Info* headers.

The *Identity* header is a signed hash of a canonical "identity string" composed of different parts of the SIP message that are separated by a "|" character. The identity string contains:

- the caller's address of record, i.e. SIP or SIPS URI or any other URI indicating the caller's identity, which is included in the *From* header;

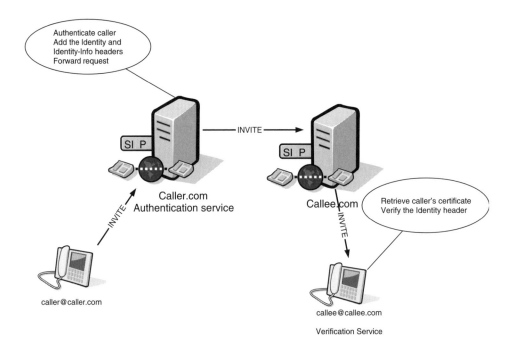

Figure 6.5   Strong identity authentication

- the callee's address of record, i.e. SIP or SIPS URI or any other URI indicating the callee's identity, which is included in the *To* header;
- the information from the *Call-Id* and *CSeq* header fields;
- the information included in the *Date* header–if a *Date* header is not included then the authentication service adds one;
- the address information indicated in the *Contact* header–if this header is empty then the corresponding field in the identity string is left empty as well;
- all the body content, i.e. the SDP part.

Using the private key of the service provider, the authentication service signs the hash calculated over the identity string and adds the *Identity* header to the SIP request. To enable the receiver of the request to check the authenticity of the request, the authentication service adds an *Identity-Info* header that contains a URI that de-references to a resource containing the certificate of the authentication service, as well as the names of the algorithms used for generating the *Identity* header. Currently only the RSA and the SHA1 (NIST 2002) algorithms are specified for the signing and the hash calculation.

On the callee side a verifier instance is used to verify the identity of the sender. The verifier can be part of a proxy or the callee's user agent itself. If the verification step is not conducted by the user agent, then the communication between the verifier and the user agent must be secured to avoid any manipulation of the SIP messages. After receiving a request with an *Identity* header, the verifier inspects the *Identity-Info* header for the location of the certificate of the signing authentication service. After retrieving the certificate the verifier checks whether it has not expired yet, whether the certificate was signed by a trusted certification authority and that it does not appear on a certificate revocation list. Further, the verification service checks whether the owner of the certificate is eligible for authenticating the caller by comparing the domain portion of the URI in the *From* header field of the request with the domain name that is specified in the subject of the certificate. If any of these checks fails then the request is rejected with a *437 Unsupported Certificate* response. Otherwise, the verifier uses the certificate to validate the signature of the request. If the signature validation fails, the request is rejected with a *438 Invalid Identity Header* response.

The *Date*, *Call-Id* and *CSeq* headers are included in the signature to avoid cut-and-paste attacks. Otherwise, the attacker could copy the *Identity* information and add it to a BYE request, for example, to terminate the caller's session.

Asserting the identity of the callee toward the caller is just as important as it is in the other direction. The caller wants to know if the reached callee is actually the one with whom he wanted to communicate. Also, when receiving a request from the callee, a BYE for example, the caller needs to verify that the termination request is actually from the callee and not from some intruder. In theory the same approach could be used for authenticating the identity of the callee. That is, an authentication service could add an *Identity* header in the responses to assert the identity included in the *To* header. However, responses cannot be authenticated. This requires either having the authentication service located on the callee's device or securing the communication between the callee and the authentication service. Even then, this may not always work because requests may be retargeted to some other user during the call establishment. For example with call forwarding, callee@callee.com might forward the call to another VoIP account under callee@callee.net. The SIP specifications in (Rosenberg *et al.* 2002b) mandate that the *To* and *From* headers in the requests and their responses must be the same. In the case of retargeting to callee@callee.net, the response would still have callee@callee.com in the *To* header which cannot be asserted by the authentication service of callee.net.

To assert the identity of the callee (Elwell 2007) proposes an approach in which the callee sends a request, UPDATE or reINVITE, to the caller once a session was established. While (Rosenberg *et al.* 2002b) mandate that the *From* header in in-dialog requests sent by a callee is to be constructed out of the *To* header that was included in the session initiating request, (Elwell 2007) suggests using the identity of the callee with whom the call was finally established, which (Elwell 2007) calls the connected identity. As illustrated in Figure 6.6, the INVITE request gets forwarded from callee.com to callee.net, and once the session is established the callee sends an UPDATE request using as the *From* header the connected identity. This identity is asserted by the authentication service of callee.net and is then verified at the caller side. For this message exchange to be successful the caller must be willing to accept the deviation from the SIP specifications and accept in-dialog requests with *From* headers that differ from the *To* header that he used for setting up the dialog. This willingness is signaled by adding the tag *from-change* in the *Supported* header in its initial INVITE request.

While the *Identity* header enables the exchange of an authenticated identity, it cannot protect itself. An attacker which acts as a man in the middle can strip out the *Identity* and *Identity-Info* headers from a request and the request would still be valid. However, the only harm done in this case would be that the callee would not be able to verify the identity of the caller, which in the worst case would lead him to reject the call.

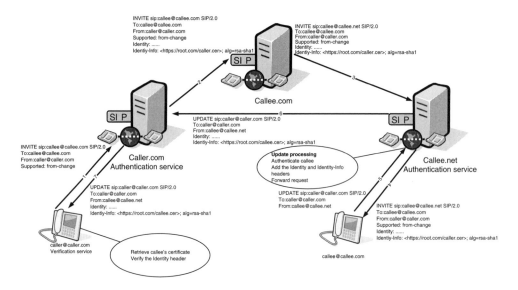

Figure 6.6 Connected identity authentication

By signing not only the SIP headers but also the SIP body, the end points of the media exchange are bound to the signed identity. Without this binding an attacker could still be able to exchange media traffic with the callee by manipulating the contact lines in the SDP parts of the SIP messages. By signing these parts as well, any manipulation of the SDP parts would be detected by the callee and the session establishment would fail.

While this binding increases the security of the communication, it also introduces some deployment issues. VoIP service providers often deploy session border controllers. SBCs are used as intermediate components between user agents or SIP components belonging to other service providers and the internal components of the VoIP provider such as proxies and application servers. Among others, SBCs are used for ensuring that the traffic entering a provider's network conforms to certain profiles, that only a certain media codecs are employed and that only a certain bandwidth is used. To provide these features, the SBCs terminate incoming transactions and initiate new ones toward the internal components. By doing so, they change different headers such as the *Contact* header, and the SDP parts (usually the "c" and "m" lines). Similarly, some of the NAT traversal technologies like TURN (Rosenberg et al. 2008c) require the manipulation of the SDP parts. If the signed SIP requests traverse only two service providers, namely origination and terminating, then the signing of the SIP request at the originating side should be done after all changes of the SIP request have taken place, i.e. after passing through any used RTP relay or SBCs. At the terminating side, the signature should be verified prior to any changes of the SIP request. To still be able to present the callee with an authenticated identity, the edge SBC in the terminating side must securely transport the request to the callee. This means, though, that the identity assertion is no longer conducted on an end-to-end basis but has become a hop-by-hop one and does not provide much more security than using the *P-Asserted-Identity*, see Figure 6.7.

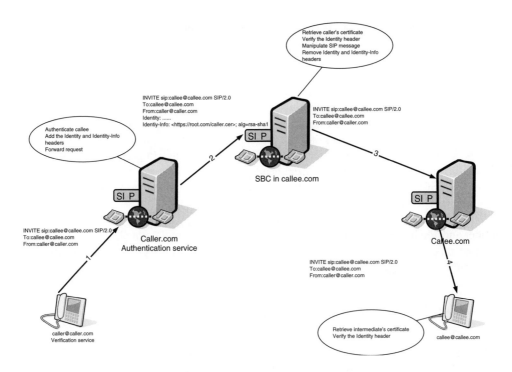

Figure 6.7   Hop-by-hop strong identity

In order to maintain the end-to-end characteristics of the authenticated identity in the presence of SBCs and media relays, an alternative SIP Identity Media mechanism may be used to ensure that the media session is established with the entity identified by the *From* header of the SIP request, which is bound to the caller by the caller's certificate subject (Wing and Kaplan 2008).

The SIP Identity Media mechanism defines a new *Identity-Media-Signature* header, which contains the authentication service's signature over the *From* and *To* address information, the SIP method, the *Date* header and certain SDP attributes such as the $a =$ *fingerprint* attribute that conveys a hash calculated over a certificate of the caller (Lennox 2006). The SDP "c" and "m" lines are not covered by the *Identity-Media-Signature* header.

The authentication service also adds an *Identity-Info* header with a reference to the location of the certificate of the authentication service. By signing the method name an attacker cannot copy the signature and add it to another request, such as BYE, and interfere with an established session. Signing the *Date* header prevents an attacker from replaying a signed request at a later point of time. An attacker could, however, copy the signed identity from one INVITE request and add it to another INVITE request but with a different SDP part. This would cause the callee to believe he is being contacted by an authenticated caller whereas the media would be exchanged with the attacker.

To prevent this, the validation of the authenticated identity in the SIP request is combined with a proof of possession of the private key. This can be achieved using different techniques such as TLS, DTLS and ZRTP (see Chapter 7); we will use DTLS (Rescorla and Modadugu 2006) for exemplification. Once the *Identity-Media-Signature* header was verified by the callee, the callee establishes a DTLS connection to the IP address indicated in the SDP "c" line. During the DTLS handshake the caller (acting as the server) sends his certificate to the callee and uses the corresponding private key to authenticate or encrypt data (depending on the actual key exchange method, see section 2.2.2.1). The callee must verify that hashing this certificate yields the same value as the one received in the $a = fingerprint$ attribute, otherwise the call is terminated.

Note that the certificate is not used for authenticating the caller but solely to provide a binding between the media path and the signaling path, by requiring the caller to prove that he owns the private key that matches the certificate the fingerprint of which was signed in the *Identity-Media-Signature* header. It is therefore enough for the caller to present a self-signed certificate and hence avoid the deployment problems of S/MIME, related to the reliance on a trusted certification authority.

While the attacker could manipulate the SDP part or copy the *Identity-Media-Signature* header and add it to another request, the attacker cannot manipulate the fingerprint attribute without this manipulation being discovered by the callee. For such an attack to succeed the attacker has to get access to private key of the caller.

6.5   Identity Theft Despite Strong Identity

The SIP specifications usually assume that users can be identified using email-like SIP URIs consisting of a textual user name and a domain name in the form of "sip:user@domain". However, as SIP is usually used for providing telephony services, users are very often identified by E.164 numbers (ITU-T Rec. E.164 2005).

Using the mechanism in (Peterson and Jennings 2006) an identity is considered to be authenticated if the signature included in the *Identity* header is successfully validated by the verifier and if the signing certificate was issued to a domain, the name of which is consistent with the domain part of the *From* header field.

When using a phone number as the user identity, the number is used as the user name part of a SIP URI in the form of "sip:+493034637170@caller.com;user=phone". In such a case, the authentication service uses a certificate owned by "caller.com" for asserting the identity. The user could also be identified using a TEL URI, in which case the identity would be "tel:+493034637170". In this case it is less obvious who can assert the identity and generate the Identity header and the verifier would not be able to assess whether the signer is actually eligible to do so. To enable a verifier to still verify the user's identity when a TEL URI is used, there must be some other way to allow a verifier to check which phone numbers are assigned to which domain, such as ENUM (Faltstrom and Mealling 2004).

As E.164 numbers are globally unique, the domain name in a SIP URI becomes of less value. Actually, users will most likely just give out their phone numbers and print these numbers on their business cards without inclusion of the domain name, and their address books will include the phone numbers but not the domain names of their contacts. Further, phones that display the caller identity also show in general only the phone number and not the complete SIP URI.

The fact that the domain name is irrelevant when using an E.164 number as the user identity provides attackers with two possibilities for falsifying their identities, even when authenticated identities are used. In the one scenario, the user indicates an E.164 number as his user name even though the number does not belong to him. The service provider adds the *Identity* header indicating that it has authenticated the user and that the user was authenticated. The request is routed to the callee. As SIP devices usually only display phone numbers but not the domain names, the callee believes that the real owner of the displayed number is calling and picks up the phone. This scenario could happen either because the provider of the attacker is malicious or simply because it does not have the means to check whether the registered phone number by the attacker actually belongs to the attacker. Actually, determining who owns an E.164 number is not always simple. Phone numbers are allocated by national authorities to telecommunications operators, who might reallocate them to other operators. Users receive their phone numbers from their operator and when changing the operator can in most countries port this number to the new operator. This makes it difficult for a VoIP service provider to check whether the phone number indicated by a user as his user name actually belongs to that user. Making things worse, there is no real need for this checking. As a malicious user cannot receive calls destined to the real owner of the phone number, the security implications are not severe. A user with the SIP URI "sip:+493034637170@caller.com" will only receive calls to +493034637170 if a caller explicitly indicated "sip:+493034637170@caller.com" as the callee's address. Calls destined to +493034627170 are routed to the owner of this number in the PSTN.

The second related scenario is illustrated in Figure 6.8. A user with the identity of "sip:+493034637170@caller.com" makes a call to callee@callee.com. The proxy at caller.com authenticates the user and adds an *Identity* header with the signature of the message. The request is then forwarded to another proxy at intermediate.com. Intermediate.com replaces the *From* header in the request with sip:+493034637170 @intermediate.com and also replaces the *Identity* header. This might happen for different reasons, such as topology hiding or routing policies. Topology hiding indicates in general the wish to hide internal information of a service provider from other providers such as the sources of the calls it is forwarding. That is, intermediate.com might not want callee.com to know that it is accepting traffic from caller.com. Routing policies could lead to a change in the *From* header as certain components in a provider's network can only accept traffic if they see its domain name in the *From* header or can only route traffic to other providers if the *From* contains its own domain name. By keeping the used phone number but changing the domain name and resigning the message, the *Identity* header no longer provides end-to-end user authentication but a hop-by-hop one similar to the *P-Assured-Identity*, see section 6.3. As the user name was not changed and assuming that the proxy at intermediate.com is trustworthy, no harm was done here. A malicious proxy could, however, do more harm and manipulate the SIP messages so that all the signaling and media data exchanged between the caller and callee traverse components controlled by it, which allows it to inject traffic and eavesdrop on the exchanged traffic.

The baiting attack (Kaplan and Wing 2008) is another approach that allows an attacker to impersonate the identity of someone else, even when authenticated identities are used. As illustrated in Figure 6.9, in the first phase of the attack, the attacker collects the bait, i.e. a signed identity of someone that is trusted by the target of the attack, say a bank. This can be achieved, for example, by asking a bank to call one back by filling in a web form. The bank calls back and includes a signed identity in the INVITE request. The attacker rejects the call but uses the signed identity to send a request to the target. The

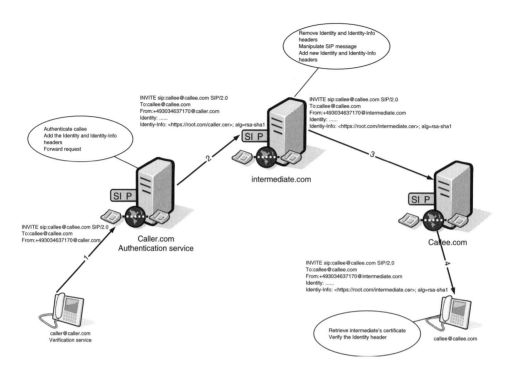

Figure 6.8 Resign attack

request sent to the target is the same INVITE received by the attacker from the bank but with an additional *Via* header included in the request. Even though the *To* header in the request sent by the attacker to the target does not include the target's URI, this will in general be ignored by user agents. The *To* address does not always have to match the user agent's address as the INVITE request might have been forwarded from some other address. After verifying the signature, the phone rings and the callee sends a positive reply including an SDP part indicating the address the callee wants to use for receiving media data. As the attacker has added itself to the *Via* list, the reply will reach the attacker, which generates an ACK request and drops the reply. At this point the attacker can start sending media data to the target, which believes the media to be coming from the bank. The baiting attack enables only a one-way communication with the target. As the attacker cannot forge the SDP part in his own INVITE request, any media sent from the target would be destined to the bank, which would drop it. However, this already enables the attacker to send a prerecorded message to the target informing him that he should call some number controlled by the attacker to confirm the account information, for example. A two-way communication can be achieved if the attacker manages to get a bait that is not bound to an SDP body. This would be the case if the attacker managed to convince the bank, for example, to send a signed NOTIFY.

Using the combination of identity authentication and proof of ownership using the media path (Wing and Kaplan 2008) can prevent the baiting attack as the target would

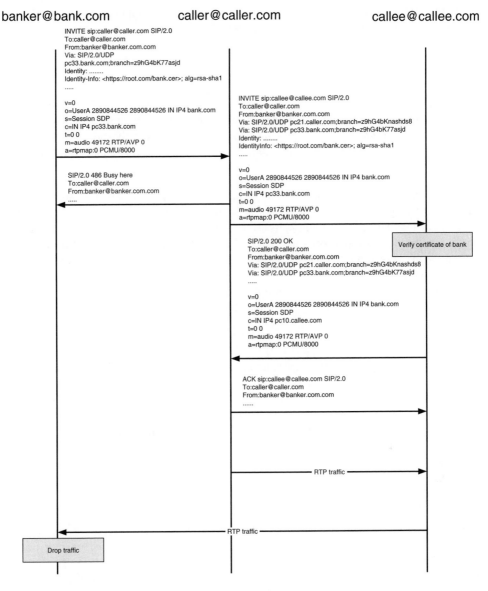

Figure 6.9   Baiting attack

not accept any media from the attacker until a DTLS connection was established with the bank first.

6.6   User Privacy and Anonymity

Alan Westin (1970) defines privacy as *the claim of individuals, groups, or institutions to determine for themselves when, how, and to what extent information about them is communicated to others*. In the context of SIP we can distinguish between different information that a user might want to keep private:

1. Personal information–personal information include the user's identity, equipment and workplace. Such information is carried in the SIP messages in headers such as *To, From, Call-Id, Contact, Organization, Server* or *User-Agent*.
2. Location information–location information indicates the IP address, used hosts, traversed networks and service provider of the user. This information is included in the *Via, Contact, Route* and *Record-Route* headers as well as the SDP part.
3. Call information–call information indicates that a user is making a call and with whom he is communicating.

A user might want to hide some or all of this information. A caller making a call to a call center might want to hide his identity so that the call center will not misuse this identity for making advertisement calls. A user that forwards his calls to another address might not want callers to know this final address so that they do not use it in the future to bypass any screening or other access control measures that might be implemented at his public address.

In terms of privacy in SIP there are four relevant parties, namely the caller, callee and the SIP service providers of the caller and the callee. The caller or callee can choose different privacy scenarios in which they hide information from each other, from the service providers or from both.

• Service-provider privacy–the user wants to reveal his identity and location information to other users but not to the service providers.
• User privacy–the user wants to reveal his identity and location information to his own service provider but not to other users or the service providers of these users.
• Absolute privacy–the user does not want to reveal information to either the service providers or other users.

These scenarios can be realized in either a user or network-provided manner

6.6.1 *User-provided Privacy*

With the user-provided privacy the user himself is responsible for hiding his personal information by using S/MIME, as described in section 6.2. Instead of placing private information in the SIP headers, the caller populates, for example, the *From* header with an anonymous value. The actual *From* identity is then revealed only to the callee after decrypting the S/MIME body part. By hiding the SDP part and *Contact* information the service providers cannot determine the addresses at which the caller wants to receive media data or SIP requests from the callee.

User provided privacy is most appropriate for the service provider privacy scenario but not for the user or absolute privacy scenarios. The caller can hide personal information from the callee, i.e. by not including *Organization* or *User-Agent* headers in the outer part of the S/MIME protected request. The callee will, however, still have access to the caller's identity that is included in the certificate used for encrypting the data.

Besides the various deployment issues related to S/MIME as described in section 6.2, this approach provides only for limited privacy. The user will not be able to hide routing related headers such as the *Via* header as otherwise it would not be possible to route responses back to the caller. As the first entry in the *Via* list includes the user's address to which responses are to be sent the service provider has access to user location information. Further, SIP components used for security, firewall and NAT traversal must sometimes inspect the SDP bodies. With an encrypted SDP part, these components will not be able to conduct their tasks and the session establishment might fail.

6.6.2 Network-provided Privacy

With user-provided privacy it is not possible to hide or encrypt routing related headers such as *Via* and *Route* as otherwise the SIP infrastructure would not be able to correctly route the SIP requests. These headers reveal information about the user's identity and location, making user-provided privacy of limited use.

To achieve a higher level of privacy and anonymity (Peterson 2002) suggests using a privacy service for anonymizing personal and location information in SIP messages. This service can be provided as part of the user's service provider SIP components or by a third-party provider. In any case, all requests that require anonymization must traverse the privacy service. The privacy service can anonymize different user-related information so that, from the point of view of the callee, the communication is between the privacy service and the callee. The privacy service can provide one or more of the following services:

- Annoymization of personal information—the privacy service anonymizes all headers that have not been already anonymized by the user agent. This includes either removing or changing some headers such as *Organization* or *User-Agent*. The content of the *From* and *Call-Id* headers might include the user's identity and is used for identifying the dialog. Hence, when sending the response back to the sender, the privacy service must substitute the anonymized values with the original values again as otherwise the user agent client would not be able to relate the responses to the proper transaction.
- Anonymization of location information—the privacy service replaces routing-related headers that include the user's location with headers that point to the privacy service. The *Contact* header pointing to the user is replaced by the address of the privacy service. The *Via* list is removed from the request and a single *Via* entry including the SIP URI of the privacy header is added.
- Anonymization of media information—the SDP part sent by the user agent includes the IP address the user is expecting to receive media data on. To hide this information the privacy service replaces the IP address and port numbers included in the SDP part with address information of a component that is under its control. This component is then responsible for routing the RTP traffic between the caller and callee.

Besides anonymizing and forwarding requests from the caller to the callee, the privacy service routes responses and requests from the callee to the caller. This requires the privacy service to be able to reconstruct any removed information. While some information is needed only for the duration of a transaction, other information might be needed for the duration of the complete dialog. Once a reply is received, the privacy service must be able to reconstruct the removed *Via* entries, add them to the reply and forward the reply based the top-most *Via* entry. As the privacy service also acts as the contact point for the caller, any requests sent by the callee will be addressed to the privacy service. The privacy service must then be capably of identifying which dialog the received requests belong to and routing the requests to the right user agent. This requires the privacy service to keep information about the established dialogs for the entire duration of the dialog. A privacy service could keep this information in one of two ways:

- The privacy service saves all information related to the established dialogs and removed headers locally. Once a reply or request is received, the privacy service maps the incoming message to the saved information, replaces any anonymized or removed information and forwards the message to the caller. This is achieved using the B2BUA concept in which a dialog is established between the caller and the B2BUA and another one between the B2BUA and the callee. Having to keep the dialog information increases the implementation complexity of the privacy service. In case of a failure of the privacy service, the dialog information will be lost and the already established dialogs cannot be terminated properly. To avoid this situation the privacy service will have to be provided in a highly available manner with the dialog information made available to a backup server that can take over the functionality of the privacy service in case of a failure.
- Instead of saving the dialog information locally, the privacy service can also include the personal and location information in an encrypted form in the forwarded requests. This could be achieved by encrypting the removed or anonymized information and adding the information as extension fields to the headers in the forwarded requests. Taking the example of the *Via* list, the privacy service would remove the *Via* list in the incoming request, encrypt the removed *Via* list using the secret key of the privacy, add a *Via* header indicating its own address and add the encrypted list to its own *Via* entry as an extension field. This field is then also included in the *Via* entry received in the response and the privacy service can decrypt it and use the resulting *Via* list for forwarding the reply. The same applies for the *Contact* header where the original contact address included in the request is replaced by the contact address of the privacy service and the user's contact address is added as an extension field in an encrypted form. While this does not require the privacy service to keep dialog state information it increases the complexity of parsing the SIP messages and fails if the encrypted information is stripped out of the messages. This could happen if some end device does not include the encrypted information in its responses or some proxy or SBC removes these parts due to some strict policy of what these components expect to see in a SIP request, for example.

Where the privacy service was provided by the service provider of the caller, the caller might have some agreement with the provider that indicates that all requests sent by the caller should be anonymized. A caller can also use a privacy service provided by a third party by routing his requests through the privacy service by indicating the address of the privacy service in a *Route* header. A callee can also use a privacy service by asking his service provider to forward all incoming requests to the privacy service or registering under his public identity a SIP URI that references the privacy service. The actual location of the callee must then be communicated to the privacy service.

To control the level of needed privacy, (Peterson 2002) extends SIP with an additional *Privacy* header in which the user can indicate one or more parameters:

- *Headers* – the user requests that the privacy service obscures only the personal and location information that cannot be anonymized by the user agent such as *Via* or *Contact*.

- *Session*–by indicating the session parameter, the user asks the privacy service to anonymize the media information as well, i.e. the IP and port numbers indicated in the SDP part.
- *User*–this is set by a SIP proxy, for example, to indicate to the privacy service that the user agent is not capable of annonymizing any headers and that the privacy service should do this on his behalf.
- *None*–with the *none* parameter the user agent indicates to the privacy service that it should not apply any privacy services.
- *Critical*–the user asserts that providing the privacy services is critical. Hence, if the required level of privacy can not be provided, the request should be rejected.
- *Id*–this parameter was added by (Jennings *et al.* 2002) and is used with the *P-Asserted-Identity*. If present, it indicates that the user does not want the asserted identity to be forwarded to nontrusted networks.

While using the privacy service as described in (Peterson 2002) provides a considerably higher level of privacy than the user-provided privacy, it still does not provide for absolute privacy. The privacy service itself has knowledge about the caller and the callee. If an attacker manages to get access to the privacy service then it can intrude on the privacy of the users. Also, by observing the traffic coming into and out of the privacy service, an attacker can collect sufficient information to guess which users are communicating with each other. Approaches such as Crowds (Reiter and Rubin 1998), Hords (Levine and Shields 2002) or Mist (Al-Muhtadi *et al.* 2002) were proposed as possible solutions for anonymizing web traffic and other sorts of Internet applications. In (Kazatzopoulos *et al.* 2008) the authors describe an approach for integrating Mist with SIP. As a result of the integration, some components have knowledge about the identity of the caller but not his location and some other components have knowledge about the location of the communicating parties but not his identity. An attacker that wishes to identify the location of a caller with whom he is communicating needs to get access to multiple components.

6.7 Subscription Theft

To be able to receive calls a user needs to bind his current location, i.e. IP address, to his SIP identity. This is achieved by sending a REGISTER request to the registrar server of his VoIP provider. Without any authentication, the attacker can assume the identity of any user and can, hence, receive all calls destined for that user. Further, if no authentication is in place, an attacker can simply issue calls on behalf of some other user by including the identity of that user in his INVITE requests.

To prevent these cases of fraud, SIP services deploy user authentication services. As explained in Chapter 3, when a SIP registrar or proxy receives a request, it challenges the issuers of the request. Without the inclusion of proper credentials the request is rejected. The attacker would need to steal the password shared between the user and the service provider in order to be able to assume the user's identity.

Besides stealing the password, an attacker can assume the identity of another user by misusing vulnerabilities in the authentication algorithm used in SIP.

```
INVITE sip:callee@voip.com SIP/2.0
  Via: SIP/2.0/UDP
client.provider.com:5060;branch=z9hG4bK74b03
  Max-Forwards: 70
  From: subscriber
<sip:subscriber@provider.com>;tag=9fxced76sl
  To: callee <sip:callee@voip.com>
  Call-ID: 3xRb9vxSit55XU8o9@provider.com
  CSeq: 1 INVITE
  Contact: <sip:subscriber@provider.com>
  Proxy-Authorization: Digest username="subscriber",
      realm="provider.com",
      nonce="wf84f1ceczx41ae6cbe5aea9c8e88d359",
opaque="",
      uri="sip:callee@voip.com",
      response="42ce3cef44b22f50c6a6071bc8"
  Content-Type: application/sdp
  Content-Length: 151

  v=0
  o=subscriber 2890844526 2890844526 IN IP4 provider.com
  s=-
  c=IN IP4 191.0.3.102
  t=0 0
  m=audio 48181 RTP/AVP 0
  a=rtpmap:0 PCMU/8000
```

```
INVITE sip:malicious@fraud.com SIP/2.0
  Via: SIP/2.0/UDP
client.provider.com:5060;branch=z9hG4bK73464
  Max-Forwards: 70
  From: subscriber
<sip:subscriber@provider.com>;tag=9fxced76sl
  To: malicious <sip:malicious@fraud.com>
  Call-ID: 3xRb9vxSit55XL98X@provider.com
  CSeq: 1 INVITE
  Contact: <sip:attacker@fraud.com>
  Proxy-Authorization: Digest username="subscriber",
      realm="provider.com",
      nonce="wf84f1ceczx41ae6cbe5aea9c8e88d359",
opaque="",
      uri="sip:callee@voip.com",
      response="42ce3cef44b22f50c6a6071bc8"
  Content-Type: application/sdp
  Content-Length: 151

  v=0
  o=attacker 2890844526 2890844526 IN IP4 fraud.com
  s=-
  c=IN IP4 192.0.4.104
  t=0 0
  m=audio 48181 RTP/AVP 0
  a=rtpmap:0 PCMU/8000
```

Figure 6.10   Replay attacks with SIP

Replay attacks in HTTP Digest authentication   With the HTTP digest authentication scheme, the user credentials are generated by calculating an MD5 hash over a secret shared between the user and his provider, a server nonce and possibly a client nonce. The nonces generated by the SIP servers are usually valid for some period of time. In a replay attack the attacker captures a SIP request that contains valid user credentials. Immediately afterwards, the attacker generates his own requests and includes the captured user credentials. As long as the nonce is still valid, the requests will be accepted. As depicted in Figure 6.10, once an attacker has managed to intercept a request including valid credentials, the attacker can use these credentials to start a session of his own while the nonce is still valid, as in Figure 6.11.

In order to eliminate the possibility of such attacks, a server can indicate its wish to use higher security measures by including a "quality of protection" (*qop*) parameter in its challenge, see section 3.9.1. In this case, a *nonce counter* is added. An authenticating server can detect a replay attack by checking the *nonce counter*. If the same *nonce counter* is used more than once or a message includes an old *nonce counter* then the server can discard the request as a replay attack. While using this approach eliminates the chances of a replay attack, it requires the authenticating server to maintain a nonce and a counter for each user for some time. An attacker can, hence, misuse this feature to launch a memory exhaustion attack by sending a large amount of requests with different user identities. Another option for reducing the dangers of reply attacks is using predictive nonces. With predictive nonces, the nonce is created based on SIP headers that do not change when resubmitting a request and the time of day. By including the time in the nonce the proxy can detect old credentials and refuse them if they are older than a certain time. Using the *To*, *From*, the SDP body and the *CSeq* methods, the attacker will only be able to initiate calls with these headers and will not be able to receive the media sent by the callee as it is directed to the owner of the credentials.

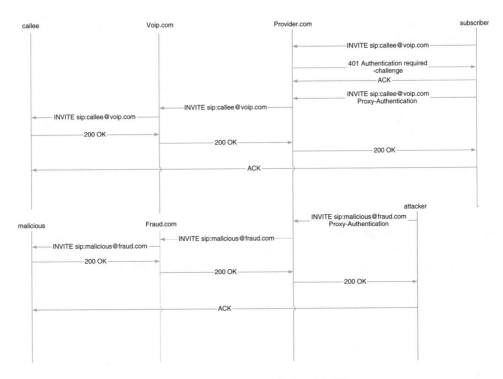

Figure 6.11   Replay attacks with SIP

Certificate validation in TLS   While using HTTP digest ensures that no entity other than the legitimate one can provide valid credentials, the scheme is still vulnerable to vital SIP headers (like for instance *Contact*, *To*, *From*) being modified in transit. As a result, S/MIME or TLS may be used to provide integrity protection and confidentiality to the SIP messages. S/MIME has been discussed in section 6.2. It provides end-to-end protection but has a number of deployment-related limitations. We will instead focus here on the checks that need to be performed on the TLS certificates.

Authentication in TLS consists of the server presenting the client with a valid certificate signed by a CA that is known to the client (that is, the client is configured with the root certificate of the respective CA). If the client is a proxy, then the server should also request a client certificate. If the client is a UA, then the client is authenticated at the SIP layer using HTTP Digest. In order to make sure that the certificate presented by the remote TLS endpoint indeed belongs to the expected participant, the following checks must be performed on the subject names contained in the certificates:

- In order to register, a UA must first establish a TLS connection to the registrar. When presented with the registrar's certificate, the UA must check that certificate has been issued to a host name that is consistent with the domain name where the UA intends to register. Similar checks must be performed when the UA establishes TLS connections with the inbound and outbound SIP proxies (which are usually functionally collocated with the registrar).

- When forwarding a SIP request to the callee's inbound SIP proxy (denoted in this context as InSP), the caller's outbound proxy (denoted OutSP and acting as a TLS client) must first establish a TLS connection to InSP (which plays the role of the TLS server). OutSP must check that the certificate presented by InSP has been issued to a host name that is consistent with the domain name from the Request-URI. Also, InSP must request a TLS client certificate from OutSP and must check that the certificate presented by OutSP has been issued to a host name that is consistent with the domain name from the *From* header of the SIP request.

Tunnel attacks on HTTP Digest over TLS   Using HTTP Digest over TLS combines the user authentication provided by the HTTP digest with the integrity protection and confidentiality of the SIP signaling provided by TLS. However, this results in a "tunneled authentication protocol", which is vulnerable to relay attacks also known as tunnel attacks. In such an attack an attacker places himself between a server and client, challenges the client to authenticate himself and uses the authentication credentials to present himself under the victim's identity to the server. The attacker can achieve this by calling the victim, requesting it to send periodically re-INVITEs using the Session Timer SIP extension directly (through handcrafted Record-Route header field), challenging the re-INVITEs, collecting credentials, and passing them on to the server. This SIP-specific attack has been described in (Abdelnur *et al.* 2008). General background on this type of attack can be found in (Asokan *et al.* 2002) and (Puthenkulam *et al.* 2003). A reasonable line of defense would be, for example, SIP clients built in a way that only sends credentials for a domain through the domain's SIP servers.

Password theft   In terms of security and access control, a SIP service is very similar to other online services such as online banking or email services. Besides the specific SIP-related theft possibilities, the general approaches for acquiring personal and secret information used in other fraud scenarios can be used as well:

- Eavesdropping on public transactions to obtain personal data. This can be done by looking over the shoulder of a person performing a transaction or by sniffing on VoIP transactions packets if encryption is not used.
- Using viruses and Trojan horses to get private data stored in computers.
- Phishing, i.e. impersonating a trusted organization or a VoIP provider and asking customers for some private information in order to fix a pretended problem.
- Stealing or hacking of equipment.
- Browsing through social networks, for instance web sites of public domains where personal data is posted. This information can then be used to guess the passwords of a user.

6.8   Fraud and SIP

According to the Communication Fraud Control Association (CFCA), in traditional telecommunication networks it is estimated that fraud accounts for annual losses of an average of 3–5 % of the operators' revenue and increasing at a rate of more than 10 % yearly. Hence, with the openness of the VoIP technology one can expect an even higher threat of fraud and higher losses of revenue. Actually, there have already been various press reports about identity theft and misuse of services, which makes fraud the biggest current threat to VoIP providers.[1] In traditional telecommunication service, fraud can take different shapes, including:

- Call selling–the fraudster acquires one or more access lines from the service provider. Using these access lines, he sells high-tariff calls below their market value. The fraudster then evades paying the bills and once the access lines used are closed by the operator, he acquires new lines from some other operator or under a different name.
- Surfing–use of another person's service by illegally obtaining access information.
- Ghosting–obtaining cheap or free service through technical means of deceiving the network, e.g. manipulating a switch or database.
- Premium rate service–with premium rate numbers, the owner of the number receives a certain sum of money from the originating provider for each call destined to that number. The provider collects the costs of the calls from his subscribers afterwards. There have been cases where office cleaners were organized to call these numbers and leave the phone off the hook all night or where the fraudster subscribed to an operator under a false identity and generated these calls himself.

While all of these fraud scenarios can be seen as direct threats to VoIP networks, the VoIP technology adds additional risks as well:

- Service plan misuse–many VoIP services are offered as flat rate services. Operators calculate these plans based on average usage scenarios. While such services are intended for personal use only, fraudsters resell the service to others, resulting in high usage and high losses to the operator.
- Credit card fraud–in this case, fraudsters use the toll-free service of a VoIP provider to find out the PIN numbers of a stolen credit card. That is, the fraudster uses the free call service of a VoIP provider to call a premium service that requires the credit card and PIN number and keeps trying this until the right number is found.

6.8.1 Theft of SIP Services

Stealing the identity of another user allows the attacker to use some service with the costs being charged to someone else. However, the attacker would be limited to the privileges of the stolen identity and all calls conducted by the attacker would have the user's identity as the originator or recipient. Fraudsters would, however, in general like to conduct fraud on a larger scale, e.g. by selling stolen services to other people and hence gaining from the fraud not only free calls but also money. This can be achieved by getting access to the infrastructure components of the SIP service, e.g. SIP proxies, databases or gateways to the PSTN. With such an access, the attacker can manipulate the authentication process so that his calls are not authenticated or are considered as legitimate or can simply ensure that no billing records are generated for his calls.

Recently there have been three patterns for conducting this kind of fraud, namely: password guessing, credential emulation and route misuse.

---

[1] http://voxilla.com/voxilla-stories/voxilla-stories/voip-fraud-the-industrys-best-kept-secret-380.html

Password guessing   SIP components usually have an administration interface that allows the administrator to configure the system, control the privileges of different users and actions and set the logging and billing criteria. This interface is usually protected through a password. Often all devices manufactured by the same company share the same password. Administrators often forget to change this password during the installation process at the provider's premises. By knowing this default password, an attacker can assume the identity of the administrator, which allows him to receive the needed privileges for misusing the service. This kind of fraud can be prevented by changing the password of the SIP components and protecting the administration interface so that it is only accessible over a trusted network link.

Credential emulation   A popular setup for VoIP services is presented in Figure 6.12. In this setup the proxy is responsible for authenticating the incoming requests and forwarding legitimate requests to the PSTN gateway. To indicate to the gateway that a request is legitimate, the proxy adds special information in the forwarded requests. This information is then used by the gateway as an indication of the legitimacy of the request and will, hence, only initiate calls to the PSTN if a request includes this information. A fraudster can detect this information either by guessing or by brute force. By including this information

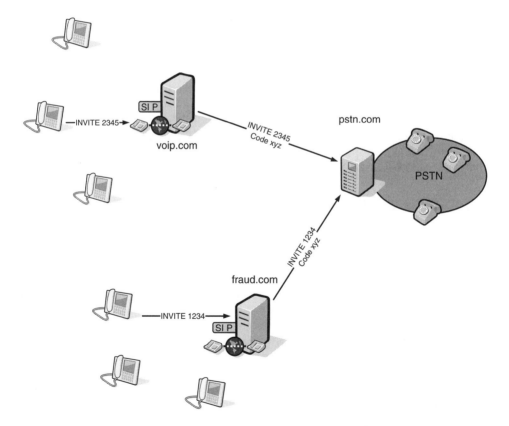

Figure 6.12   Credential emulation attacks with SIP

in his own requests, a fraudster can fool a gateway into believing that his requests are legitimate. By running his own proxy server and adding this information to the requests of his customers, the fraudster will receive access to the PSTN without having to pay for it.

In general this kind of attack is more complex. The fraudster needs to detect gateways that accept SIP signaling requests directly from the Internet and use this kind of authentication approach. Further, to cover their traces, fraudsters first get access to a VoIP server of an enterprise or a university with a wideband Internet access and then route the calls through these servers.

To protect against such fraud, the communication between the proxy and the gateway must be secured. This can be achieved by having the gateway reject all SIP requests arriving from any other IP address than those of a set of trusted proxies. This could, however, be circumvented by having the fraudster spoof the IP addresses of his requests. Higher security can be achieved by establishing a secure tunnel, e.g. using IPSec or TLS, between the proxy and gateway and rejecting all SIP traffic not arriving over this secured link.

Route misuse   The authentication process is usually triggered at SIP proxies when a user initiates a session to a restricted destination. A restricted destination can be a phone number that is reached through a PSTN gateway. The address of the restricted destination is presented in the *Request-URI* and the *To* headers. Only if the caller can authenticate himself is the request forwarded to the called destination. To bypass the authentication process but still reach the restricted destination, the caller adds the IP address or the DNS name of the PSTN gateway as a *Route* header or as an *maddr* field in the *Request-URI*. When seeing a *Route* header or an *maddr* field in the *Request-URI*, the proxy should use this information for forwarding the request. If the SIP proxy did not recognize that the address belongs to the PSTN gateway, then the request would reach the gateway without being authenticated first. To prevent this kind of fraud, the proxy should check the addresses in the *Route* and *maddr* fields and decide whether they belong to restricted destinations. This is not always possible as the proxy might not know the addresses of the PSTN gateways.

7

Media Security

Industry, service providers and ultimately end users expect a security architecture that secures not only the signaling messages but also the exchanged media traffic. While protocols such as IPsec, see section 2.2.2.1, provide a solid basis for securing IP traffic, media-related security protocols should also address the specific characteristics of media traffic and signaling protocols which make them very application-specific.

An important issue related to the security of the multimedia communications is the feasibility of securing the traffic in an end-to-end manner. There are legal as well as technical aspects that put this paradigm in question. In the first category there is the support for lawful interception, which requires that conversations can be made available to legal agencies under specific circumstances. This means that it must be possible for specific encrypted conversations be decrypted by a third party on the basis of an interception warrant and without the involvement of the original parties. In the latter category there are middleware boxes such as NATs and firewalls that need to perform various processing on the media and therefore need to be able to access it in unencrypted form. Therefore, an end-to-middle security model that involves one or more trusted intermediate entities is necessary in many service scenarios.

A significant amount of effort has been invested by the IETF in designing a security architecture for the media plane, with some of the resulting solutions implemented and publicly available [minisip, available from minisip[1] zfone, available from Zfone[2]].

3GPP is following this work and some of these protocols are being considered for adoption in IMS, with emphasis on fulfilling the security requirements of the various IMS services and scenarios (33.828 TR 2008).

This chapter is devoted to describing the most significant protocols that have resulted so far from this effort. In section 7.1 we first give a brief overview of Real-time Transport Protocol (Schulzrinne *et al*. 2003) used for transporting the media traffic with IMS and SIP. Section 7.2 describes the security extension added to RTP. The mechanisms for exchanging the keys needed for securing the traffic are described in Section 7.3.

---

[1] http://www.minisip.org/
[2] http://zfoneproject.com/

SIP Security   Dorgham Sisalem, John Floroiu, Jiri Kuthan, Ulrich Abend and Henning Schulzrinne
© 2009 John Wiley & Sons, Ltd

7.1 The Real-time Transport Protocol

The Real-time Transport Protocol (Schulzrinne *et al.* 2003) designed within the Internet Engineering Task Force is the most widely used application layer protocol for real-time communication over the Internet. SIP as well as H.323 (ITU-T Rec. H.323 2006) define RTP as the application-level transport protocol for data.

RTP is an end-to-end protocol that is often used together with other transport protocols, in particular UDP. RTP has no notion of a connection; it may operate over either connection-oriented (say, TCP) or connection-less lower-layer protocols (typically, UDP). It does not depend on particular address formats and only requires that framing and segmentation are taken care of by lower layers. RTP offers no reliability mechanisms. It is typically implemented as part of the application or as a library rather than integrated into the operating system kernel.

RTP sessions consist of two streams, namely a data stream for audio or video data packets and a stream of control packets using the sub-protocol called RTP Control Protocol (RTCP). In general, data and control streams use separate ports; however, they may be packed into a single lower-layer stream as long as RTCP packets precede the data packets within the lower-layer frame.

As illustrated in Figure 7.1, RTP adds to the data an additional header, which is sent as part of the data payload of the transport layer. RTP data headers include the following fields:

- **Version field (V)**–this field indicates which version of RTP is being used; version 2 is the common one.
- **Extension bit (X)**–the X bit indicates whether an extension header follows the fixed header or not.

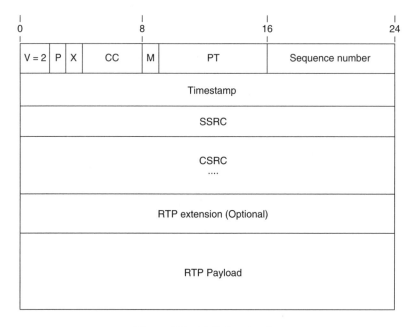

Figure 7.1   RTP data header

- **CSRC count field (CC)**–the CC field contains the number of CSRC identifiers following the header.
- **Payload type field (PT)**–this field is used to dynamically distinguish between different audio and video encodings.
- **Marker bit (M)**–the marker bit is used to delineate application level data units such as video frames and audio talk spurts.
- **Sequence number (SEQ)**–the sequence number is incremented by one for each RTP data packet and can be used by the receiver for detecting losses.
- **Timestamp**–the timestamp reflects the sampling instance of the first octet in the RTP data packet.
- **SSRC, CSRC**–these fields contain the identity of the sending source.

For monitoring the QoS of a session or tracing the number and identities of the members in a session, each session member periodically sends RTCP control packets to all other session members using the same distribution mechanism used for the data packets.

Each RTCP packet begins with a fixed part similar to that of the RTP header followed by one or more parts carrying a variety of information such as:

- **RR**–receiver reports (RR) consist of several entries with each corresponding to one active sender. Each entry contains information about the percentage and number of the lost packets observed in the stream from the corresponding sender, the highest sequence number seen and timing information.
- **SR**–sender reports (SR) include information about the amount of sent data and when this report was generated. Additionally, when the sender is also a receiver, it includes reports about the received data in an identical way to the receiver reports.
- **SDES**–the source description packets (SDES) include the identification information about the session member.
- **Bye**–this packet indicates the end of a member's participation.
- **APP**–the application packets (APP) contain application specific information and can be used for experimental purposes.

7.2 Secure RTP

The media exchanged during sessions established using SIP include audio, video, text and application data (e.g. fax, whiteboard). This data is exchanged by means of the Real-time Transport Protocol (Schulzrinne *et al.* 2003). For protecting the RTP datagrams, one may consider IPsec (see section 2.2.2.1) as well as TLS (see section 2.2.2.1) or DTLS (see section 2.2.2.2) as possible candidates. However, the exchange of real-time data introduces a few subtle requirements, the fulfilling of which have a great impact on the performance of transporting and processing the media. These requirements are:

1. There must be no or a minimal increase in the payload size. Therefore the use of additional headers and payload fields to carry per-packet information such as initialization vectors, sequence numbers, security association identifiers and padding must be avoided. Instead, fields that already exist in the RTP payload must be used to derive this information.

2. Header compression mechanisms must not be hampered. These compression mechanisms are based on the idea that the values of many of the fields contained in the IP, UDP and RTP headers remain unchanged across the datagrams that make up a given data flow. Also, some other fields change in a predictable way. As a result, the values of all those fields can be conveyed only once when the data flow is established and need not be transmitted in the subsequent packets. At the other end, a packet is identified by a "connection ID" and the headers are restored. On wireless and slow point-to-point links, these techniques [see (Bormann *et al.* 2001) and (Koren *et al.* 2003)] reduce the size of the headers from 40 bytes (on IPv4) to just a few bytes. As a result, the headers up to the RTP header must be left unencrypted.

3. The decryption of the encrypted payloads must be possible despite datagrams being re-ordered or lost.

4. The decryption process must be able to start before the entire data is received. Also, pipelining and parallelization of the cryptographic operations must be enabled to the largest possible extent, in order to achieve real-time performance. This has resulted in making the counter mode of operation for the cryptographic algorithm the primary design choice (see section 2.1.1);

5. The hardware (in terms of number of gates) required by the cryptographic algorithms must be minimized because it will be typically be used on low power mobile terminals. This has to do not only with the design of the cryptographic algorithm (which must be fast both in hardware and software), but also with the mode in which the algorithm is used: for instance the stream ciphers use only the encryption operation of the algorithm for both the enciphering and deciphering, which results in reducing the silicon footprint by half.

The Secure Real-time Transport Protocol (SRTP) (Baugher *et al.* 2004) has been specifically designed to be suitable for real-time data transmission and specifies a new RTP profile (RTP/SAVP) that offers confidentiality, integrity protection, data origin authentication in case of point-to-point communications and replay protection to the RTP and RTCP traffic.

We will introduce SRTP by describing its functionality from the perspective of the three-component model introduced in section 2.2.1.1, which consists of a data plane, a policy and a signaling component:

- In the data plane, SRTP defines modifications of the RTP payload that enable datagrams containing encrypted media and a MAC field to be transported between SRTP endpoints (see section 7.2.2).
- In the policy plane, SRTP defines "security associations" (called "cryptographic contexts" in SRTP terminology) that describe the type of protection (confidentiality, integrity protection), the cryptographic algorithms and the session keys used to secure the RTP and RTCP streams (see section 7.2.1). An SRTP cryptographic context is associated with one RTP or RTCP stream, therefore the "traffic selector" that describes the protected data flow is identified by the SSRC and the destination transport address of the RTP/RTCP stream.
- SRTP does not provide a signaling component, instead a number of key management protocols such as SDES, MIKEY, ZRTP and DTLS-SRTP (see section 7.3) may be used

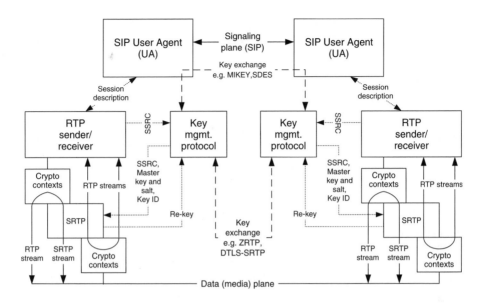

Figure 7.2   The main elements of the media plane security framework

to negotiate the security parameters of the SRTP cryptographic contexts and provide the SRTP endpoints with a pair of shared secrets (denoted as "master key" and "master salt"). SRTP uses the master key and master salt obtained in this way as input into a key derivation mechanism to generate the session keys necessary to the SRTP cryptographic contexts (see section 7.2.4).

Figure 7.2 illustrates the conceptual interaction between the main components involved in setting up a multimedia session between two participants. The incoming and outgoing RTP and RTCP datagrams are matched against the SRTP cryptographic contexts as the last step before sending them out and as the first step following their reception. If a match occurs, then the respective SRTP cryptographic context provides all the necessary parameters to define the type of processing that needs to be performed.

The key exchange takes place either over the signaling path or the media path, as described in section 7.3. Some key management protocols need to interact with the RTP application in order to retrieve the SSRC values for setting up the SRTP cryptographic contexts (while some others perform this binding dynamically). Also, key management protocols should allow SRTP to trigger re-keying, that is, the re-negotiation of fresh keys. Because the old and new keys coexist for a certain amount of time, the key management protocol must provide the possibility for the SRTP endpoints to announce the key identifiers.

7.2.1   The SRTP Cryptographic Context

An SRTP cryptographic context consists of the following parameters:

- A rollover counter (ROC), which counts the number of times the RTP sequence number carried in the RTP packet has rolled over. In this way, the ROC extends the 16-bit RTP sequence number to a 48-bit "SRTP packet index". The ROC need not be explicitly exchanged between the SRTP endpoints because in all practical situations a rollover of the RTP sequence number can be detected (unless 2^{16} consecutive RTP packets are lost, which amounts, for instance, to twenty minutes of voice traffic).

- A sequence number s_1 indicating the highest RTP sequence number received, which assists in detecting when the RTP sequence number rolls over and hence when the ROC needs to be incremented.
- A replay list (when authentication and replay protection are provided) to protect against replay attacks (see section 7.2.3).
- The encryption algorithm, which may be one of: NULL, AES in counter mode (AES-CTR) or AES in f8 mode (see section 2.1.1.2).
- The message authentication algorithm. Only HMAC-SHA1 (Krawczyk *et al.* 1997) has been specified so far.
- The master key(s), which *must* be random and kept secret.
- A master key usage counter, which counts the number of datagrams that have been protected with the current master key. The same master key should be used to protect at most 2^{48} RTP datagarms or 2^{32} RTCP datagrams (as subsequently explained in section 7.2.3). The key management must be triggered to provide a new master key whenever the lifetime of the current one is close to expiry.
- An optional *key derivation rate* (which is a power of 2 in the range $0-2^{24}$) that enables fresh sets of session keys to be periodically generated during the lifetime of the current master key.
- A master salt, which is used to provide additional entropy to the key derivation process in order to enhance the robustness of the master key against off-line key collision attacks (see section 2.1.1.2);
- The length of the session keys. The session keys are derived from the master key and master salt and are the keys effectively used by the encryption and MAC algorithms (see section 7.2.4).
- Information related to the presence in the SRTP/SRTCP packets and the current value of a Master Key Identifier (MKI). The MKI identifies the master key and salt that the sender used to protect a certain datagram and is necessary in order to support re-keying and the replacement of the old master key and master salt with fresh ones (e.g. as result of the key usage counter of the old keys approaching the maximum limit). The MKI needs to be signaled and retrieved from the key management protocol (see Figure 7.2). This information consists of:
    - an MKI flag indicating whether an MKI is being used (and therefore carried in the SRTP/SRTCP datagrams);
    - the length of the MKI field, if the MKI field is present;
    - the current value of the MKI field used by the sender and a list of MKI values and their corresponding master keys that are still considered valid by the receiver.
- ⟨From, To⟩ values, specifying the lifetime for a master key in terms of an RTP extended sequence numbers interval, as an alternative (but less flexible) mechanism to the MKI.

The cryptographic context of an SRTCP stream shares by default most of the above parameters with its associated SRTP stream, except that no ROC counter is used for RTCP and separate master key usage counter and replay lists are maintained by the SRTCP cryptographic context. Note, however, that some key management schemes provide a distinct ⟨master key, master salt⟩ pair for the SRTP and respectively the SRTCP cryptographic contexts.

An SRTP cryptographic context is identified by a ⟨SSRC, dst. IP, dst. port number⟩ tuple containing the RTP Synchronization Source (SSRC), the destination IP address and the destination port number. An RTP session is identified on each participating host by one destination IP address and a pair of destination port numbers (for exchanging the RTP and respectively the RTCP traffic). Therefore one separate cryptographic context will be created for each particular RTP stream (identified by a distinct SSRC value) of an RTP session. Also, a separate cryptographic context will be created for the RTCP stream that corresponds to an RTP stream.

In order to facilitate the interoperability, SRTP defines a minimal set of mandatory cryptographic algorithms, which are AES-CTR for key derivation, AES-CTR and NULL for encryption and HMAC-SHA1 for integrity protection. Optionally AES in f8 mode may be used for encryption. Also, in some restricted circumstances (see section 7.2.2) NULL integrity protection may be used. The introduction of additional cryptographic algorithms in the future is not precluded.

Also, SRTP specifies default values for all the cryptographic context parameters so that in most scenarios the key management protocols need not explicitly negotiate any of them besides the master key, master salt and the SSRC values.

7.2.2   The SRTP Payload Structure

SRTP encrypts the RTP payload including the RTP padding, if present, and appends to the RTP datagram an optional Master Key Identifier field and a recommended authentication tag (see Figure 7.3). The RTP header is not encrypted in order to allow header compression to be applied to the SRTP traffic and to enable the SSRC to be seen in clear by the receiving end(s) so that the cryptographic context can be identified. The authentication tag contains a MAC applied over the RTP header and the RTP payload up to the MKI field.

SRTCP encrypts the payload of the compound RTCP packet and appends an encryption flag, an SRTCP index, an SRTCP MKI and an authentication tag (see Figure 7.4). The encryption flag indicates whether an SRTCP packet has been encrypted or not and allows individual RTCP packets in a compound RTCP packet to be re-grouped into one unencrypted RTCP compound packet and one encrypted RTCP compound packet (Schulzrinne *et al.* 2003). The unencrypted RTCP compound packet would typically contain information (e.g. receiver report packets) that needs to be available to third-party entities, while the encrypted RTCP compound packet contains confidential information (e.g. the source description packet prepended by an empty receiver report packet).

The RTCP header is not encrypted (for the same reasons as for the RTP header).

The authentication tag contains a MAC applied over the RTCP header, the RTCP payload, the encryption flag and the SRTP packet index, up to the MKI field.

Some wireless access technologies exhibit significant losses of the transmission efficiency when the size of the RTP datagrams increases, while others do not allow any increase of the RTP datagrams size at all. This involves an authentication tag shorter than

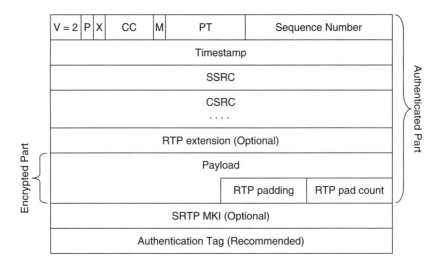

Figure 7.3   The structure of an SRTP packet

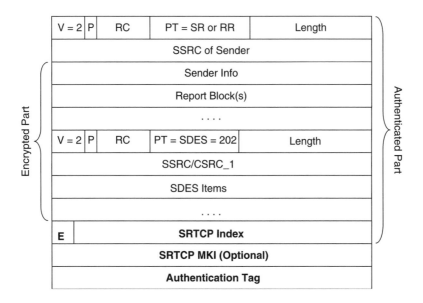

Figure 7.4   The structure of an SRTCP packet

the default 80 bits or no authentication tag at all being used. In the absence of authentication, stream ciphers are very vulnerable because an attacker that somehow guesses the plaintext can alter the ciphertext so that it decrypts to a plaintext of his choice (see section 2.1.1.2). Also, even when the attacker injects arbitrary datagrams into the media stream, the decryption of random data may lead some RTP applications to crash. Therefore it is recommended that SRTP is never used with weaker (shorter authentication tag) or no authentication, unless the threat of such attacks can be completely ruled out.

7.2.3   Sequence Numbering

The RTP header contains a 16-bit sequence number, whose original purpose was to enable the reconstruction of the media stream at the receiving end.

In the context of SRTP, sequence numbers have two additional useful functions:

- replay protection, using a mechanism similar to that employed by AH and ESP (see section 2.2.1.1);
- generation of per-packet encryption keystreams (see section 7.2.4).

This has resulted in the extension of the range of the RTP sequence numbers by means of the ROC to a 48-bit SRTP packet index and in including a 31-bit SRTCP packet index in the SRTCP packets.

7.2.4   The Key Derivation Procedure

The key derivation in SRTP is illustrated in Figure 7.5. SRTP uses the master key and the master salt provided by an external key management protocol (such as SDES, MIKEY, ZRTP, DTLS-SRTP, see section 7.3) as input to a PRF (see section 2.1.1.3) to derive a set of session keys consisting of an SRTP encryption key, an SRTP salting key and an SRTP authentication key.

The SRTP encryption and salt keys are used to generate the keystreams that are used for the encryption and decryption of the SRTP and SRTCP packets. The SRTP authentication key is used to calculate and validate the MAC of the SRTP and SRTCP packets.

The PRF used for the session keys derivation is based on AES-CTR encryption algorithm. The master key is used as the AES encryption key and the Initial Value is generated using concatenation, shift and XOR operations from the following parameters:

- A *label*, which takes values in a set of six predefined constants, which identify the six distinct keys that belong to the two separate sets of authentication, encryption and salting session keys, used by the SRTP and SRTCP streams.
- The SRTP and SRTCP packet indexes and the *key derivation rate*, which are used to refresh the session keys during the lifetime of the current master key. This is achieved by feeding the result of the integer division of the packet index by the *key derivation rate* into the key derivation process. A *key derivation rate* of zero (which is also the default value) is interpreted as no key derivation taking place.
- The master salt.

For data encryption and decryption a distinct keystream is generated for each packet, enabling each individual SRTP and SRTCP packet to be decrypted independently of the reception of the previous ones. The keystream is generated using the encryption session key as the AES encryption key. In the case of AES-CTR, the IV is calculated using concatenation, shift and XOR operations from the following parameters:

- SRTP/SRTCP packet index, which results in the keystream being unique for each SRTP/SRTCP packet;

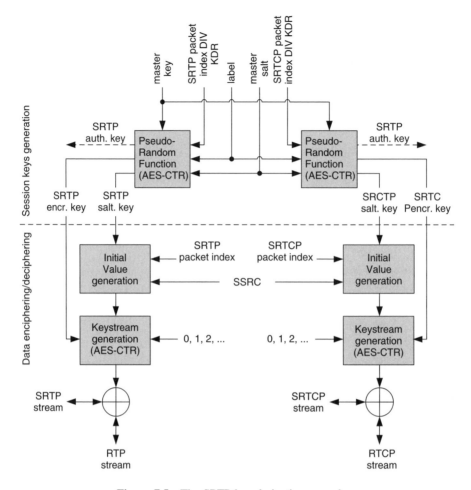

Figure 7.5 The SRTP key derivation procedure

- SSRC, which ensures that distinct keystreams are used by different RTP streams in an RTP session;
- the session salting key.

In AES using the f8 cryptographic mode, the IV is calculated slightly differently, in that RTP/RTCP header fields are also included and the salting session key is used to encrypt the IV to obtain a "masked" IV, which is used in the subsequent enciphering process.

The most important problem the key derivation and the key management must solve is related to the "two-time pad" (see section 2.1.1.2), which may occur if a keystream is reused. Therefore the use and the sharing of a master key must be driven by the following rules:

- Sharing the master key and master salt between a SRTP stream and the corresponding SRTCP stream is safe because different sets of keys are generated for SRTP and SRTCP respectively, using different label values.

- Sharing the master key and master salt between multiple SRTP streams is safe provided that the SSRCs are different, which is true for SRTP streams belonging to the same SRTP session. The danger of "two-time pad" occurring in such a case is limited because possible SSRC collisions are likely to be quickly eliminated and the probability of the packet indexes also overlapping during this time is very small. It is, however, recommended that each SRTP/SRTCP stream uses its own master key and in fact some key management protocols negotiate different master keys for the SRTP streams in each direction. This approach is especially useful for key management protocols that do not provide support for the exchange of the SSRC values.
- Because the SRTP and the SRTCP packet indexes are used in the process of generating the keystreams, duplicate keystreams will be produced in case they roll over. For this reason, re-keying (i.e. generation of a new master key and master salt) is necessary whenever an SRTP/SRTCP stream exhausts its index space (2^{48} for SRTP streams and 2^{31} for SRTCP streams), or the SRTP session must be terminated. When a master key is shared between several SRTP/SRTCP streams in an RTP session, then the re-keying must be triggered when any of the several index spaces is exhausted.

For re-keying to be possible, a mechanism is necessary to indicate which set of session keys has been used to protect a certain SRTP/SRTCP datagram. This task is accomplished by the MKI field introduced earlier.

Alternatively, SRTP defines a ⟨From, To⟩ mechanism that may be used to indicate the range of SRTP packet indexes when a certain master key is used. The re-keying of the associated SRTCP stream occurs at the same time as the re-keying of the SRTP stream. If an RTP session consists of multiple SRTP streams that share the same master key, one single "master SRTP stream" will be chosen to be the one that dictates when the re-keying is initiated, based on its SRTP packet index. The ⟨From, To⟩ mechanism has the advantage that it does not require any master key identification information to be explicitly carried in the SRTP datagrams (as is the case with the MKI). It has, however, the disadvantage that the SRTP streams other than the "master SRTP stream" may experience brief master key de-synchronizations between the sending and receiving ends at the re-keying time.

The session key derivation procedure may be invoked several times during the lifetime of a master key by setting the *key derivation rate* to a nonzero value. For instance, if the *key derivation rate* is set to 2^{10}, then a fresh set of session keys will be generated after every 2^{10} SRTP datagrams are exchanged. This has the effect of limiting the volume of data encrypted with the same session key, which results in a smaller volume of data available for cryptanalysis and less data exposed in case a session key is compromised.

Early versions of SRTP used for the purpose of integrity protection an algorithm called Truncated Multi-Modular Hash (TMMH; McGrew 2001), which has been obsoleted in favor of HMAC-SHA1. The TMMH algorithm belongs to a class of Message Authentication Code functions that require longer keys. A relic of this requirement is that SRTP defines a keystream prefix, which represents the length at the beginning of the keystream used by the TMMH. When HMAC-SHA1 is used, the keystream prefix must be zero.

7.2.5 The SRTP Interaction with Forward Error Correction

Forward Error Correction (FEC) is an error control mechanism that increases the reliability of the data transmission over unreliable transmission channels. FEC works by the sender adding redundant information into the data that he sends, which enables the receiver to completely or partially recover the lost data. This results in increased bandwidth usage; however FEC is an attractive alternative when mechanisms based on retransmissions cannot be used. One example is the real-time applications, the delay requirements of which cannot accomodate retransmissions.

FEC algorithms may be used in order to protect the RTP streams against packet loss (Li 2007). The use of FEC is relevant to SRTP in two ways. First, the order of applying FEC and SRTP to a RTP stream must be consistent between the sender and the receiver. If not otherwise agreed by the two parties (e.g. by means of the key management protocol), SRTP (Baugher *et al.* 2004) specifies that the FEC processing must be performed before SRTP processing on the sender and in the reverse order on the receiver.

The second aspect regards the securing of the data produced by the FEC algorithm. It must be noted in the first place that, if the RTP stream must be encrypted, then the FEC data associated to it must be encrypted too, because the FEC data reveals information about the plaintext data used to produce it.

The FEC data may be transported in two ways:

- As RTP payload in a separate RTP stream. This RTP stream shares the same SSRC value as the original, error-protected RTP stream, but has a different destination transport address. In this case, the RTP stream carrying the FEC data must be encrypted and integrity protected using a different pair of SRTP master key and master salt than the original RTP stream. SRTP does not make any provision about deriving keying material for securing the FEC data, therefore a separate set of SRTP master key and master salt needs to be provided by the key management protocol.
- As a secondary codec in a redundant encoding format (Perkins *et al.* 1997). In this case the FEC data is carried in the RTP payload of the same RTP packets as the error-protected media and is therefore secured along it.

7.3 Key Exchange

In order to be operational, the SRTP framework needs to interact with a key management protocol that facilitates the participants in a multimedia session to generate the master keys as well as negotiate and agree on the parameters of the SRTP cryptographic contexts.

In order to negotiate security parameters on behalf of multimedia sessions, a key management protocol must be able to address one or several scenarios that are typical of multimedia sessions and differ in terms of the roles the various participants have in setting up the security parameters of the media streams:

- One-to-one (unicast), where the security parameters of the media sessions are either established by mutual agreement or each party sets the security parameters for its own outgoing streams.
- Simple one-to-many (multicast), where the sender is in charge of setting the security parameters of the outgoing stream.
- Many-to-many without a centralized control unit (like in case of multicast) where each party sets the security parameters for its own outgoing media.

- Many-to-many with a centralized control unit (like in a conference bridge scenario) where the participants in a multimedia session retrieve the security parameters from a group controller.

Several key management protocols have been proposed and in this section we will focus our discussion on those that have received the most consideration and either resulted in RFCs or are under active development at the time of this writing. They are:

- SDP Security Descriptions for Media Streams (SDES);
- Multimedia Key Exchange (MIKEY) and the Key Management Extensions for SDP and RTSP (key-mgmt);
- ZRTP;
- DTLS-SRTP.

A generic set of requirements for the multimedia key management protocols has been specified in (Wing *et al.* 2008), which also contains a comprehensive collection of key management protocol proposals, including some of historical value.

The first two key management protocols may be classified as signaling plane key management protocols because they use SDP attributes to carry their payloads. Therefore, their message exchange sequence must be compatible with the SDP offer/answer model (Rosenberg and Schulzrinne 2002a), which means that it must complete in at most one round trip. In order to ensure that the key management protocol payloads are integrity protected, the SIP signaling itself must be integrity protected.

The last two key management protocols take a different approach: they multiplex their messages onto the RTP ports and the message exchange takes place concurrently with the actual media exchange. They may therefore be characterized as media plane key management protocols. Even though the media plane key management protocols are decoupled from the signaling plane, they indirectly rely on it to transport peer authentication information and provide a binding between the media plane endpoints and the signaling plane endpoints, which is necessary to protect against connection hijacking attacks. Therefore, in this case integrity protection for the SIP signaling is also desirable.

The following mechanisms may be used to protect the SIP signaling:

- S/MIME (see section 6.2)–can provide end-to-end confidentiality for the body of a SIP message and partial confidentiality for the SIP headers, integrity protection for the body and the immutable headers of the SIP message and identity authentication for the sender of the SIP message. The main drawback of S/MIME is that encrypting the SDP payloads end-to-end hinders intermediate signaling entities that need to access and/or update SDP attributes. Also, S/MIME does not provide protection against replay attacks.
- TLS–does not offer end-to-end protection to the SIP messages but only hop-by-hop confidentiality and integrity protection. This is an acceptable security model for those scenarios where a trust relationship has been established between the terminals and the SIP proxies, as well as between the SIP proxies operating in different administrative domains. In such a scenario it is assumed that the SIP proxies do not act maliciously and that the outbound SIP proxies grant access to the SIP infrastructure only to authenticated users.

- Strong authenticated SIP identity (see section 6.4)–provides identity authentication for the initiator of a SIP dialog. This results in a "single-sided verification" model that enables the callee to verify the integrity of the SIP request, validate the identity of the caller and the authenticity of the key exchange. This model can be considered secure provided that the callee cooperates in the sense that he performs these checks and takes appropriate measures in case they fail. The major limitation of the SIP identity mechanism is that no protection is offered to the SIP responses, and therefore the callee's identity is not authenticated towards the caller. The SIP connected identity (see section 6.4) solves this problem by requiring the callee to send an additional request (usually an UPDATE) following the reception of an invitation. This, however, occurs after the actual key exchange has taken place, which happens during or concurrently with the initial SDP offer/answer exchange.

These mechanisms require support from PKIs. As an alternative to using a PKI, multimedia key management protocols may take advantage of the interactive nature of the multimedia communication and use a Short Authentication String (SAS) to authenticate the key exchange and achieve identity authentication. Using the SAS mechanism requires that the users participate in bootstrapping the security relationship between the communicating endpoints by one participant reading aloud the SAS value displayed by his terminal so that the other participant can compare it with the SAS displayed by his own terminal. This scheme, however, does not completely eliminate the danger of user impersonation, particularly when the other party is an automaton or the voice of the interlocutor is not *a priori* known or is unrecognizable, e.g. due to poor quality transmission. It also requires the capability of the terminals to display such a short string.

Finally, it must be pointed out that a number of scenarios are particularly challenging to the key management protocols; they include forking, retargeting and early media. In case of retargeting, the identity of the UAS terminating a dialog-initiating request will be different from the one originally intended by the UAC. This is a problem if the UAC is required to know the identity of the UAS in advance, for instance for the purpose of sending him confidential information encrypted with the UAS's public key already in the initial handshake message. The same applies to forking, where the identity of the terminating UAS (or UASs) may be different from the originally intended callee.

When forking occurs, the UAC may receive responses from multiple UASs. As a result, multiple key exchange sessions will be established and multiple cryptographic contexts will be created. It is therefore important that mechanisms are available to enable the key exchange handshakes to be correctly mapped to the media sessions on behalf of which they negotiate the security parameters and generate the keying material.

Early media designates the media sent by the callee as soon as she receives the invitation. The caller may therefore start receiving media before the SDP answer. If the media is encrypted and the caller requires attributes from the SDP answer in order to derive the necessary keying material to decrypt it, media clipping will occur. This affects all the key exchange schemes that require a full round trip to take place in the signaling plane, which in particular is the case of SDES and some of the MIKEY exchange modes. Early media is better supported by the key management protocols that operate in the media plane because they enable the media to be first sent unencrypted and then switch to SRTP after both endpoints have finished setting up the cryptographic contexts. Security preconditions (Andreasen and Wing 2007) may be used to delay the sending of the early media (and media in general) until both ends finish setting up the cryptographic contexts, which ensures that early media is in no case sent unencrypted and also allows for the identity of the sender to be authenticated.

The support for retargeting, forking and early media will be discussed on a case-by-case basis throughout the following sections.

7.3.1 SDP Security Descriptions for Media Streams

The SDP Security Descriptions for Media Streams (Andreasen *et al.* 2006) enables the peers in a two-party unicast communication to exchange security parameters and keys that allow them to set up SRTP cryptographic contexts.

SDES is not an authenticated key exchange mechanism because the security parameters and the keys are carried in cleartext in form of SDP attributes. SDES relies instead on the SIP signaling to be protected or encrypted by other means, like for instance S/MIME or TLS. Therefore, SDES may be regarded as an enhancement of the "*k*=" SDP parameter (Handley *et al.* 2006), which was deemed insufficient because the establishment of data security associations in general and SRTP cryptographic contexts in particular require far more parameters than just a key.

7.3.1.1   Negotiation of the SRTP Cryptographic Parameters

SDES introduces a new SDP attribute, named *crypto*, with the following structure:

$$a = \text{crypto:} \langle \text{tag} \rangle \langle \text{crypto-suite} \rangle \langle \text{key-params} \rangle [\langle \text{session-params} \rangle]$$

One or more crypto attributes may accompany a media line in an SDP offer. A crypto attribute should not be specified at the session level. Multiple crypto attributes indicate that the sender of the SDP offer (the offerer) supports multiple security schemes, in the given order of preference; this enables the answerer to select one of them in the subsequent SDP answer.

The fields in a crypto definition carry the following information:

- The $\langle \text{tag} \rangle$ field contains a number and is used in the SDP offer/answer process to uniquely identify the crypto attributes associated with a certain media line.
- The $\langle \text{crypto-suite} \rangle$ field encodes the suite of encryption and authentication algorithms used to protect the respective media stream.
- The $\langle \text{key-params} \rangle$ field specifies one or more sets of keying material that is used to protect the traffic in the sending direction. One such set has the following structure:

$$\text{inline: base64}(\langle \text{key} \rangle \parallel \langle \text{salt} \rangle)[\text{``} \parallel \text{''} \langle \text{lifetime} \rangle][\text{``} \parallel \text{''} \langle \text{MKI} \rangle \text{`` : ''} \langle \text{MKI-length} \rangle]$$

and contains: the SRTP master key ($\langle \text{key} \rangle$) concatenated with the salting key ($\langle \text{salt} \rangle$) and encoded in base64, followed by an optional maximum lifetime of the key ($\langle \text{lifetime} \rangle$) represented in number of packets protected with the respective keys, and an optional MKI value accompanied by the MKI field length. If missing, the lifetime of the key defaults to the 2^{48} SRTP packets and 2^{31} SRTCP packets (whichever occurs first, see section 7.2.3). Multiple key sets identified by different MKI values can be defined for a media stream to enable MKI-based re-keying. On the other hand, no support for the From−To type of re-keying (see section 7.2.4) is offered in SDES.

The ⟨session-params⟩ field provides a number of additional SRTP cryptographic context parameters. When missing, the default values defined by SRTP are used. These parameters are:

- KDR–the *key derivation rate*, which is 2 power the number specified by this parameter;
- UNENCRYPTED_SRTP, UNENCRYPTED_SRTCP, UNAUTHNTICATED_SRTP –indicating which type of protection the SRTP/SRTCP packets should (not) be afforded;
- WSH–the window size hint, is used as a hint for the size of the SRTP replay protection window in the absence of other sources of information like for instance SDP bandwidth modifiers;
- FEC_ORDER–indicating the order in which Forward Error Correction and SRTP are applied to the media packets (see section 7.2.5);
- FEC_KEY–the master key and salt to be used for the FEC stream where the FEC stream is sent as part of a different RTP session (e.g. to a different port number than the media stream).

There are two further parameters that characterize an SRTP cryptographic context: the SSRC and the ROC. Because SDES operates exclusively in the signaling plane, no explicit correlation between the negotiated SRTP cryptographic contexts and the SSRC values of the actual media streams can be established at the signaling time. For this reason a "late binding" mechanism is necessary (see section 7.3.1.2), which maps the SSRC values to the SRTP cryptographic contexts created by SDES when media traffic is actually received.

The ROC is always initialized to zero because in the peer-to-peer unicast scenario that characterizes SDES, both participants join at about the same time. Also, the initial SQN value should be generated in the range $0-2^{15} -1$ (instead of the full range of $0-2^{16} -1$). This reduces the chances of the sender's and receiver's ROCs becoming out of sync to those cases when at least 2^{15} consecutive RTP packets are lost at the beginning of the media session. In such a case the sender will increment the ROC whereas the receiver will not. However losing 2^{15} consecutive RTP packets corresponds, for instance, to several minutes of conversation, which is unlike to occur in practical scenarios.

Figure 7.6 illustrates the SDP payloads in a SDES exchange where the offer contains two crypto attributes for the media, one of which specifies two MKIs. The answerer picks the first proposal (tag = 1) and responds with the same cipher suite and the SRTP master and salting keys for his own sending direction.

As already mentioned, the SRTP master and salting keys provided by each endpoint in the $a = crypto$ attribute of the SDP are used to derive the SRTP session keys employed by the respective endpoint to encrypt and generate the MAC for the RTP and RTCP streams that it sends. This differs conceptually from most of the other SDP attributes (like for instance transport address and codecs), where the entity providing them offers information related to the media that it receives. Providing the SRTP keys for the sending–rather than receiving–direction is meant to avoid a two-time pad situation occurring in case of forking scenarios when the SSRC from distinct UASs collide: if the SRTP keys for the receiving

---

SDP offer

v=0
o=alice 2890844526 2890842807 IN IP4 10.47.16.5
s=Confidential conversation
i=Security parameters exchanged with SDES
u=http://www.example.com/abc.html
e=alice@example.com
c=IN IP4 168.2.17.12
t=2873397496 2873404696
m=audio 49170 RTP/SAVP 0
a=crypto:1 AES_CM_128_HMAC_SHA1_80
inline:WVNfX19zZW1jdGwgKCkgewkyMjA7fQp9CnVubGVz|2^20|1:4
⟨session-params⟩
a=crypto:2 F8_128_HMAC_SHA1_80
inline:MTIzNDU2Nzg5QUJDREUwMTIzNDU2Nzg5QUJjZGVm|2^20|1:4;
inline:QUJjZGVmMTIzNDU2Nzg5QUJDREUwMTIzNDU2Nzg5|2^20|2:4
⟨session-params⟩

SDP answer

v = 0
o=bob 25690844 8070842634 IN IP4 10.47.16.5
s=Confidential conversation
i=Security parameters exchanged with SDES
u=http://www.example.com/abc.html
e=bob@example.com
c=IN IP4 168.2.17.11
t=2873397526 2873405696
m=audio 32640 RTP/SAVP 0
a=crypto:1 AES_CM_128_HMAC_SHA1_80
inline:PS1uQCVeeCFCanVmcjkpPywjNWhcYD0mXXtxaVBR|2^20|1:4

Figure 7.6   Example of an SDES offer/answer

direction were provided by the UAC instead, then all the UASs would have used them, which combined with the SSRC collision would have led to two-time pad. With the current design, each UAS will pick its own keys.

As a final observation, an endpoint must provide SRTP keys even if it only acts as a media receiver; in this scenario the receiver's key will be used to protect the RTCP receiver reports that she sends.

7.3.1.2   Late Binding of the SRTP Cryptographic Contexts

Figure 7.7 describes the mapping between a SDES exchange and the resulting SRTP cryptographic contexts. The SRTP cryptographic contexts are identified by ⟨SSRC, port number, IP address⟩ triplets, where the "_O" and "_A" suffixes indicate the offerer and the answerer, respectively.

The SRTP cryptographic contexts that require "late binding" (in fact those located on the receiving end of each media stream) are marked with a "*" symbol instead of an actual SSRC value. This indicates that the SSRC parameter of the cryptographic context is assigned with the SSRC value carried in the first SRTP packet received that fulfills the

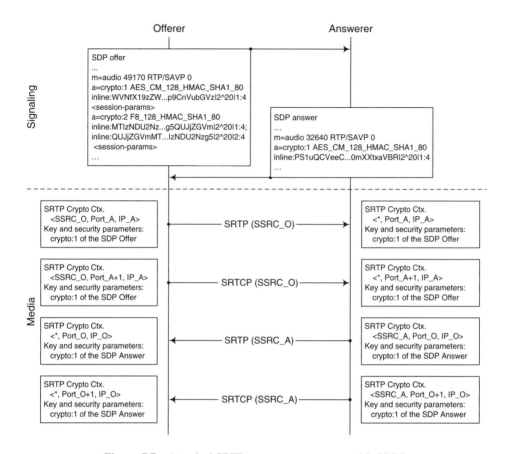

Figure 7.7 A typical SRTP crypto context setup with SDES

following conditions: (i) the destination transport address of the SRTP packet matches that of the SRTP cryptographic context and (ii) the SRTP packet is successfully authenticated with the SRTP authentication key of the SRTP cryptographic context.

The presence of the SRTP authentication tag is necessary when the "late binding" is involved because otherwise the mechanism can be easily exploited by DoS attacks to create false bindings. The "late binding" mechanism is discussed in more detail in section 7.3.4.2.

7.3.1.3 Support for Forking, Retargeting and Early Media

When SIP forking occurs, the UAC may receive SIP responses from multiple UASs, as well as multiple SRTP streams. Because SDES is exclusively a signaling plane key exchange protocol, it does not provide any mechanism to allow the UAC to map multiple SRTP sources (SSRCs) to their corresponding SIP responses. Therefore, the "late binding" mechanism will not be able to correctly associate the incoming SRTP/SRTCP streams to the SRTP cryptographic contexts received in the SDP answers. There are two possible solutions to this problem:

- The UAC updates the media transport parameters in the SDP offer (for instance by sending an UPDATE) to a different value, for each new dialog established by a forked call. In this way, the transport parameters will uniquely identify the SRTP cryptographic contexts for the purpose of the late binding.
- The UAC will try to authenticate an incoming SRTP/SRTCP datagram using all the "unbound" SRTP cryptographic contexts and will bind the SSRC in the incoming SRTP/SRTCP datagram to that SRTP cryptographic context for which the authentication was successful. This approach requires additional MAC validations until all the SRTP cryptographic contexts are "bound"; however the computational burden incurred is negligible. Nevertheless, it must be observed that this try-and-fail approach offers a limited window of opportunity for a DoS attack until the "late binding" process completes and all the SRTP cryptographic contexts are "bound".

When retargeting occurs, the exchange of cipher suites and keys will take place between the UAC and the final target. However, the keying material contained in the SDP offer will also be seen by the intermediate targets, which represents a security weakness. This issue equally affects the SIP forking. Therefore, achieving end-to-end security for the media traffic requires first determining the identity of the final target e.g. by means of (Elwell 2007) and then updating the *crypto* attribute in the SIP offer and sending it encrypted end-to-end e.g. using S/MIME.

In early media scenarios the UAS starts sending encrypted media as soon as she receives the invitation from the UAC. This involves encrypted media possibly reaching the UAC before the SDP answer that contains the master and salting keys necessary for the UAC to decrypt the incoming media and will result in media clipping.

7.3.2 Multimedia Internet Keying

The Multimedia Internet Keying (MIKEY; Arkko *et al.* 2004) is a key management protocol designed to perform key exchange and negotiate cryptographic parameters on behalf of multimedia applications. SRTP constitutes the main use case of MIKEY; however by design MIKEY is independent of the secure data transport protocol on behalf of which it performs the key exchange. The MIKEY messages are transported in the SDP payloads, encoded in base64 (Josefsson 2006), using the SDP key management extensions (see section 7.3.2.5).

A MIKEY exchange consists of an I_MESSAGE, sent in the SDP offer by the initiator of the session invitation (denoted as initiator), and an optional R_MESSAGE, sent by the receiver of the invitation who responds with the SDP answer (responder). A MIKEY message is composed of one header and a number of payloads. Some of the payloads are used to define the parameters of the SRTP cryptographic contexts (see section 7.3.2.1), while some others are used by the key exchange itself (see section 7.3.2.3).

7.3.2.1 The Crypto Session Bundle and the Crypto Sessions Map

A MIKEY exchange enables the parties to agree upon or update the security parameters of a set of Crypto Sessions (CSs), that form a Crypto Session Bundle (CSB). A CSB is identified by a CSB ID, which must be unique per initiator–responder pair and is carried in the MIKEY header. In case of SRTP, a CS corresponds to an SRTP cryptographic context and its associated SRTCP cryptographic context.

The MIKEY message header also contains a "CS ID map" that consists of a linear array of ⟨Policy_#, SSRC, ROC⟩ tuples. Each tuple corresponds to one CS and is identified by an implicit CS ID that represents the index of the tuple (starting with 1) in the array.

The Policy_# indicates which security policy applies to the respective CS. The same security policy may apply to more or all CSs in the CSB. The security policies are carried in Security Policy (SP) payloads. Each SP payload encodes in a Type–Length–Value (TLV) format a number of parameters that are specific to the underlying secure data transport protocol. In case of SRTP, they include:

- the authentication and encryption algorithms used;
- whether SRTP encryption and/or authentication should be used;
- whether SRTCP encryption should be used;
- authentication/encryption/salting key lengths;
- key derivation rate.

Figure 7.8 illustrates on a theoretical example how the parameters of three CSs are structured across the MIKEY payloads. The key derivation procedure is described in section 7.3.2.2. In our example the CS ID 1 and 2 share the same security policy Policy_1,

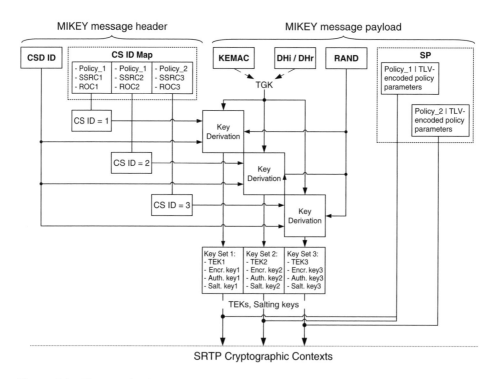

Figure 7.8 The mapping between a MIKEY crypto session and SRTP cryptographic contexts

while CS ID 3 uses Policy_2. The SSRC and ROC from the CS ID map, the security parameters of the corresponding policy and the corresponding key set provide all the information necessary to set up the SRTP cryptographic contexts.

In case of bidirectional media streams, the initiator provides in the CS ID map the SSRC values of the media streams that it sends and sets the SSRC values of the media streams that he expects to receive from the opposite direction to zero, expecting that the responder will fill them appropriately in the response message.

7.3.2.2   The Key Derivation

Besides the security policies and their mapping to the crypto sessions, the MIKEY exchange also enables the parties to negotiate a key set for each CS. The key set consists of a Traffic Encrypting Key (TEK), an authentication key, an encryption key and a salting key, which the underlying secure data transport protocol may use according to its own needs. For instance, SRTP only needs the TEK (for the SRTP master key) and the salting key because it uses its own key derivation procedure to obtain encryption, authentication and salting keys for the SRTP and SRTCP streams.

Central to the key derivation process is the TEK Generating Key (TGK), which depending on the key exchange method (see section 7.3.2.3) may be carried in encrypted form in a key data transport payload (KEMAC) or may be obtained following an exchange of Diffie–Hellman public keys carried in Diffie–Helmann data payloads (DHi and DHr) (see Figure 7.8).

The KEMAC payload can in fact transport not only one but several TGKs, as well as TEKs optionally accompanied by salting keys and key validity data (KV). The MIKEY key derivation procedure is based on a Pseudo-Random Function (see section 2.1.1.3) that uses the HMAC-SHA1 (Krawczyk *et al.* 1997) algorithm and the TGK as input key. The PRF is applied to a concatenation of the following parameters:

- a constant, which identifies the type of the key–TEK, authentication key, encryption key or salting key;
- the CS ID;
- the CSB ID;
- a random number carried in the RAND payload.

The inclusion of the CSB ID and CS ID guarantees that the key sets are unique for each CS shared between two given parties.

Alternatively, instead of a TGK, the parties may exchange in the KEMAC payload a TEK. In this case no key derivation takes place and, if the keys are negotiated on behalf of SRTP, then the TEK must be accompanied by a salting key.

Exchanging a TEK involves all the SRTP streams in the CSB sharing the same master and salting keys. It must therefore be ensured that no SSRC collision occurs among the CSs that belong to the same CSB (this condition is necessary in order to avoid the "two-time pad" problem, see section 2.1.1.2). This condition is satisfied if, for instance, a CSB corresponds to one single RTP session.

The KV payload that accompanies a TGK/TEK key is used to specify either the MKI value or the From–To packet index validity interval for the SRTP-related keys.

7.3.2.3 Key Exchange Methods

MIKEY offers limited support for security parameter negotiation. The main reason is that the number of round trips required by the key exchange must be reduced to as few as one single message in some scenarios and must complete in at most one round trip in all other scenarios. If for some reason the responder cannot accept the SRTP policies offered by the initiator, then it must respond with an error message that indicates the supported policies, encoded into MIKEY SP payloads. The initiator then has the choice of accepting one of the proposed policies and starting a new exchange. The other reason is that, in multicast scenarios, for instance, there must be one single security policy for all the receivers. Therefore the sender is the one that pushes the SRTP policies to the receiver(s).

The MIKEY base specification (Arkko *et al.* 2004) specifies three key exchange methods:

- pre-shared key (MIKEY-PSK);
- public key encryption (MIKEY-RSA);
- Diffie–Hellman (MIKEY-DHSIGN).

Two more exchange methods have been subsequently specified in order to cover a number of additional use cases. They are:

- HMAC authenticated Diffie–Hellman method (MIKEY-DHMAC) (Euchner 2006);
- a reverse public key encryption method (MIKEY-RSA-R) (Ignjatic *et al.* 2006).

The five key exchange methods are illustrated in Figure 7.9–7.13. We will start by describing a number of payloads that carry general information and are used in many of the key exchange methods:

- Timestamp (T)–this is necessary to protect against replay attacks. This involves the participants having loosely synchronized clocks.
- Random number (RAND)–the random number is used in the process of generating the TEKs (see Figure 7.8). It offers protection against off-line precomputation attacks on the key derivation process by increasing the entropy of the TGKs.
- Initiator/responder identity (IDi/IDr)–this contains a NAI or a URI that allows the peer to retrieve or identify the appropriate PSK or public key that has to be used to validate the MIKEY message. Identities need not be sent in the MIKEY messages if they can be inferred by other means, like for instance from the signaling protocol onto which the MIKEY exchange is piggybacked. It is, however, recommended that the initiator includes the responder's identity in the I_MESSAGE, in order to counter DoS attacks that consist of redirecting to the victim a valid I_MESSAGE intended for other recipients. This will trigger the victim to create states for the cryptographic contexts and depending on the key exchange method may involve a signature verification or an exponentiation to calculate a Diffie–Hellman shared secret. Also, the response messages in case of the Diffie–Hellman methods should include the initiator's identity in order to ensure that the secret is established with the intended party. MIKEY does not address user privacy issues, since the management of the identity is largely the resort of the signaling protocol that transports the MIKEY messages.

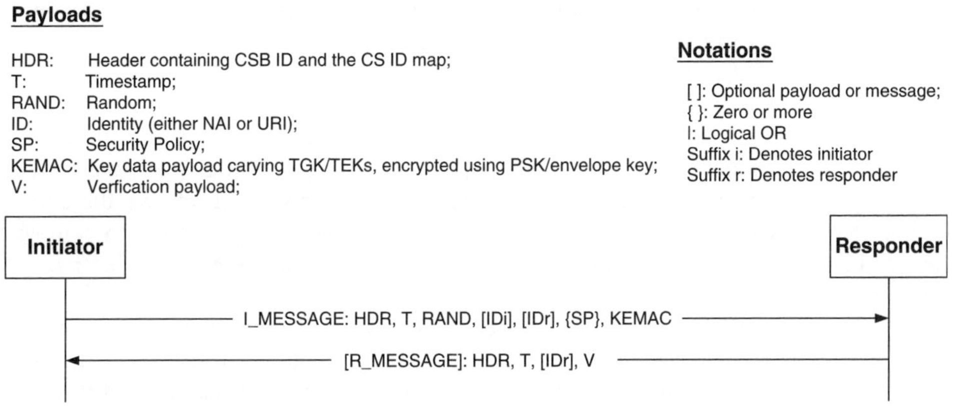

Payloads

HDR: Header containing CSB ID and the CS ID map;
T: Timestamp;
RAND: Random;
ID: Identity (either NAI or URI);
SP: Security Policy;
KEMAC: Key data payload carying TGK/TEKs, encrypted using PSK/envelope key;
V: Verfication payload;

Notations

[ ]: Optional payload or message;
{ }: Zero or more
I: Logical OR
Suffix i: Denotes initiator
Suffix r: Denotes responder

```
┌──────────────┐                                                    ┌──────────────┐
│  Initiator   │                                                    │  Responder   │
└──────────────┘                                                    └──────────────┘
       │                                                                   │
       │────── I_MESSAGE: HDR, T, RAND, [IDi], [IDr], {SP}, KEMAC ────────▶│
       │                                                                   │
       │◀──────────── [R_MESSAGE]: HDR, T, [IDr], V ───────────────────────│
       │                                                                   │
```

Figure 7.9   MIKEY-PSK

- Certificate (CERT)–this contains a certificate or an URL from where the certificate can be retrieved. A certificate chain is encoded as a sequence of CERT payloads.

The preshared key method   The preshared key (MIKEY-PSK) exchange method is illustrated in Figure 7.9. This method is applicable when the initiator and the responder are preconfigured with a PSK. The TGK(s)/TEK(s) are chosen by the initiator and sent to the responder in encrypted form in the KEMAC payload carried in the I_MESSAGE. The KEMAC payload also contains a MAC calculated over the entire message.

For the encryption and integrity protection of the message payloads, the MIKEY endpoints generate a set of shared keys consisting of an authentication, an encryption and a salting key, which is distinct from those generated on behalf of SRTP (see section 7.3.2.2). These keys, subsequently denoted MIKEY message keys, are obtained by applying the same PRF function, using PSK as the input key, to a concatenation of the following parameters:

- a constant that identifies the type of the key;
- the CSB ID;
- the random number contained in the RAND payload.

For the encryption of the KEMAC payload, AES-CTR is used with the MIKEY message encryption key and an Initialization Vector calculated using XOR and concatenation operations from the following parameters: the MIKEY messsage salting key, the CSB ID and the timestamp sent by the initiator in the T payload.

The MAC carried in the KEMAC payload is generated using the HMAC-SHA1 algorithm and the MIKEY message authentication key. The MAC provides integrity protection as well as data origin authentication, since only the entity that shares the same PSK with the responder can create a valid MAC.

The initiator has the choice to request a verification message from the responder, which carries a verification payload (V) that contains the result of the HMAC-SHA1 function calculated using the MIKEY message authentication key over the entire verification message concatenated with the identities and the timestamp (as they appear in the respective payloads). This provides mutual authentication of the MIKEY exchange.

In MIKEY-PSK the initiator may refrain from requesting a verification message, for instance, in order to avoid having to validate too many verification messages that would be received in the case of one-to-many scenarios. In such a case the authentication of the responder is only implicit, provided that only an entity that possesses the same PSK as the initiator will be able to decrypt the content of the KEMAC and retrieve the correct TGK/TEK. A MIKEY verification message is, however, necessary in order to complete the CS ID map whenever the responder is expected to send media or at least RTCP receiver reports.

MIKEY-PSK may be used in one-to-one as well as small group communication scenarios where the sender of the media chooses the security policy and the TGK/TEK key. MIKEY-PSK exclusively involves symmetric cryptographic computations and therefore it has a reduced computational complexity. On the downside, MIKEY-PSK is affected by the typical scalability and out-of-band PSK distribution issues that affect all the PSK-based schemes. It also does not offer Perfect Forward Secrecy because once the PSK is compromised, all past keys used to protect the MIKEY messages and the data traffic are compromised as well.

The Public Key Encryption method   The Public Key Encryption exchange method is illustrated in Figure 7.10. This method, subsequently referred to as MIKEY-RSA, is meant to eliminate the scalability problems that arise when PSKs are used in conjunction with a large number of participants.

MIKEY-RSA requires that the initiator *a priori* knows or obtains, using methods outside of the MIKEY protocol, a public key that belongs to the responder. If multiple public keys are available for the responder, the initiator indicates which one it has used by providing a hash (carried in the CHASH payload) over the responder's certificate containing the respective public key.

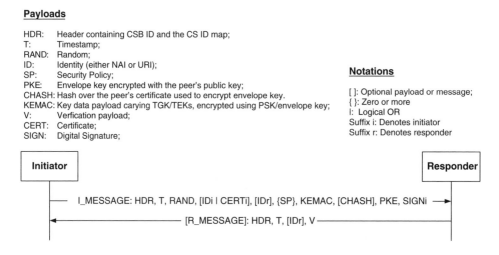

Payloads

HDR:	Header containing CSB ID and the CS ID map;
T:	Timestamp;
RAND:	Random;
ID:	Identity (either NAI or URI);
SP:	Security Policy;
PKE:	Envelope key encrypted with the peer's public key;
CHASH:	Hash over the peer's certificate used to encrypt envelope key.
KEMAC:	Key data payload carying TGK/TEKs, encrypted using PSK/envelope key;
V:	Verfication payload;
CERT:	Certificate;
SIGN:	Digital Signature;

Notations

[ ]: Optional payload or message;
{ }: Zero or more
|: Logical OR
Suffix i: Denotes initiator
Suffix r: Denotes responder

Initiator		Responder

I_MESSAGE: HDR, T, RAND, [IDi | CERTi], [IDr], {SP}, KEMAC, [CHASH], PKE, SIGNi →

← [R_MESSAGE]: HDR, T, [IDr], V

Figure 7.10   MIKEY-RSA

The initiator uses the responder's public key to encrypt a so-called envelope key, which it sends in an envelope data payload (PKE). The envelope key is used to generate the MIKEY message keys (i.e. the authentication, encryption and salting keys that are used to protect the message payloads), in a similar way to how PSK is used in MIKEY-PSK. The MIKEY message keys are used in MIKEY-RSA to encrypt and integrity protect the KEMAC payload.

The initiator digitally signs the I_MESSSAGE using its own certificate, which it may either send in a CERT payload (CERTi), or expect the responder to retrieve it based on the identity information (IDi) or just know it (by means outside of the MIKEY protocol). The digital signature is carried in the SIGNi payload and provides integrity protection and authentication. Depending on how the PKI is organized, a certificate chain may need to be provided.

The choice of encrypting an envelope key using the responder's public key (and then using the envelope key to protect the KEMAC payload) rather than the KEMAC payload itself is motivated by the fact that large KEMAC payloads would require more than one asymmetric encryption/decryption operation. Conversely, encrypting just the envelope key requires only one such operation, at the expense of multiple but much faster symmetric encryption/decryption operations necessary to authenticate and decrypt the KEMAC.

The exchange requires one round trip if the initiator requires explicit authentication of the responder (which takes place in a way similar to the MIKEY-PSK scheme, by means of the verification message). Otherwise, the exchange consists of one single message. In this latter case, the authentication of the responder is implicit because only the entity that possesses the private key that matches the public key used by the initiator to encrypt the envelope key will be able to decrypt it correctly.

A verification message is necessary in order to complete the CS ID map whenever the responder is expected to send at least RTCP receiver reports or media.

MIKEY-RSA may be used in one-to-one as well as small group communication scenarios where the sender of the media chooses the security policy and the TGK/TEK key. Because it relies on public-key cryptography, MIKEY-RSA involves operational complexity (due to interaction with PKI) as well as significant computational effort. The impact of the latter may be diminished by using the envelope key as a PSK in subsequent key exchanges that take place between the two parties. MIKEY-RSA does not offer PFS because, as soon as the private RSA encrypting key is compromised, all past secrets encrypted with that key can be recovered by the attacker.

The Diffie–Hellman method The Diffie–Hellman (MIKEY-DHSIGN) method is illustrated in Figure 7.11. It enables the initiator and responder to exchange ephemeral Diffie–Hellman keys and derive a shared secret, which is then used as a TGK to derive keying material for SRTP.

The DH method always requires one round trip. The peers mutually authenticate the MIKEY exchange by digitally signing the messages they send (SIGNi and SIGNr payloads). This method inherits the general properties of the ephemeral Diffie–Hellman exchange, which are: (i) enabling both the initiator and the responder to contribute entropy to the process of deriving the shared key and (ii) providing PFS. This may be perceived as an advantage over the key exchange methods where the responder has to trust the secret key provided by the initiator (such as in MIKEY-PSK and MIKEY-RSA) and which allow

Payloads

HDR: Header containing CSB ID and the CS ID map;
T: Timestamp;
RAND: Random;   **Notations**
ID: Identity (either NAI or URI);
SP: Security Policy;   [ ]: Optional payload or message;
DH: Public Diffie-Hellman key;   { }: Zero or more
KEMAC: Key data payload carying TGK/TEKs, encrypted using PSK/envelope key;   l: Logical OR
CERT: Certificate;   Suffix i: Denotes initiator
SIGN: Digital Signature;   Suffix r: Denotes responder

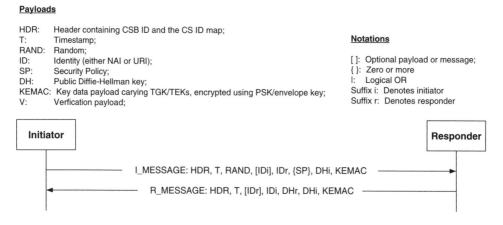

Figure 7.11   MIKEY-DH

Payloads

HDR: Header containing CSB ID and the CS ID map;
T: Timestamp;
RAND: Random;   **Notations**
ID: Identity (either NAI or URI);
SP: Security Policy;   [ ]: Optional payload or message;
DH: Public Diffie-Hellman key;   { }: Zero or more
KEMAC: Key data payload carying TGK/TEKs, encrypted using PSK/envelope key;   l: Logical OR
V: Verfication payload;   Suffix i: Denotes initiator
                                                          Suffix r: Denotes responder

Figure 7.12   MIKEY-DHMAC

an attacker to retrieve all the shared keys ever created or exchanged using a compromised PSK or a compromised private encryption key.

On the downside, exactly this property prevents MIKEY-DHSIGN from scaling to group communication scenarios, because the data must be protected with a different set of session keys for each recipient. MIKEY-DHSIGN also requires interaction with a PKI.

A variation of the MIKEY-DHSIGN method is the MIKEY-DHMAC method (Euchner 2006), illustrated in Figure 7.12. The MIKEY-DHMAC method uses PSK for the purpose of mutual authentication, instead of digital signatures as in the MIKEY-DHSIGN method. The mutual authentication of the MIKEY exchange is achieved by each party calculating a MAC over the message and including it in the KEMAC payload. The MAC is calculated using the MIKEY message authentication key, which is derived from the PSK in a similar way to the MIKEY-PSK. MIKEY-DHMAC reduces the complexity associated with the public-key cryptography at the price of reducing the scalability.

Payloads

HDR: Header containing CSB ID and the CS ID map;
T: Timestamp;
RAND: Random;
ID: Identity (either NAI or URI);
SP: Security Policy;
PKE: Envelope key encrypted with the peer's public key;
GenExt: Generalized Extension;
KEMAC: Key data payload carying TGK/TEKs, encrypted using PSK/envelope key;
CERT: Certificate;
SIGN: Digital Signature;

Notations

[ ]: Optional payload or message;
{ }: Zero or more
I: Logical OR
Suffix i: Denotes initiator
Suffix r: Denotes responder

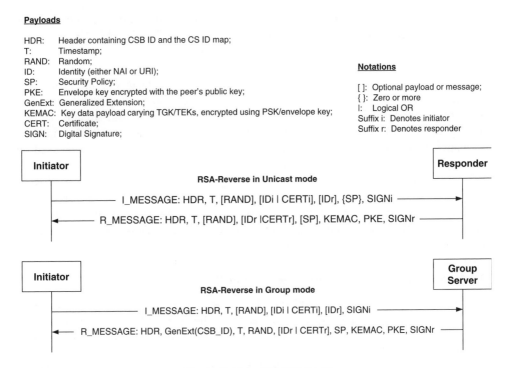

Figure 7.13   MIKEY-RSA-R

The MIKEY-RSA reverse method   The MIKEY-RSA reverse method (MIKEY-RSA-R) (Ignjatic *et al.* 2006) is illustrated in Figure 7.13. It addresses a number of scenarios that are not supported by the MIKEY-RSA method, like for instance:

- Scenarios when the initiator cannot determine the identity or obtain a certificate of the responder. Call forwarding is a typical scenario when the identity of the ultimate responder cannot be determined in advance. These scenarios are addressed by the MIKEY-RSA-R in unicast mode.
- Group communication scenarios that require the participants to download the TGK and the security policies from a group key server. These scenarios are addressed by the MIKEY-RSA-R in group mode.

The major difference from the MIKEY-RSA is that in MIKEY-RSA-R it is the responder who specifies the security policies and the key.

In MIKEY-RSA-R in unicast mode the initiator specifies the security policies and provides the CS ID map of the RTP streams it originates. The keys (envelope key and TGK/TEKs) are, however, generated by the responder. The responder uses the public key from the initiator's certificate to verify the signature SIGNi and to encrypt the envelope key carried in the PKE payload in the R_MESSAGE. This involves the initiator's certificate being valid for both signing and encrypting. Similarly to MIKEY-RSA, the envelope key is then used to derive the MIKEY message authentication, encryption and salting keys that protect the KEMAC payload that transports the TGK/TEKs.

The mutual authentication of the MIKEY exchange is completed by the initiator verifying the signature provided by the responder over the R_MESSAGE (SIGNr).

The MIKEY-RSA-R method in group mode may be used in the context of a group management architecture where the participants receive an invitation to join a conference and then contact a group server to download the security parameters of the media sessions. In this scenario, the I_MESSAGE is sent by the receivers to the group server. The CS ID map advertised in the I_MESSAGE will typically contain zero streams and the main use of the I_MESSAGE is to provide the receiver's certificate (CERTi) or identity information that can be used to retrieve his certificate (IDi). The CSB ID to be used by the group, together with CS ID map, security policies and keying material are provided by the group server in the R_MESSAGEs. The group's CSB ID is sent in a MIKEY generalized extension (GenExt) payload (Carrara *et al.* 2006) and it is the one used in the key derivation process (so that all the receivers end up with the same set of session keys), whereas the CSB IDs provided by the initiators in the I_MESSAGEs are solely used to identify each individual MIKEY exchange.

7.3.2.4 Support for Forking, Retargeting and Early Media

Forking is supported in MIKEY-PSK and MIKEY-RSA only under the provision that all the UASs that terminate the forked branches share the same PSK and respectively the same RSA private key. This condition might be fulfilled in some particular scenarios (e.g. all UASs belong to the same user), but it is not generally true.

Retargeting is not supported in MIKEY-PSK and MIKEY-RSA because the identity of the ultimate target cannot be determined prior to the SIP session establishment. However, MIKEY-PSK and MIKEY-RSA require the peer's PSK and public RSA key to encrypt and authenticate the payload containing the secret key, which is sent in the I_MESSAGE.

MIKEY-DHSIGN and MIKEY-RSA-R support forking and retargeting. In case of forking, different TGKs will be established with each UAS terminating the forked branches. MIKEY-DHMAC uses PSK for the purpose of mutual authentication and therefore it exhibits the same limitations with regard to forking and retargeting as MIKEY-PSK, because the initiator needs to know the identity of the UAS(s) in order to determine the appropriate PSK to authenticate the KEMAC payload sent in the I_MESSAGE.

Because the SSRC map carries in the MIKEY messages, the SSRC collisions that could occur in case of forking when two or more UASs that terminate the forked branches pick the same SSRC value can be detected when the R_MESSAGEs are received by the initiator.

The MIKEY DH-based methods and MIKEY-RSA-R produce media clipping in early media scenarios because the UAC needs to receive the R_MESSAGE carried in the SDP answer in order to be able to derive the necessary session keys to authenticate and decrypt the incoming media traffic.

In MIKEY-PSK and MIKEY-RSA the keying material is exclusively provided by the initiator and therefore the responder can start sending encrypted early media. However the responder's SSRC is only known after the R_MESSAGE has been received by the initiator, which involves the encrypted early media received by the initiator up to that moment not being associated with the cryptographic context, unless some sort of "late binding" is provided.

7.3.2.5 Key Management Extensions for SDP

The key management extensions for SDP[3] (Arkko *et al*. 2006) consist of a new SDP attribute *key-mgmt* that enables the messages of a key management protocol to be transported along with the other media parameters, provided that the key management protocol is suitable for an SDP offer/answer model, i.e. the key exchange occurs in at most one round trip. These extensions do not provide any additional security properties to the key management protocol or any kind of protection to the *key-mgmt* attributes themselves.

A *key-mgmt* attribute contains a key management protocol identifier (e.g. "mikey") followed by the actual message encoded in base64 (Josefsson 2006). A *key-mgmt* attribute may be defined in an SDP payload either at the SDP session level or at the media level. In the former case, the key management protocol exchanges the keys and negotiates the other cryptographic parameters on behalf of all the media streams that belong to the respective multimedia session. Conversely, a media level *key-mgmt* attribute applies only to the respective RTP session and overrides any *key-mgmt* session level attribute. Multiple *key-mgmt* attributes may be specified for the same media line or multimedia session, indicating that the offerer supports multiple key management protocols; following that, the answerer selects one of them in the subsequent SDP answer. Figure 7.14 illustrates an example of an SDP offer negotiating a key management protocol at the session level (on behalf of an audio and a video RTP session) and indicating support for MIKEY and another hypothetical key management protocol.

Indirectly, the *key-mgmt* attributes are protected if the SIP signaling itself is protected end-to-end or hop-by-hop between trusted SIP servers. If this is not the case, then the key management extensions may be manipulated by a man in the middle even if the key management protocol itself is secure against such attacks. The manipulation consists of the man in the middle removing some or all *key-mgmt* attributes. Thus, if several *key-mgmt* offers are made, a man in the middle may choose to leave the "weakest" one and remove the others. In order to protect against this type of attack, each individual key management protocol must take additional measures to enable the secure exchange of the full list or offered protocols.

```
v=0
o=alice 2891092738 2891092738 IN IP4 host.example.com
s=Secured session
t=0 0
c=IN IP4 host.example.com
a=key-mgmt:mikey HRoYXQgYnkgYSBw...
a=key-mgmt:hypotheticmp mRlZmF0aWdhYmxl...
m=audio 39000 RTP/SAVP 98
a=rtpmap:98 AMR/8000
m=video 42000 RTP/SAVP 31
a=rtpmap:31 H261/90000
```

Figure 7.14 Example of an SDP offer containing a *key-mgmt* attribute

---

[3] And RTSP (Schulzrinne et al. 1998); however we will restrict our discussion to SDP.

Moreover, the man in the middle may remove all of the *key-mgmt* attributes. In order to eliminate such a possibility, local policies may be enforced to forbid the setup of a session if one or more *key-mgmt* attributes have been offered but a *key-mgmt* attribute is missing from the answer. This also involves the key exchange protocol always requiring an answer from the responder (remember that some MIKEY exchange modes can choose not to request a response).

7.3.3 ZRTP

ZRTP (Zimmermann *et al.* 2008) is a key agreement protocol that enables the participants in a one-to-one multimedia session to establish a shared secret and agree on the security parameters for setting up SRTP sessions. ZRTP is essentially an authenticated ephemeral Diffie–Hellman exchange, with the mutual authentication being performed by means of a Short Authentication String (SAS). In this way ZRTP can operate without support from a PKI. It may however take advantage of the existence of a PKI by providing an alternative option for the key exchange to be mutually authenticated using digital signatures. ZRTP also combines a number of security mechanisms that reduce its vulnerability to man-in-the-middle attacks, even if the key exchange goes unauthenticated.

The ZRTP exchange takes place over the same port numbers used by the multimedia session for the RTP traffic; the initial messages are sent as each participant learns the port number of the peer. There will be as many distinct ZRTP exchanges as RTP sessions are defined in the multimedia session.

Figure 7.15 illustrates the structure of a ZRTP packet. The ZRTP header uses a common format with the RTP header. It specifies an RTP version number of zero to differentiate it from the regular RTP traffic (which uses the version number 2) and includes a specific ZRTP magic cookie to differentiate it from STUN packets (Rosenberg *et al.* 2003).

The ZRTP packet header also contains:

- A sequence number that is initialized to a random value and is incremented with each ZRTP packet sent. The sequence number is used to estimate losses and detect out-of-order message arrival.

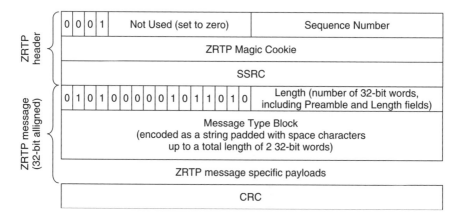

Figure 7.15 The format of a ZRTP packet

- The SSRC value of the RTP stream with whom the ZRTP packet shares the transport endpoints.

A ZRTP packet carries a ZRTP message, which starts itself with a preamble, followed by the message length, message type and the payloads that are specific to that particular ZRTP message type. The ZRTP packets end with a 32-bit CRC to detect transmissions errors as a supplementary mechanism to the UDP checksum (Stone *et al.* 2002).

The ZRTP exchange, illustrated in Figure 7.16, has been designed to be robust to packet losses because the practice has shown that the initial media packets that follow the setup of a session are affected by higher loss rates. Therefore, ZRTP employs an aggressive retransmission scheme (20 *Hello* retransmisions are performed in about 4 seconds). Acknowledgment messages (*HelloACK*, *Confirm2ACK*) are used to stop the retransmisions of a message when no explicit response to that message is received.

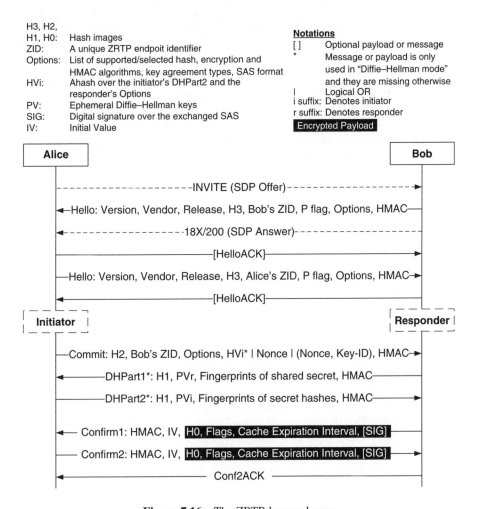

Figure 7.16 The ZRTP key exchange

ZRTP implements three key agreement modes:

- **Diffie–Hellman mode**–an ephemeral Diffie–Hellman exchange is performed in order to derive a "ZRTP master secret" (see section 7.3.3.2). The public Diffie–Hellman keys are carried in the *DHPart1* and respectively *DHPart2* messages. The *Commit* message carries a hash commitment of the initiator's public Diffie–Hellman key (*HVi*, described in section 7.3.3.7).
- **Preshared mode**–uses cached shared secrets to derive the ZRTP master secret for the current session. *DHPart1* and *DHPart2* messages are not exchanged in the Preshared mode and a *Nonce* and a *KeyID* are sent in the *Commit* message instead of the hash commitment (*HVi*). The *KeyID* carries a "fingerprint" that enables the responder to identify the cached shared secrets used for the derivation of the ZRTP master secret. If the fingerprint does not match, then the responder either returns an error message or reverts to the Diffie–Hellman mode, where it becomes the initiator. Because the Preshared mode derives the entire keying material from cached shared secrets, which are stored on the device across multimedia sessions, it does not offer Perfect Forward Secrecy. For this reason, the use of the Preshared mode to generate keying material for a new multimedia session must be restricted to ZRTP endpoints having very limited computational resources and in general should be avoided whenever possible.
- **Multistream mode**–as soon as a multimedia session has been established and a ZRTP master secret has been generated for one media session, using either the Diffie–Hellman or the Preshared mode, ZRTP master secrets for each of the other media streams in the multimedia session may be derived based on the existing ZRTP master secret, without having to perform a new Diffie–Hellman exchange. *DHPart1* and *DHPart2* messages are not exchanged in the Multistream mode and a *Nonce*, which is used in the key derivation process, replaces the *HVi* in the *Commit* message. The multistream mode is also used to resume encryption during the lifetime of a multimedia session after previously switching the session to plaintext using the *GoClear/ClearACK* exchange.

7.3.3.1 Discovery and Capabilities Negotiation

The ZRTP discovery phase consists of the ZRTP endpoint exchanging *Hello* messages. This enables each participant to discover whether the peer implements ZRTP and which cryptographic algorithms, key agreement types and SAS formats are supported. *Hello* messages may be sent at any time after the peers learn each other's RTP transport address from the SDP payloads. The sender of the first *Hello* message is in general the one that learns the transport parameters of his peer first, namely the receiver of the SDP offer.

A *Hello* message (see Figure 7.16) has the following structure:

- the ZRTP protocol version implemented, the vendor and the release;
- the ZRTP identifier of the sender *ZID*; the *ZID* is a 96-bit unique identifier of a ZRTP endpoint; a ZRTP endpoint may use multiple *ZID*s, up to one per distinct peer;
- a hash image H_3 (described in section 7.3.3.5);
- a P (Passive) flag that enables an endpoint to indicate that he expects the other side to initiate the ZRTP exchange;

- a list of supported hash, cipher and HMAC algorithms (to be used by the SRTP crypto-graphic context), ZRTP key agreement modes and SAS encodings, denoted as Options.

After the discovery phase, the actual ZRTP key agreement is triggered by one of the parties sending a *Commit* message and thus becoming the initiator. The *Commit* message contains the selected hash, cipher and HMAC algorithms as well as the ZRTP key agreement mode and SAS format. Depending on the selected key agreement mode, the *Commit* message contains:

- a hash commitment *HVi* (described in section 7.3.3.7) in Diffie–Hellman mode;
- a *Nonce* in multistream mode;
- a *Nonce* and a shared secret fingerprint *KeyID* in Preshared mode.

In the situation when both ZRTP endpoints simultaneously send a *Commit* message, the initiator is decided by assigning a higher precedence to the one that has selected the Diffie–Hellman key agreement mode, and a higher precedence to the higher *HVi* (respectively *Nonce*) value where both peers have selected the same key agreement mode.

7.3.3.2   The Diffie–Hellman Key Exchange and the Key Derivation

The *DHPart* messages exchange ephemeral Diffie–Hellman keys, which are used to derive a shared secret (denoted as *DHResult*). ZRTP also implements a key continuity mechanism (to be described in section 7.3.3.4) that uses cached shared secrets generated during previous sessions established with a certain peer as input for deriving the ZRTP master secret of the current session with the same peer. In order to enable the peers to check the consistency of their shared secrets cache, fingerprints calculated over the cached shared secrets are exchanged in the *DHPart* messages.

The ZRTP master secret s0 calculation is done according to the following formula:

$$s0 = \text{hash}(\text{counter} \mid DHResult \mid \text{"ZRTP-HMAC-KDF"} \mid \text{ZIDi} \mid \text{ZIDr} \mid total\_hash \mid$$
$$\text{len}(s1) \mid s1 \mid \text{len}(s2) \mid s2 \mid \text{len}(s3) \mid s3)$$

where the "i" and "r" suffixes denote the initiator and the responder, and

- *counter* is always set to 1;
- *DHResult* represents the Diffie–Hellman secret which is calculated by each party after receiving the peer's public key;
- *total_hash* represents a hash calculated over the concatenation of the responder's *Hello*, the initiator's *Commit* and the two *DHPart* messages of the ZRTP exchange (see Figure 7.16);
- *s1*, *s2* and *s3* represent placeholders for the cached shared secrets, which are either derived from the ZRTP master secrets of previous sessions established between the two parties or represent preconfigured shared secrets set up using an out-of-bound mechanism;

- the lengths of the *s1*, *s2* and *s3* keys–a length of zero indicates that the corresponding key is missing from the *s0* calculation.

The *s1*, *s2* and *s3* are obtained in the following way:

- *s1* represents one of the "retained secrets" *rs1* and *rs2*. The retained secrets and their use are described in section 7.3.3.4.
- *s2* is an optional auxiliary shared secret *auxsecret* that may be established or pre-configured using any available out-of-bound mechanism. It may be a hash calculated over the concatenation of the SRTP master and salting keys when such keys are available from the signaling layer (e.g. from SDES or MIKEY). Alternatively, *s2* may represent a hashed password that was previously set up on the ZRTP peers.
- *s3* represents an optional *pbxsecret* typically used in the "trusted man-in-the-middle" scenario (described in section 7.3.3.9).

The fingerprints exchanged in the *DHPart* messages are calculated by both the initiator and the responder for each of the *rs1*, *rs2*, *auxsecret* and *pbxsecret* by applying a HMAC function, using the respective shared secret as the key, to the character string constants "Initiator" and "Responder". Each endpoint then calculates the expected value of each fingerprint (by using the same string constant as the peer used for generating it) and compares it with the received one. A match between the received and the calculated fingerprints indicates that both the peers have the correct shared secret. In case of a mismatch, the respective shared secret will be omitted from the calculation of *s0* and its length will be set to zero.

By using different string constants as input to the HMAC function, the fingerprints calculated by the initiator and the responder for the same shared secret will be distinct. Also, if any of the cached shared secrets is missing (for instance *auxsecret* or *pbxsecret*), it will be replaced with a random number. The fingerprint verification will fail in this case and the ZRTP endpoints will therefore omit the random number from the *s0* calculation. However, this mechanism makes it impossible for an eavesdropper to detect if and which shared secrets are in fact missing. In particular, this conceals the situation when the ZRTP peers do not share any secret yet, which is when they are the most vulnerable to a man-in-the-middle attack.

Figure 7.17 illustrates the key derivation procedure used by the Diffie–Helman key agreement mode. The ZRTP master secret *s0* is used as input to a HMAC function, which is applied to a set of distinct character string constants to derive the following session keys:

1. The SRTP master and salting keys: *srtpkeyi*, *srptkeyr*, *srtpsalti* and *srtpsaltr*. The initiator uses the ⟨*srtpkeyi*, *srtpsalti*⟩ keys to derive the SRTP session keys to encrypt and integrity protect the outgoing SRTP/SRTCP streams, while the responder uses the same keys to validate the authentication tag and decrypt the incoming streams. The ⟨*srtpkeyr*, *srtpsaltr*⟩ keys are used in a similar manner for the SRTP/SRTCP streams flowing in the opposite direction.
2. The shared keys *zrtpkeyr* and *zrtpkeyi*, used by the responder to encrypt the *Confirm1* message and by the initiator to encrypt the *Confirm2* message. The same keys are used by the ZRTP peer to decrypt the respective messages.

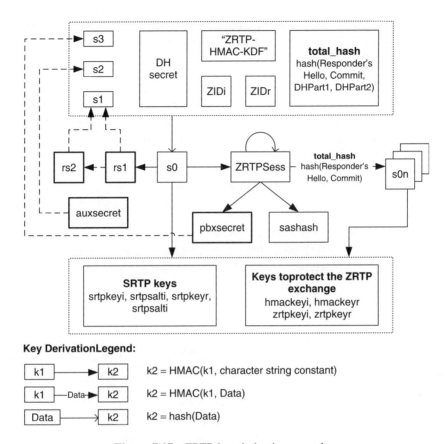

Key DerivationLegend:

k1 ➝ k2	k2 = HMAC(k1, character string constant)
k1 —Data➝ k2	k2 = HMAC(k1, Data)
Data ➝ k2	k2 = hash(Data)

Figure 7.17 ZRTP key derivation procedure

3. The shared keys *hmackeyr* and *hmackeyi*, used by the responder to integrity protect the *Confirm1* message and by the initiator to integrity protect the *Confirm2* message. They are also employed to integrity protect the *GoClear* messages. The same keys are used by the ZRTP peer to validate the authentication tag of the respective messages.
4. A ZRTP session key *ZRTPSess*, which is used to further derive the following session keys:

   (a) the *sashash*–used to generate the SAS;
   (b) the *pbxsecret*–used in trusted man-in-the-middle scenarios (see section 7.3.3.9);
   (c) a set of keys, one per media stream used to generate the set of session keys described at points 1, 2 and 3 above in multistream mode.

If the encryption is turned off during a session's lifetime, a fresh *ZRTPSess* key is generated by applying a hash function to itself. This ensures that former *ZRTPSess* secret keys cannot be easily compromised retroactively if a certain *ZRTPSess* is compromised at some time. The new *ZRTPSess* key will be used to generate the new ZRTP master secret using the multistream mode in case the encryption is turned on again during the session's lifetime.

Also, after each Diffie–Hellman calculation, a new retained secret $rs1$ is derived from $s0$ and the old $rs1$ is stored in $rs2$. The retained secrets themselves are fed back into the calculation of the next $s0$ for the purpose of providing key continuity (discussed in section 7.3.3.4).

The $s0n$ denotes the ZRTP master secret of the nth media stream and applies to the cases when the multimedia session contains more than one RTP session and hence a corresponding number of distinct ZRTP exchanges. The $s0n$ is derived using the ZRTP multistream mode from the *ZRTPSess*, rather than performing another Diffie–Hellman key exchange.

In order to achieve Perfect Forward Secrecy, all the shared keys illustrated in Figure 7.17, except for the retained secrets, must be erased as soon as they are no longer needed for deriving other keys (e.g. $s0$), or the protected media session requiring them is switched to cleartext or terminated.

The retained secrets are stored across sessions; therefore they alone cannot ensure PFS (which is the case of the ZRTP preshared mode), unless they are combined with the result of an ephemeral Diffie–Hellman exchange (as in the case of the ZRTP Diffie–Hellman mode).

7.3.3.3 Key Confirmation

The *Confirm* messages (*Confirm1*, *Confirm2* and *Confirm2ACK*, see Figure 7.16) conclude the ZRTP exchange. They provide key confirmation between the ZRTP peers and allow them to exchange encrypted and integrity protected data.

The *HMAC*s are calculated over the encrypted parts of the *Confirm1* and *Confirm2* messages using *hmackeyr* and respectively *hmackeyi* (see section 7.3.3.2).

The encryption uses AES in Cipher Feedback mode (section 2.1.1.2), the encryption keys *zrtpkeyr* and *zrtpkeyi* and the *IV*s that are sent along with the encrypted data.

The encrypted information carried in *Confirm1* and *Confirm2* consists of:

- the cache expiration interval–indicates the time interval after which the last created retained secret may be removed;
- the optional signature block–enables the ZRTP peers to exchange a digital signature over the SAS, whenever the ZRTP peers can use a PKI to obtain each other's signing key; if the SAS are authenticated using digital signatures, there is no need for the users to verify them interactively;
- H_0, a random number representing the initial value in a hash chain, used to protect the ZRTP exchange against packet injection (see section 7.3.3.5);
- a number of miscellaneous bits that indicate:
    - E (Enrollment) bit–indicates whether an enrollment procedure aimed at bootstrapping a trust relationship with a trusted third party is taking place (see section 7.3.3.9).
    - V (Verification) bit–indicates whether the SAS has been explicitly validated by the users on the ZRTP endpoints during the previous call. If this was the case, then, thanks to the key continuity mechanism (see section 7.3.3.4), there is no need for SAS verification in the subsequent calls.
    - A (Allow Clear) bit–indicates whether the ZRTP endpoint allows the communication to be switched from SRTP to cleartext RTP.

– D (Disclosure) bit–indicates whether the ZRTP endpoint has the ability to disclose the secrets that it shares with the peer to a third party. This is not a protocol feature but rather a measure to ensure that wiretapping-friendly ZRTP-enabled products behave in a transparent way.

Besides the actual key exchange, ZRTP also defines three additional exchanges to handle specific situations:

• The *Error/ErrorACK* exchange–allows a ZRTP exchange to be terminated if an error is detected.
• The *GoClear/ClearACK* exchange–allows the ZRTP endpoints to switch from SRTP to plain RTP at the user's request. Later on the communication may be re-encrypted by triggering a ZRTP handshake through some GUI element, such as a "Go Secure" button.
• The *SASrelay/RelayACK* exchange–enables ZRTP to support scenarios where a trusted intermediate entity terminates the ZRTP session from one participant and sets up a second ZRTP session to the other participant (see section 7.3.3.9).

7.3.3.4 Key Continuity

As soon as a ZRTP exchange is confirmed to have not been tampered with (flag V is set, see section 7.3.3.3), a key continuity mechanism is used to ensure that no man in the middle can sneak into any subsequent conversation between the respective ZRTP endpoints. The key continuity mechanism is based on the idea of using shared secrets generated in previous sessions in the process of deriving the shared secrets for the current session. For this purpose, ZRTP generates a new retained secret $rs1$ each time a new $s0$ shared secret is established, using the following formula:

$$rs1 = \text{HMAC}(s0, \text{``retained secret''})$$

and stores the previous $rs1$ in $rs2$ (see Figure 7.17).

Storing two consecutive retained secrets enables two ZRTP endpoints to recover from the situation when a ZRTP exchange is terminated abruptly so that, for example, the initiator derives a new $s0$ after receiving the *DHPart1* message but *DHPart2* never reaches the responder. In such a situation the $rs1$ of one peer will match the $rs2$ of the other peer during the shared secrets fingerprints comparison (see section 7.3.3.2). As a result both parties will set $s1$ to the matching retained secret and will resynchronize their caches.

The key continuity mechanism is similar to the "baby duck security model" (Gutmann 2008) employed by the Secure Shell (SSH) application. The "baby duck security model" ensures that any subsequent key exchange between two peers is secure based on the proof or assumption that no man in the middle was present at the time of the initial session establishment, when the endpoints cache the peer's credentials. In the case of SSH the credentials consist of public signing keys while in the case of ZRTP they consist of the retained shared secrets.

Because in ZRTP the retained secrets are updated with each ZRTP exchange, the window of opportunity for a man-in-the-middle attack where a retained secret is compromised is limited to the very next call. From this point of view the ZRTP model provides better protection than the SSH model, where a compromised private signing key will affect all subsequent sessions.

7.3.3.5 ZRTP Packet Injection Mitigation

Packet injection represents a significant class of DoS attacks that can be mounted by attackers that are not located on the media path.

One mechanism to protect a ZRTP exchange against packet injection is the use of *total_hash* (see Figure 7.17) in the calculation of *s0*. A mismatch, however, can only be detected after performing the calculation of the Diffie–Hellman shared secret, which is a computationally expensive operation.

ZRTP also makes use of a hash chain mechanism to protect a ZRTP exchange against malicious packet injection on a message-by-message basis, as the exchange proceeds, with minimal computational effort.

The hash chain elements are denoted as H_0, H_1, H_2 and H_3, where H_0 is a 256-bit random number independently chosen by each party and $H_{i+1} = \text{hash}(H_i)$, for $i = 0$–2, where the hash function is SHA-256 (SHS 2007). The hash values are provided in successive messages in the inverse order, starting with H_3 (see Figure 7.16). As a result, an eavesdropper (who does not posses H_0, which is sent last and encrypted) cannot provide a correct H_{i-1} value by only knowing H_i, which he saw in the previous message.

By verifying that $H_{i-1} = \text{hash}(H_i)$ the receiver can make sure that the message in the signaling sequence has been sent by the same ZRTP endpoint as the previous one. Finally, H_0 is sent encrypted in the *Confirm* messages and therefore only the H_0 produced by the peer of the Diffie–Hellman exchange will successfully validate the entire hash chain.

7.3.3.6 Man-in-the-middle Attack Mitigation

The SAS represents the primary mechanism used by ZRTP to authenticate the ephemeral Diffie–Hellman exchange and therefore protect against man-in-the-middle attacks. It consists of the validation of a short human-readable string over a secure out-of-bound channel, with the purpose of ensuring that the peers of the ZRTP exchange have obtained the same ZRTP master secret and therefore no man-in-the-middle attack occurred.

The SAS value is a human-readable string that usually consists of four "human-oriented" base32 symbols.[4] The SAS value is obtained by truncating and converting to base32 the *sashash* (see Figure 7.17). The SAS may be validated in the following ways:

- Interactively–the two persons that established the call verify the SAS by reading it aloud. This method takes advantage of the interactive nature of the real-time multimedia communications and has the merit of completely eliminating the need for a PKI. On the other hand, it requires that the end devices possess the capability to display the

---

[4] https://zooko.com/repos/z-base-32/base32/DESIGN

SAS so that the user can read it. Additional considerations regarding this method are addressed in Section 7.3.3.7.

- Automatically, over the media plane–if each of the ZRTP endpoints possesses a signing key of the peer, then the SAS may be signed and exchanged in the *Confirm* messages. This method, however, requires PKI support.

ZRTP provides two additional mechanisms to authenticate the ephemeral Diffie–Hellman exchange without performing an SAS validation:

- The key continuity mechanism (see section 7.3.3.4).
- The *a=zrtp-hash* media level attribute (described in section 7.3.3.8), which may be used to convey the protocol version and a hash value calculated over the *Hello* message. Provided that the signaling is integrity-protected end-to-end, H_3, which is carried in the *Hello* message, will be integrity protected as well (see Figure 7.16). In the next step, the *Hello*, *Commit* and *DHPart* messages are integrity protected by the *HMAC*s payloads, which use H_2, H_1 and respectively H_0 as keys. As a result, the ephemeral Diffie–Hellman exchange is authenticated between the signaling endpoints and an SAS validation is not necessary in this case. This method requires that ZRTP and the signaling layer interact so that the hash value calculated over the *Hello* message can be passed to the signaling plane.

7.3.3.7 Interactive SAS Validation

In order to be convenient for users to read and compare it, the SAS has a relatively small size (for instance four base32 symbols). This increases the chances of a man in the middle succeeding in finding two Diffie–Hellman keys that produce an SAS collision (see Figure 7.18a). In order to counter this possibility, a hash commitment is sent by the initiator of the ZRTP exchange in the *Commit* message. The hash commitment is carried in the *HVi* payload and represents a hash calculated over the responder's *Hello* message and the initiator's subsequent *DHPart2* message, which contains the initiator's public Diffie–Hellman key. Not only must a potential man in the middle now produce an SAS collision, but the generated public Diffie–Hellman must also yield the same hash value as the one sent in the hash commitment. This mechanism is illustrated in Figure 7.18b.

The interactive SAS validation has the considerable advantage that it does not rely on PKI. However, it also has a number of limitations. First, having the SAS spoken aloud by one of the participants requires that the phones of both participants are fitted with a display that the SAS can be read from. Also, a voice impersonation attack can be mounted if the participants do not *a priori* know each other's voices. The attacker, however, must carry on with the entire conversation[5] because passing it on to the legitimate endpoint any time after the SAS validation will be easily noticed by a change in the interlocutor's voice. The attack is therefore much more laborious as compared, for instance, with substituting the fingerprint of a public key in a session initiation message. Also, the legitimate participants have at their disposal means beyond the protocol itself to help them become more confident

---

[5] As well as subsequent ones, basically until he obtains the valuable information he was looking for and decides to drop off at the price of the victim immediately detecting the forgery.

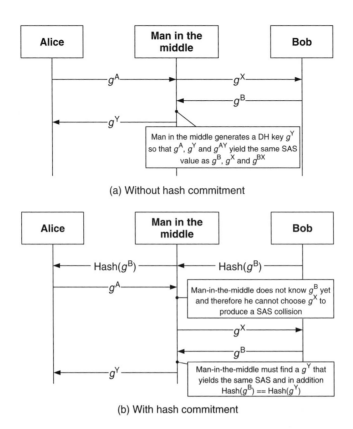

(a) Without hash commitment

(b) With hash commitment

Figure 7.18   A Diffie–Hellman exchange without and with a hash commitment

or more suspicious that they are or are not talking to the right person. However, SAS verification has little relevance when the callee is an automaton, such as an IVR system or a voicemail box. Finally, SAS verification cannot be performed where centralized media mixing is taking place, as for instance when a conference bridge is used to interconnect a number of participants.

There are concerns that, even when the parties are both humans and know each other's voices in advance, impersonation is still possible because a man in the middle only needs to be able to impersonate the victim during the SAS verification. In order to do so, the attacker could record the victim saying many or all of the 32 symbols that can appear in a base32 SAS during previous calls, and replay them accordingly during the SAS verification in a subsequent call. The difficulty for the attacker, however, is that he cannot determine in advance when, how and in what context the participants will read the SAS. The participants may, for instance, validate the SAS in a puzzle-solving fashion like: "In the word 'puzzle' the letter that appears twice is the same as the third symbol of the SAS". Moreover, thanks to the key continuity mechanism (see section 7.3.3.4), a thorough SAS validation need not be done for every session; therefore the participants can afford to do it in a quite elaborate manner.

```
v=0
o=bob 2890844527 2890844527 IN IP4 client.biloxi.example.com
s=
c=IN IP4 client.biloxi.example.com
t=0 0
m=audio 3456 RTP/AVP 97 33
a=rtpmap:97 iLBC/8000
a=rtpmap:33 no-op/8000
a=zrtp-hash:1.00fe30efd02423cb054e50efd0248742ac7a52c8f91bc2df881ae642c371ba46df
```

Figure 7.19   Example of an SDP offer containing a *zrtp-hash* attribute

7.3.3.8   ZRTP Signaling Plane Attribute

Figure 7.19 illustrates the use of the $a=zrtp$-$hash$ media level attribute, which is used to:

- Indicate the Protocol Version.
- Authenticate the ZRTP exchange provided that the signaling is integrity-protected end-to-end (as described in section 7.3.3.6).
- Provide a binding between the endpoints in the media plane (and the SAS resulting from the ZRTP exchange) and the endpoints in the signaling plane.
- Indicate support for ZRTP in the signaling plane and hence allow the parties to negotiate a key management protocol when more than one of them is supported by at least one of the parties. While exchanging *Hello* messages is the primary ZRTP discovery mechanism, not explicitly announcing the ZRTP support in the signaling plane may lead to a duplication of security functions when MIKEY or SDES are also supported by the endpoints.

Multiple $a=zrtp$-$hash$ attributes may be present if a ZRTP endpoint implements multiple protocol versions. In absence of the $a=zrtp$-$hash$ attribute, two ZRTP endpoints will negotiate the protocol version by exchanging multiple *Hello* rounds, during which the ZRTP endpoint implementing a higher protocol version will fall back to a lower protocol version (provided that it supports one) until the peers agree or abort the ZRTP exchange.

7.3.3.9   Trusted Man-in-the-middle Scenarios

ZRTP provides support for scenarios when an end-to-end ZRTP exchange between the participants is not possible. In such a case, an intermediate entity (e.g. a PBX) acting on behalf of one of the participants may be registered as a trusted man in the middle and relay the ZRTP exchange between the two participants. Figure 7.20 illustrates such a scenario, where Bob is located behind a PBX. In the first step Bob must enroll his phone with the PBX, which results in Bob's phone and the PBX sharing a secret, namely *pbxsecret* (see Figure 7.17), which is used to provide key continuity. The enrollment may take place, for instance, by having Bob initiate a call to a special extension of the PBX or any other mechanism that enables *pbxsecret* to be installed on the two parties. When Alice calls Bob, the PBX terminates the ZRTP session with Alice (step 2 in Figure 7.20) and

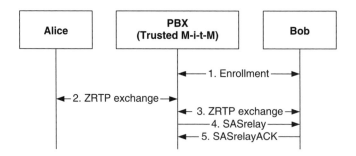

Figure 7.20 A ZRTP trusted man-in-the-middle scenario

establishes another one with Bob (step 3). This results in Alice and Bob ending up with different SAS values but the PBX facilitates the SAS validation by relaying to Bob the value of the SAS established between Alice and the PBX using an *SASrelay/RelayACK* exchange (steps 4 and 5).

7.3.3.10 Support for Forking, Retargeting and Early Media

Retargeting does not raise any specific issues because the ZRTP exchange and SAS verification will take place between the UAC and the ultimate UAS. Intermediate entities are not supposed to interfere with the ZRTP signaling, except for the trusted man-in-the-middle scenario.

When forking occurs, a distinct ZRTP session will be established with each of the UASs that terminate a forked branch. Each ZRTP session will have a distinct SAS and each ZRTP session will negotiate security parameters and will generate keying material for a distinct media session.

No media clipping occurs where early media is being sent by the UAS because the early media is sent over RTP and then switched to SRTP after the peers have completed the derivation of the SRTP session keys, which occurs after the Diffie–Hellman exchange has been completed and hence after the UAC had received the SDP answer. By the time the early media starts to be protected, the identity of the sender can also be validated except for the case where the interactive SAS validation is necessary and can therefore only take place after the establishment of the multimedia session.

7.3.4 DTLS-SRTP

DTLS-SRTP (McGrew 2008) is an extension of the DTLS protocol that enables the participants in a point-to-point media session to agree on the cryptographic parameters and derive keying material necessary to establish SRTP sessions.

The DTLS-SRTP handshake takes place in the media plane and is multiplexed on the same UDP ports as the media itself. DTLS-SRTP inherits the TLS model that a DTLS association (the correspondent of a TLS connection) protects the data exchanged between one pair of local and remote UDP port numbers. As a result, one distinct DTLS association will correspond to each SRTP or SRTCP stream. A bidirectional media session will require the setup of four DTLS-SRTP sessions that will protect the SRTP and the SRTCP streams in each direction. This overhead can, however, be significantly reduced in the following ways:

- When symmetric RTP (Wing 2007) is being used, the number of DTLS-SRTP sessions will be reduced to two, one protecting the SRTP streams and one protecting the SRTCP streams, in both directions.
- Also, RTP and RTCP can in fact be multiplexed on the same port (Perkins 2007), which also reduces the number of DTLS-SRTP sessions by a factor of 2.
- When multiple DTLS handshakes must be performed within the same multimedia session, the DTLS session resumption mechanism should be used to reduce the amount of signaling and cryptographic calculations in all subsequent DTLS handshakes that follow the initial one.

The DTLS-SRTP handshakes are initiated as soon as a DTLS endpoint (which becomes the DTLS client for the respective DTLS association) learns the media port numbers advertised by the DTLS peer (which becomes the DTLS server) in the SDP payload.

7.3.4.1 Negotiation of the SRTP Cryptographic Parameters

A DTLS-SRTP handshake starts with the DTLS client sending a *Client Hello* message containing an *use_srtp* extension. The *use_srtp* extension carries a list of SRTP Protection Profiles and an optional MKI value. The server chooses one of the SRTP Protection Profiles proposed by the client and returns it in an *use_srtp* extension in the *ServerHello* message. If none of the SRTP Protection Profiles are acceptable to the server, no *use_srtp* extension is returned, in which case the DTLS handshake may be aborted.

The SRTP Protection Profiles are (in a similar manner to the TLS cipher suite specifications) constant code values (e.g. SRTP_AES128_CM_SHA1_80), each of them defining one combination of: (i) the encryption algorithm, block size and associated cryptographic mode (any of NULL or AES on 128 or 256 bits in Counter Mode) and (ii) the hash algorithm (HMAC-SHA1) and SRTP authentication tag length (32 or 80 bits).

For all other parameters DTLS-SRTP uses the SRTP default values for: (i) the length of the keys that correspond to the selected cryptographic algorithms; (ii) the *key derivation rate* (which is 0); (iii) the lifetime of the keys (which is 2^{31}); (iv) the relative order of applying Forward Error Correction and SRTP processing (FEC and then SRTP on the sender); and (v) the SRTP replay window size (64).

The default values of some parameters, such as the relative order of applying Forward Error Correction and SRTP processing, or the SRTP replay window size, may be overridden by explicitly providing them over the signaling path in a similar manner to the ⟨session-params⟩ parameter of the *a=crypto* SDP attribute in SDES (see section 7.3.1.1). These parameters are declarative (they are specified by the sender of the media and are not subject to negotiation) and are carried along with the sender's media description; therefore they are known at the time the DTLS-SRTP handshake is initiated.

When the *use_srtp* extension has been negotiated between two DTLS endpoints, the application data sent by the local application will be delivered to SRTP rather than sent through the DTLS record layer (see section 2.2.2.1). On the receiving end, datagrams arriving at the media transport address are differentiated between RTP, DTLS handshake messages and STUN datagrams based on the first octet, which takes values in disjoint intervals for the three protocols.

7.3.4.2 Late Binding of SRTP Cryptographic Contexts

It must be observed that, even though in DTLS-SRTP the security parameters of a given SRTP cryptographic context are established in the context of a DTLS association identified by the local and the remote transport addresses (consisting of the IP address and the port number), the SRTP streams are matched against the cryptographic context based on the destination transport address and SSRC value. Because DTLS-SRTP associations are agnostic of the SSRC parameter, a "late binding" mechanism is necessary to enable the receiving endpoints to bind the SRTP cryptographic contexts to the actual SSRC values.

Figure 7.21 illustrates the relationship between the DTLS-SRTP sessions and the SRTP cryptographic contexts for a bidirectional media session that does not use symmetric RTP and RTP-RTCP multiplexing. The arrows representing the DTLS handshakes indicate

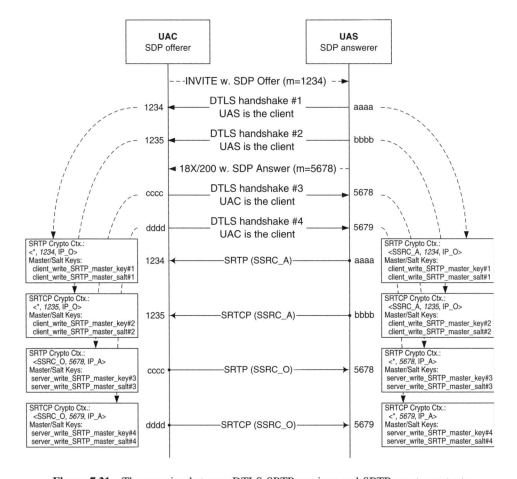

Figure 7.21 The mapping between DTLS-SRTP sessions and SRTP crypto contexts

the initiating direction (client→server). The "*" symbol is used to mark the "unbound" SRTP/SRTCP cryptographic contexts. The "late binding" mechanism works according to the following algorithm:

1. The destination transport address and SSRC of the media packet received are checked against the "bound" SRTP cryptographic contexts. If a match is found then the media packet is authenticated and decrypted accordingly. If no match is found then the algorithm proceeds with step 2.
2. The destination transport address of the received media packet is matched against all the existing SRTP cryptographic contexts that share the respective transport address. If no match is found then the packet is discarded. Otherwise, the algorithm proceeds with step 3. This check needs to be performed even after all the SRTP cryptographic contexts have been bound to SSRC values because the SSRC values may change during the lifetime of a media session. Also, multiple SSRC values may appear in the media packets received at a certain transport addresss (e.g. the destinations of a forked call sending media to the caller).
3. The SRTP packet is authenticated using the SRTP authentication key of each of the matching SRTP cryptographic contexts. If successful, a new instance of the respective SRTP cryptographic context is created and bound to the SSRC value of the received packet and the algorithm terminates. If the authentication fails, then the packet is discarded. This trial and error approach leads to an amplification of a DoS attack by a factor equal to the number of DTLS associations set up for a given transport address (as in the case of a forked call). The effect on the CPU usage is, however, limited because only authentication operations using symmetric cryptographic algorithms are involved. Additional protective measures may be taken by prioritizing the processing of packets having known SSRC values. Also, the source transport address of the received media may be checked to match that of a DTLS association established with the local transport address.

7.3.4.3 Key Derivation

DTLS-SRTP uses the TLS keying material extractor mechanism (Rescorla 2008) to derive, in addition to the keying material used by the DTLS record layer itself, one distinct ⟨SRTP master key, SRTP salting key⟩ key pair for each direction (see Figure 7.22). For this purpose DTLS-SRTP uses a predefined label value of "EXTRACTOR-dtls_srtp" as input to the TLS key derivation function (see section 2.2.2.1).

When RTP and RTCP are not multiplexed on the same ports, the RTP stream and its corresponding RTCP stream will be associated with different DTLS-SRTP associations and as a result distinct SRTP master and salting keys will be used by the SRTP and SRTCP cryptographic contexts. This represents a slight deviation from the SRTP model, where the SRTP and the corresponding SRTCP cryptographic contexts share most of the security parameters, including the master and salting keys. A DTLS-SRTP implementation must, however, ensure that the same cipher suites are negotiated for the SRTP and its corresponding SRTCP cryptographic contexts. Perfect Forward Secrecy of the keying material can be ensured by using one of the ephemeral public key exchange modes of DTLS.

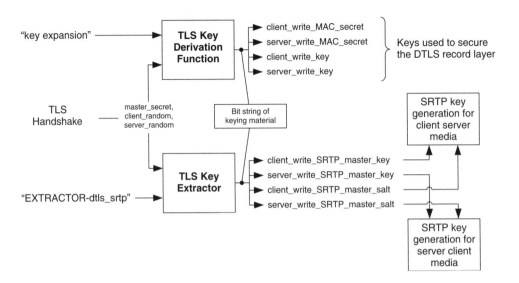

Figure 7.22   DTLS-SRTP key derivation

7.3.4.4   Authentication of the Key Exchange and Interaction with the Signaling Plane

The DTLS-SRTP handshake exchanges public keys, which are transported in the DTLS messages in the form of self-signed certificates. Both the client and the server must send a public key, meaning that, when *use_srtp* is present, the server must request a certificate from the client and the client must send a *Certificate Verify* message. This ensures that the handshake is mutually authenticated between the DTLS peers.

In order to protect the DTLS-SRTP handshake against man-in-the-middle attacks, it must also be verified that a binding exists between the endpoints that established the DTLS-SRTP association and those that signaled the setup of the multimedia session. This is achieved by means of the *a = fingerprint* SDP attribute (Lennox 2006), which carries a hash calculated over the self-signed certificate that the DTLS endpoint sent during the DTLS-SRTP handshake.

Using the *a=fingerprint* attribute to provide identity authentication for the DTLS-SRTP handshake requires that either (i) the signaling path is end-to-end integrity protected or (ii) the endpoints have previously cached each other's fingerprint and hence bootstrapped a "baby duck security model". Caching a fingerprint should be done only when the signaling is integrity-protected end-to-end; subsequent DTLS-SRTP associations can then be securely established in absence of signaling plane protection. However, caching public key fingerprints does not eliminate the danger of user impersonation, which becomes possible if the corresponding private key is compromised.

7.3.4.5   Support for Forking, Retargeting and Early Media

DTLS-SRTP supports retargeting because the DTLS-SRTP session will be established between the UAC and the ultimate UAS. Forking scenarios are accordingly also supported;

the DTLS associations are mapped to the corresponding SIP sessions by means of the $a = fingerprint$ attribute (included in the SDP payload).

DTLS-SRTP enables the early media to be encrypted because the setup of the SRTP cryptographic contexts that protect the early media takes place as soon as the UAS receives the INVITE request. However, the identity of the sender of the early media can only be validated after the caller receives the SDP answer carrying the callee's fingerprint. This provides a window of opportunity to an attacker that eavesdrops the SDP offer to establish a DTLS association instead of the legitimate callee and send early media before the caller receives the callee's fingerprint and has the chance to authenticate the public key used by the peer to establish the DTLS association.

7.3.4.6   DTLS-SRTP Extension for Group Communications

DTLS-SRTP was originally designed to address one-to-one communication scenarios. It therefore does not scale well to one-to-many scenarios because the sender has to authenticate and encrypt the media stream individually for each receiver.

The DTLS-SRTP Key Transport (Wing 2008) defines a new upper layer DTLS protocol that runs on top of a protected DTLS Record Layer and enables a sender to push the SRTP keying material and SSRC value to a receiver with whom he has already established a DTLS association. This allows the DTLS-SRTP to address small group one-to-many scenarios, where the sender performs the functions of a key server that distributes the same key to all the receivers.

7.3.5   The Capability Negotiation Framework

Putting in place a security mechanism for the media traffic requires the participants to negotiate, in the first place, the following two aspects:

1. The RTP profile–this aspect is related to achieving "best-effort SRTP", which means using SRTP (RTP/SAVP profile) with those peers that support it and plain RTP (RTP/AVP profile) with the others. ZRTP allows the participants to discover whether they support SRTP; however if this capability negotiation is missing from the key management protocol, the participants need to announce and negotiate during the session establishment the use of one of the RTP/AVP or RTP/SAVP profiles.
2. The key management protocol–negotiating between different key management protocols that are transported within the $a = key\text{-}mgmt$ attribute is straightforward; however, currently only MIKEY is transported in this way and in fact each key management protocol defines a different subset of SDP attributes.

The Session Description Protocol (SDP; Handley *et al.* 2006) was originally specified for the purpose of announcing the media parameters of the multimedia sessions and therefore it offered very limited negotiation capabilities. The widespread use of SDP led to an increasing demand to address more complex scenarios that required a certain degree of capabilities negotiation between the participants in a multimedia session. As a result, the SDP offer/answer model (Rosenberg and Schulzrinne 2002a) was proposed.

However SDP and its existing extensions still do not offer a convenient way to negotiate one out of several alternative RTP profiles and combine it with one out of several sets of

SDP attributes. The SDP capability negotiation framework proposed in (Andreasen 2008) provides a backward-compatible way to address these limitations.

In brief, the SDP capability negotiation framework defines a number of new attributes:

- A supported capability attribute *a=acap* that defines one SDP attribute as a capability. The *a=acap* attribute is followed by a numeric handler and the actual SDP attribute. The handler enables the SDP attribute to be subsequently combined with other *a=acap* attributes and RTP profiles to complete a media description. An *a=acap* attribute may be specified at both session or media level.
- A transport capability attribute *a=tcap* that specifies a list of supported RTP profiles.
- A potential configuration attribute *a=pcfg* that specifies a combination of supported and transport capabilities (contained in the *a=acap* and respectively *a=tcap* attributes). For this purpose the *a=pcfg* attribute uses the numeric handler associated with the *a=acap* attributes and the index of the RTP profiles listed in the *a=tcap* attribute. The *a=acap* attributes may be combined inside the same *a=pcfg* attribute by enumerating *a=acap* handlers, using "or" constructors (|) or marking then as optional by including the handlers in brackets ([,]). The *a=pcfg* attributes are followed by a number indicating the preference level assigned by the offerer to the resulting media description and can only be specified at the media level.
- An actual configuration attribute *a=acfg* which is used by the answerer to specify which (if any) of the provided potential configurations has been selected.

Figure 7.23 provides an example of an SDP offer that provides a total of nine possible choices for establishing a multimedia session that consists of one audio and one video session. Each of the two media sessions may be use either "plain" RTP or SRTP (indicated by the presence of the RTP/SAVP profile in the transport capabilities a=tcap attribute).

A callee that does not support SRTP will select "plain" RTP for both media sessions. On the other hand, if SRTP is used, either of MIKEY (session level attribute) or SDES

```
v=0
o=alice 2891092738 2891092738 IN IP4 lost.example.com
s=
t=0 0
c=IN IP4 lost.example.com
a=acap:1 key-mgmt:mikey AQAFgM0XflABAAAAAAAAAAAAAAAAsAyO...
a=tcap:1 RTP/SAVP RTP/AVP
m=audio 59000 RTP/AVP 98
a=rtpmap:98 AMR/8000
a=acap:2 crypto:1 AES_CM_128_HMAC_SHA1_32
inline:NzB4d1BINUAvLEw6UzF3WSJ+PSdFcGdUJShpX1Zj|2^20|1:32
a=pcfg:1 t=1 a=1|2
m=video 52000 RTP/AVP 31
a=rtpmap:31 H261/90000
a=acap:3 crypto:1 AES_CM_128_HMAC_SHA1_80
inline:d0RmdmcmVCspeEc3QGZiNWpVLFJhQX1cfHAwJSoj|2^20|1:32
a=pcfg:1 t=1 a=1|3
```

Figure 7.23 An SDP offer using the SDP capability negotiation framework

```
v=0
o=- 24351 621814 IN IP4 192.0.2.2
s=
t=0 0
c=IN IP4 192.0.2.2
m=audio 54568 RTP/SAVP 98
a=rtpmap:98 AMR/8000
a=crypto:1 AES_CM_128_HMAC_SHA1_32
inline:WSJ+PSdFcGdUJShpX1ZjNzB4d1BINUAvLEw6UzF3|2^20|1:32
a=acfg:1 t=1 a=2
m=video 55468 RTP/SAVPF 31
a=rtpmap:31 H261/90000
a=crypto:1 AES_CM_128_HMAC_SHA1_80
inline:AwWpVLFJhQX1cfHJSojd0RmdmcmVCspeEc3QGZiN|2^20|1:32
a=acfg:1 t=1 a=3
```

Figure 7.24   An SDP answer using the SDP capability negotiation framework

(media level attributes) may be employed to set up the cryptographic contexts for each of the protected media streams: $a = 1|2$ means either *acap:1* or *acap:2*, and $a = 1|3$ means either *acap:1* or *acap:3* will be used.

If $a = pcfg:1$ $t = 1$ $a = 1$ is selected by the answerer for both media sessions, then the MIKEY handshake will provide keying material for both media streams, since the key-mgmt attribute ($a = acap:1$) is specified at the session level. It is also possible, although there is no obvious use case for it, that MIKEY is used for one media stream and SDES for the other.

Figure 7.24 illustrates an SDP answer where the answerer has selected SDES for both media streams (as indicated by the actual configuration $a = acfg$ attributes). In this case, the RTP/SAVP profile is specified in the media lines of the SDP answer and the answer will provide its own SRTP keys in the *a=crypto* attributes.

The use of the SDP capability negotiation framework may be equally used to announce support for the DTLS-SRTP and ZRTP key management protocols by listing the $a = fingerprint$ and respectively the *a=zrtp-hash* attributes as supported capabilities.

7.3.6   Summary

We have presented so far a number of key management protocols that enable participants in multimedia sessions to negotiate the security parameters and the keying material necessary to protect the media exchange with SRTP. We have also discussed aspects related to forking, retargeting and early media as well as their applicability to small group communications. In this section we will summarize the security dependencies and vulnerabilities of these key management protocols (see Table 7.1) based on a number of criteria.

The first criterion is the dependency of the key management protocol on the security of the signaling plane. This is in fact a weakness of the key management protocol because signaling cannot be easily protected (and in particular encrypted) end-to-end, primarily because intermediate SIP entities involved in the routing of the SIP messages need access to various SDP attributes (besides the SIP headers) and also because in many situations a PKI to enable the terminals to mutually authenticate end-to-end is not available. As

Table 7.1 A comparison of the key management protocols

Criteria	SDES	MIKEY	ZRTP	DTLS-SRTP
Dependency on signaling plane security	End-to-end encryption	End-to-end integrity protection	End-to-end integrity protection during initial contact. None, if SAS verification can be performed	End-to-end integrity protection during initial contact
Vulnerability due to compromised long-term credentials	Yes	Yes	Limited to next session, can be detected in subsequent sessions	Yes
DoS attacks on the key exchange protocol	Flood with INVITEs	Flood with INVITEs	Flood with INVITEs provided that the victim sends back SIP responses	Flood with INVITEs, provided that the victim responds with *ClientHello* messages. Also, an eavesdropper in either the signaling or media plane can force the caller to perform bogus DTLS handshakes until he receives the client's fingerprint
Amplification factor for DoS attack on the media plane due to "late binding"	Yes	No	No	Yes

a result, in most scenarios the signaling security must rely on a transitive hop-by-hop trust model (see section 3.9.1.3). Alternatively, an authentication service may be used to authenticate the identity of the sender of a SIP request and offer integrity protection to security-related SDP attributes. This mechanism, however, works in an unidirectional fashion, therefore it needs to be applied in both the direct and the reverse ("connected identity") directions (see section 6.4). ZRTP and DTLS-SRTP are the least dependent on the signaling plane being protected; they make use of key continuity mechanisms, which require the signaling to be end-to-end integrity protected only when the communicating parties make their first contact. In addition to that, ZRTP can also bootstrap a secure relationship between two parties without any security support from the signaling in the case when the participants can rely on the SAS validation (the participants are humans and are able to recognize or "trust" each other's voice).

The second criteria concerns the vulnerability of the key management protocol in case the credentials used for the identity authentication (e.g. shared secrets, private signing keys), including those used for key continuity, are compromised. In such a situation, an attacker can impersonate the victim in all subsequent calls. ZRTP offers the best protection, by reducing the window of opportunity for an attack to take place to the very next call (provided that a SAS verification is not performed) and enabling the legitimate entities to subsequently observe that an attack has occurred.

The third criteria is related to the possibility of mounting DoS attacks on the key management protocols by initiating bogus handshakes or by injecting faked packets into the handshake. This would force the communicating parties to perform useless asymmetric key computations or abort the session. In case of DTLS-SRTP, an attacker that eavesdrops either the signaling or the media plane can inject bogus DTLS handshakes. The window of opportunity for such attacks is, however, limited to the time interval between the initial INVITE and DTLS *ClientHello* message until the fingerprint of the client's public signing key is received by the caller in the SDP answer.

Another possible attack in this category is flooding the victim with INVITE messages. In the case of SDES, this will result in the creation of many SRTP cryptographic contexts; MIKEY will also trigger a corresponding number of asymmetric key computations. In DTLS-SRTP the INVITEs themselves do not trigger any cryptographic operations, however the attacker can respond with a sequence of *ServerHello ... ServerHelloDone* to each *ClientHello* the victim sends, forcing the DTLS client to perform asymmetric key computations. In ZRTP the attacker has to wait for the SDP answer before being able to continue with the handshake and trigger asymmetric key computations on the victim's side.

The last criterion concerns the ability of the key management protocol to signal the SSRC values of the protected media streams. If this capability is missing, then a late binding mechanism must be used (see section 7.3.4.2). This results in an SRTP endpoint trying to authenticate SRTP packets carrying unknown SSRC values against all SRTP cryptographic contexts (including the bound ones, since SSRC values may change during the session's lifetime). This facilitates and amplifies the DoS attacks because the attacker need not use a valid SSRC value and also one single bogus SRTP packet triggers MAC verification operations against multiple SRTP cryptographic contexts.

8

Denial-of-service Attacks on VoIP and IMS Services

8.1 Introduction

The goal of denial-of-service attacks is to prevent legitimate users from using the attacked victim host. This can be achieved by flooding the victim host with a high number of useless packets that deplete one or more resources of the host, such as bandwidth, memory or CPU, so that the victim no longer has sufficient resources left for serving legitimate requests. Another type of attack is based on misusing certain vulnerabilities of the victim host's software. Using only a small number of malicious data packets, an attacker can take complete control of the host or cause it to be unavailable for legitimate requests. To increase the difficulty of discovering and tracing the source of the attack back, denial of service attacks are usually conducted in a distributed manner. With this approach, a large number of systems on the Internet are manipulated by the attacker and caused to send traffic to the victim. The first of such distributed denial of service (DDoS) attacks were already being observed in the summer of 1999 (Criscuolo 2000). The first high profile attack targeted Yahoo.com and cost Yahoo a significant loss in revenue (Wired 2000).

Security threats, especially denial-of-service attacks, are considered minimal in current circuit-switched telecommunication networks. This is achieved by using a closed networking environment dedicated to a single application, namely voice. The end devices are usually rather simple in terms of required hardware and software and are, hence, difficult to manipulate and to misuse for launching denial of service attacks. Furthermore, the access to the network is restricted and closely monitored, making the detection of suspicious activities simple.

Compared with PSTN networks, VoIP services are much more open. When looking at a typical VoIP setup, see Figure 8.1, we can distinguish between three main parts, namely, the user premises, the VoIP infrastructure and a number of supporting services.

SIP Security   Dorgham Sisalem, John Floroiu, Jiri Kuthan, Ulrich Abend and Henning Schulzrinne
© 2009 John Wiley & Sons, Ltd

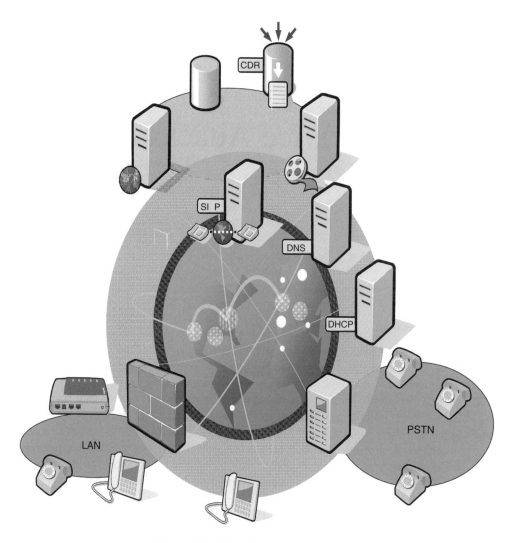

Figure 8.1   Overview of a VoIP service

User premises   The user premises include all the devices and applications needed by a private user or an enterprise to use a VoIP service. This might include any of the following:

- **User devices**–often called user agents (UA), user equipment (UE) or customer premises equipment (CPE). The user devices represent the interface of the users to the VoIP service. User devices can be a VoIP phone that connects to the Internet through an Ethernet or wireless LAN interface, an IAD that allows the connection of a PSTN phone to a VoIP service or a software application running on a PC or a mobile phone.
- **NAT and firewalls**–firewalls are used to protect a network from attacks and misuse. Network address translators hide the internal network structure from the outside and

enable the sharing of one or more public IP addresses among a larger number of hosts. These devices often have extensions for supporting VoIP, with the goal of blocking unwanted traffic and enabling only legitimate users to access VoIP services.

- **Gateways and PBX**–PBXs are often deployed in enterprise environments and offer advanced telephony services such as call forwarding or conferencing. Increasingly, IP-based PBX systems are replacing traditional PBX systems and offer the same services. Additionally, such systems often act as gateways between the VoIP service and the PSTN network.
- **DHCP**– for any device to be able to send and receive data packets over the Internet it requires an IP address. The Dynamic Host Configuration Protocol (Droms 1993) reduces the configuration overhead required for allocating an IP address to each VoIP component. This is done by enabling a host that requires an IP address to communicate with the DHCP server and to retrieve an IP address without any manual interaction by the administrator of the device.

SIP-based VoIP infrastructure   The SIP-based VoIP infrastructure includes components needed by the service provider to be able to offer VoIP and other SIP-based services. This includes the signaling components such as the SIP proxy and registrar, session border controllers, application servers such as conferencing or voicemail servers and gateways to traditional networks.

- **SIP servers**–SIP servers are responsible for processing the SIP messages, routing the SIP messages to the correct destination and possibly executing additional services such as user authentication and PSTN-like services such as call forwarding. In the terminology of the SIP specifications (Rosenberg *et al.* 2002b), these servers are usually called proxy or redirect servers and in the IMS specifications such servers are called the call session control functions. Session border controllers, which also deal with routing of SIP messages and with the provision of services, are also considered as SIP servers here.
- **PSTN gateways**–Often also called media and signaling gateways or interworking units (IWU), PSTN gateways connect the Internet to the PSTN and enable calls from the Internet to the PSTN and vice versa. These components have the following functionalities:
  - termination of PSTN signaling;
  - transcoding functionality (voice encoding from G.711 to other media encoding such as G.727 or G.729);
  - splitting voice samples into RTP packets.
- **Application servers**–application servers are components that enhance the VoIP service with additional services such as conferencing, voicemail or integration with other applications, such as calendars or media players.
- **NAT-traversal support**–SIP messages carry addressing information of the communicating end parties as part of the message content. A user device that is behind a NAT has only a private IP address that cannot be contacted from the public Internet. To enable calls to and from such user devices, additional mechanisms are needed. Technologies such as STUN (Rosenberg *et al.* 2003), TURN (Rosenberg *et al.* 2008c) or

ICE (Rosenberg 2007) require the service provider to maintain additional servers that can themselves become the targets of DoS attacks.

Supporting services For a VoIP service to actually function, a number of supporting services are needed. These include DNS servers for resolving names into IP addresses, databases for collecting user information and provisioning systems for managing user profiles and system configuration.

- **Address resolution servers**–SIP addresses are usually described as URIs in the form of "sip:user@domain" or as E.164 (ITU-T Rec. E.164 2005) numbers such as "+49303030". As with any Internet service, there is a need to translate between the high-level service-specific names and IP addresses. For this there are two major components in the VoIP architecture:
  - DNS servers–domain name servers (Mockapetris 1989) constitute a distributed database enabling the mapping between host or service names and IP addresses.
  - ENUM severs–Telephone Number Mapping, ENUM (Peterson *et al.* 2004) is a DNS-based service that enables the mapping between an E.164 number and a SIP URI. This service allows a VoIP user to have an E.164 number that can be dialed from the PSTN and that still gets routed to a VoIP device.
- **AAA servers**–authentication, authorization and accounting servers contain the necessary information to authenticate a user (e.g. password) as well as the user profile. The user profile indicates in general the user's specific services, such as white and black lists or call forwarding specification. AAA servers can be considered as general databases that are accessed by the SIP servers directly using application programming interfaces or communication protocols such as RADIUS (Rigney *et al.* 1997) or DIAMETER (Calhoun *et al.* 2003).
- **Provisioning systems**–provisioning systems allow the service provider to configure the offered services, add or delete subscribers from the database and query the logging information and call data records generated by the VoIP infrastructure. Provisioning systems also enable the subscribers to configure their profile or configure call forwarding, for example, or check their call logs using a web interface.
- **UA configuration servers**–to properly function in a VoIP environment, user agents require certain information. This information includes the user's domain address, the addresses of the VoIP servers the device should be contacting and other information related to NAT traversal. Configuration servers automate this procedure and relieve the user from having to enter this information manually into the user device.

From a technical point of view, a SIP-based voice over IP service resembles an email service more than a traditional telecommunication service. VoIP components use the same physical infrastructure as email services whether fixed or wireless, the same transport protocols, e.g. TCP/IP, and very often the same operating systems and hardware, e.g. PCs. Hence, it is only natural to expect a VoIP system to suffer from the same security threats as an email system or any other system on the Internet. Further, attackers can use the specific nature of VoIP systems, i.e. the used VoIP protocols and components, to launch DoS attacks on VoIP infrastructures. We can distinguish three distinct attack methods that can lead to complete or partial VoIP service interruption:

- Flooding attacks–these attacks aim at depleting the resources of VoIP components so that the VoIP components become unavailable for processing legitimate VoIP requests or make the processing time considerably longer.
- Misuse attacks–with this kind of attacks, an attacker misuses SIP messages in order to illegally use the VoIP service, terminate sessions of other users or forward calls to malicious destinations. While these kinds of attacks do not cause the complete service to become unavailable, they can deny one or more users access to the VoIP service.
- Indirect attacks–For a VoIP service to function properly, a number of supporting services are needed. These services are responsible for address resolution, provisioning and billing. By attacking these systems or interrupting the communication between the VoIP components and these supporting servers, an attacker can misuse the service and considerably reduce the availability of the VoIP service.

The effects of a DoS attack on a VoIP system depend very much on the nature of the victim. On the one hand, targeting a DoS attack on a user agent only results in preventing a user from using her phone. On the other hand, targeting a SIP proxy prevents all users served by this proxy form receiving or initiating any calls. While we will be addressing attack scenarios targeting different components of a VoIP service, our main concentration will be on attacks targeting the VoIP infrastructure. On the one hand, infrastructure is the most obvious target for attackers. On the other, the effects of a failed infrastructure component are much more devastating both in terms of service availability and loss of revenue and reputation to the provider.

8.2 General Classification of Denial-of-service Attacks

Denial-of-Service attacks try to prevent a legitimate user from accessing a service or a network resource, or bring down the servers offering such services. While such attacks might target hosts belonging to individuals or small enterprises, often the targets of denial of service attacks are servers offering some sort of services to a larger user community, e.g. web or mail servers of ISPs or enterprises. In the last few years DoS attacks have increasingly become a major problem for computer security and have increased in frequency, severity and sophistication. According to a 2007 CSI/FBI survey report, 25 % of respondents reported that they have suffered DoS attacks directed against them (Richardson 2007).

In communication networks, denial-of-service attacks can target protocol processing functions at different layers of the protocol architecture. That is, attacks can be launched on the network, transport or application layers and can be roughly classified as illustrated in Figure 8.2. With this classification, we mainly distinguish between nonintentional attacks and intentional or malicious attacks. Whereas nonintentional attacks are usually the result of an implementation or configuration error, malicious attacks are initiated intentionally by attackers. Intentional attacks can be further subdivided into flooding and protocol misuse attacks. Flooding attacks, also referred to as exhaustion or depletion attacks, aim at depleting one or more resources of the victim and making it incapable of conducting its tasks and processing incoming requests. These resources are mainly CPU, bandwidth and memory. Protocol misuse attacks aim at taking the victim down or taking control of the victim by misusing certain implementation vulnerabilities.

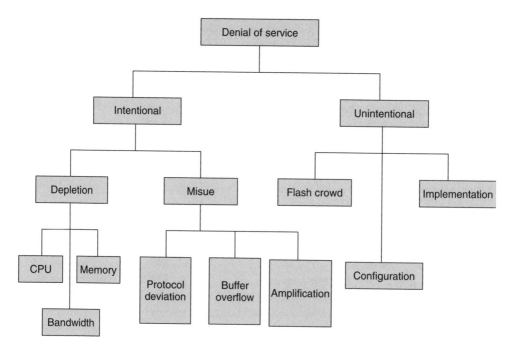

Figure 8.2 Classification of DoS attack types

Another approach for classifying attacks is to distinguish between single source and distributed denial-of-service attacks. In contrast to single source attacks with a DDoS attack, the attacker only initiates the attack. The actual attack is conducted by a large number of manipulated hosts. This not only makes the detection of the attacker more difficult, but also the resources the attacker needs to commit, in terms of bandwidth and CPU, are much lower.

A further classification possibility is to differentiate between direct and indirect attacks. With direct attacks, the attack is targeted directly at the victim by sending it a large amount of requests or misusing certain vulnerabilities of the victim's implementation to make it cease to work properly. With an indirect attack, the attacker targets some supporting services like DNS or the database used by the victim. Hence, the victim might still be working properly and not showing any signs of being attacked, but can still not provide the services it is supposed to offer.

8.3 Bandwidth Consumption and Denial-of-service Attacks on SIP Services

The basic goal of a denial of service attack is to cause the server to become overloaded so as not to be able to serve all of the incoming requests. This causes the server to start dropping incoming requests without distinguishing between legitimate and malicious requests. While this is already bad enough, SIP is usually transported over UDP and if a request is dropped the request is retransmitted a number of times. If an overloaded server is receiving a constant number of new requests over UDP, the total amount of traffic arriving at the server will actually increase over time as at least the legitimate users will retransmit lost requests. This will further worsen the server's overload situation and increase the drop rate. In the following a theoretical model's presented for determining the total amount of bandwidth that will be used when a SIP server is overloaded and is receiving more requests than it can process.

When used with UDP, SIP uses an exponential retransmission behavior to enable reliable message delivery. If after sending a request a SIP sender does not receive a response, either because the request itself or the response to it was lost, the sender retransmits the request after $T1$ seconds (Rosenberg *et al.* 2002b), and the timeout value is increased. After a maximum number of retransmissions have been sent, the sender will give up. For the case of INVITE requests, user agents use an exponential retransmission behavior up to the so-called TimerB. That is, a request is retransmitted at time points T1, 3T1, 7T1, 15T1 and up to TimerB. This can be represented as a series in the form of:

$$(2^1 - 1)T1, (2^2 - 1)T1, (2^3 - 1)T1, \ldots, (2_i^N - 1)T1 \tag{8.1}$$

with $[(2_i^N - 1)T1 = \text{TimerB}]$.

Non-INVITE requests use the exponential retransmission behavior up to a timer called T2 and then every T2 seconds up to TimerF.

In its simplest form a SIP server can be modeled as a processing entity of certain capacity and a queue for the incoming requests. Each incoming request requires certain processing capacity and occupies the processing entity for a certain period of time. Requests arriving while the processing entity is occupied by a request are queued. If new requests arrive at a higher rate than the rate at which the processing entity empties the queue, then the queue will become full and incoming requests will be dropped.

With only a processing capacity of C requests per T1 seconds and an incoming rate of $(r > C)$ INVITE requests per T1 seconds, $(r - C)$ requests are lost each T1 seconds. These are retransmitted T1 seconds later. The retransmitted packets also suffer from a loss and will have to be retransmitted later. Hence, the sender's transmission rate (R_i) can be depicted in Table 8.1.

Table 8.1 Transmission behavior of INVITE requests

Time	R_i	Lost
0	r	$r - C$
1 T1	$2r - C$	$2r - 2C$
2 T1	$2r - C$	$2r - 2C$
3 T1	$3r - 2C$	$3r - 3C$
4 T1	$3r - 2C$	$3r - 3C$
....	....	...
7 T1	$4r - 3C$	$4r - 4C$
....	....	...

Table 8.2 Transmission behavior
of non-INVITE requests

Time	R_o	Lost
0	r	$r - C$
1 T1	$2r - C$	$2r - 2C$
2 T1	$2r - C$	$2r - 2C$
3 T1	$3r - 2C$	$3r - 3C$
4 T1	$3r - 2C$	$3r - 3C$
....	....	...
7 T1	$4r - 3C$	$4r - 4C$
....	.....	....
15 T1	$5r - 4C$	$5r - 5C$
....	.....	....
23 T1	$6r - 5C$	$6r - 6C$
....	.....	....

The amount of INVITE requests (R_i) received by the server at any time point can then be determined as:

$$R_i(n) = r + (r - C) \times \left(\left\lfloor \frac{\ln(n + 1)}{\ln 2} \right\rfloor \right) \tag{8.2}$$

Note that the maximum value of n is

$$N = \frac{Timer\,B - T1}{T1}.$$

When reaching the steady state, e.g. $n = N$, the loss rate (l_i) can be determined as:

$$l_i = \frac{(r - C) \times \left(\left\lfloor \frac{\ln(n+1)}{\ln 2} \right\rfloor + 1 \right)}{r + (r - C) \times \left(\left\lfloor \frac{\ln(n+1)}{\ln 2} \right\rfloor \right)} \tag{8.3}$$

For non-INVITE requests, the transmission behavior is slightly different, see Table 8.2 with the value of T2 set to 4 seconds and T1 set to 0.5 seconds.

Here, the amount of non-INVITE requests (R_o) received by the server at any time point can be determined as:

$$R_o(n) = \begin{cases} r + (r - C) \times \left(\left\lfloor \frac{\ln(n+1)}{\ln 2} \right\rfloor \right) & n \leq (T2 - T1)/T1 \\ r + (r - C) \times \left(\left\lfloor \frac{\ln(M+1)}{\ln 2} \right\rfloor + \left\lfloor \frac{n-M}{M} \right\rfloor \right) & \text{otherwise} \end{cases} \tag{8.4}$$

with $M = (T2 - T1)/T1$ and assuming that M is an integer.

If a server can only process a certain number of requests, then under a denial-of-service attack, the loss of legitimate requests will further amplify the attack. Table 8.3 depicts

Table 8.3 Effects of overload on total number of requests

r	1500	2000	2500	3000	3500	4000
$R_o(N)$	5000	10000	15000	20000	25000	30000

the expected rate of requests per second $[R_o(N)]$ when flooding a server with a maximal capacity ($C = 1000$) with different rates (r) with $T1 = 0.5$, $T2 = 4$ and TimerB set to 32 seconds. From these results we can identify that, due to overload, the amount of retransmitted requests might exceed the original amount of sent requests several fold. Note that in the above calculation we assume that the attacker does not retransmit lost requests. In such a case the ratio of amplification is even higher.

Note that Equations (8.2) and (8.4) describe a worst case scenario. In the discussion it is assumed that all lost requests are retransmitted. An attacker will most likely not implement a standard conforming SIP client and maintain a complete SIP state machine but will just send SIP requests and not bother about retransmissions. In this case, only legitimate requests will be retransmitted. The term ($r - C$) used to indicate the number of to be retransmitted requests should be set to $[(r_{att} + r_{leg} - C) \times r_{leg}/r_{att} + r_{leg}]$ with r_{att} as the rate with which the attacker is generating traffic and r_{leg} as the rate of the legitimate traffic. This is, however, a best case scenario. In general, as some of the traffic from the attacker passes other proxies that act in transaction-state manner and retransmit dropped requests, the actual rate with which requests arrive at a proxy will be between the best and worst case scenario calculations.

8.4 Bandwidth Depletion Attacks

The access link to the Internet is a host's only means for communicating with other hosts and receiving and sending data packets. Such access links have a limited bandwidth capacity that can range from a few kilobits per second to several gigabits per second depending on the host's task, e.g. whether the host is for private usage or is providing some service for a large ISP.

With bandwidth depletion attacks, the attacker generates a large number of data packets that cause the link connecting the attacked victim to the Internet to become overloaded. Legitimate traffic will only have a low chance of reaching the victim and getting served. A popular kind of these attacks is the so-called "UDP flood attacks". The User Datagram Protocol (UDP) (Postel 1980) is a connectionless protocol. A sender can transmit a UDP packet to a receiver without any previous connection setup and the receiver is required to receive the packet and process it. By sending a large number of UDP packets, the attacker can saturate the link to the victim. Further, the victim receives the packets and checks if they are destined for some application. If that is the case then the packets are forwarded to that application that tries to process these packets and either serve them if they contain meaningful content or to drop them otherwise. If there is no application listening to the ports on which the packets are received, then the host drops them and sends an ICMP (Internet Control Message Protocol) message (Postel 1981a) packet to the sender of the packets. In both cases the host also requires considerable CPU resources for processing and dealing with this high number of packets. Often the sender also spoofs the source address in the packets. On the one hand this helps to hide the identity of the attacker;

on the other it also causes the ICMP packets generated by the victim to be sent to some other host, inflicting a denial-of-service attack on that host on the one side and reducing the bandwidth usage of the attacker on the other.

A similar kind of attack is the ICMP flood attack. ICMP is used for management purposes on the Internet, such as for measuring the round trip delay or number of routers traversed on the path between two hosts. For example, ICMP echo requests allow the user to send a request to a destination and receive a response within a round trip time. By sending a large number of such packets, an attacker can congest a victim and make it unreachable.

Bandwidth deletion attacks can be targeted at any host that has a TCP/IP stack regardless of the applications running on the host. Hence, while these attacks do not target any SIP-specific characteristics they can be launched at any SIP component.

8.5 Memory Depletion Attacks

Various Internet protocols and services can be modeled as sequences of transactions. That is, a client sends a request to a server. The server processes the request and conducts some actions such as asking the client for further information or routing the request to yet another server. The transaction is then considered to be terminated once the server receives the requested information from the client or a response from another server. Throughout the duration of the transaction the server maintains some state information describing the request and the status of the transaction. The memory dedicated to such state information is usually dimensioned so as to allow the server to serve a certain number of requests with the transactions having an expected average lifetime. An attacker can deplete the memory of a server either by generating a higher number of requests than the victim's memory was dimensioned for or by prolonging the duration of the transactions.

A SIP component can be the victim of various memory depletion attacks that can either target the TCP/IP stack of the host or specifically target the SIP implementation of the host.

8.5.1 General Memory Depletion Attacks

A prominent example of memory depletion attacks is the TCP-SYN attacks (McClure *et al.* 2003a). To establish a TCP session, a three-way handshake is conducted between the client and the server (see Figure 8.3). That is, the client sends a SYN (synchronize) request. This request is acknowledged by the server with an ACK message, which includes its own SYN request (ACK-SYN). To complete the session establishment, the client acknowledges the server's ACK-SYN request by sending an ACK message. If the server does not receive an ACK message it will resend its ACK-SYN message and keep the state information describing this transaction until the expiry of a timeout.

By sending a high number of SYN requests and not acknowledging the server's ACK-SYN requests, an attacker can force the server to maintain a large amount of state information for a duration much higher than the average value the server's memory was designed for. Thereby the server's memory becomes depleted and the server soon runs out of memory needed for serving legitimate requests. Similar to the UDP flooding attacks, the attacker can also spoof the source address in its SYN requests. This causes the ACK-SYN

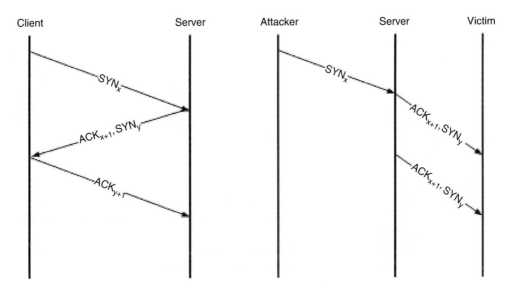

Figure 8.3   TCP-SYN attack scenario

messages generated by the victim to be sent to some other host and, hence, reduces the bandwidth usage of the attacker and inflicts a denial of service attack on another host. TCP-SYN attacks can be launched on any server willing to establish TCP connections with other hosts. As TCP is one of the transport protocols supported by SIP, a SIP server can become the target of such an attack just like any other web or email server.

Another type of memory depletion attack uses the possibility of fragmenting an IP packet in multiple fragments (Postel 1981b). In this case the attacker sends a sequence of fragments and withholds the first (or any other) fragment. The IP processing entity of the victim then waits for the arrival of all fragments as it is not able to reassemble the fragments and pass them up to the next layer. After the expiry of a timeout the IP fragments that could not be reassembled are discarded. Until this point these fragments occupied a part of the memory available to the victim. Hence, by generating a high number of such fragments, the memory of the victim can be depleted.

8.5.2   Memory Depletion Attacks on SIP Services

When a SIP component receives a request it generally maintains some state information about the transaction created by this request until the transaction is terminated. The available memory allows the component to process some maximal number of transactions with the transaction state information requiring on average a certain memory size and residing in memory on average for a certain period of time. An attacker can deplete the available memory by generating a large number of requests, by prolonging the time period during which the transaction state resides in memory, by increasing the amount of memory needed for maintaining the transaction state information, or by a combination of these.

8.5.2.1 Memory Depletion Attacks on SIP Proxies

To be able to process an incoming message, SIP proxies copy the message into their internal buffers. The amount of buffered data and the time period a proxy is supposed to keep the buffered data can vary depending on whether the server is working in a stateful or stateless mode. In any case, the server will at least need to maintain the buffered data while contacting other entities such as AAA or DNS servers or a database. Depending on the message type, headers used, the number of Via header fields and the body of the message, the size of a SIP message might range from a few hundreds of bytes up to a few thousand.

- **Stateless servers**–stateless SIP servers only need to maintain a copy of the received messages while processing those messages. As soon as the destination to which a message is to be sent to is determined and the message is sent out, the server can delete the buffered data.
- **Stateful servers**–in general we can distinguish between two types of states in SIP:
  - **Transaction state**–this is the state that a server maintains between the start of a transaction, i.e. receiving a SIP request and the end of the transaction, i.e. receiving a final response for the request plus some additional waiting time to make sure that the response has actually been correctly delivered. Typically, transaction context consumes a few kilobytes of memory depending on message size, on whether a message is forked to different destinations and memory management overhead. The duration of the transaction depends very much on the type of request. A session initiating an INVITE transaction usually involves user interaction and can last for several seconds. Other transactions last only for the time needed to process the request at the different components it traverses as well as the network propagation delay. The processing time includes the time needed to process the message itself as well as the time needed for conducting any other services such as checking the user profile in the database or resolving an address using DNS. In case of losses, the proxy keeps and retransmits the message up to TimerB (respectively TimerF).
  - **Dialog state**–in some scenarios servers may need to maintain some information about the session throughout the lifetime of the session. This is especially the case for communication involving firewall or NAT traversal or for special accounting and security reasons, as is the case for the IMS components such as the S-CSCF. Further, various components, such as session border controllers, maintain session state as well.

In the following text different attack scenarios are described with SIP proxies referred to as the attack targets. These attacks also apply to SIP components that are based on the B2BUA concept such as SBC or SIP ALGs, which terminate incoming sessions and start new sessions to the destinations. However, as these components need to processes two transactions, namely incoming and an outgoing, for each request such components require a larger amount of memory than proxies, memory exhaustion attacks can have even more negative results on them.

Brute force attacks   A SIP session is identified mainly by the *From* and *To* tags as well as the *Call-ID* and the *Via* branch field. Hence, the simplest method for mounting an attack on the memory of a SIP server is to initiate a large number of SIP sessions with different session identities, i.e. with different *To*, *From* tags, *Call-ID* or *Via* branch fields. If the number of generated calls is larger than the level for which the memory of the SIP proxy was designed, then the memory of the proxy will be depleted after some time and the proxy will start dropping incoming requests due to lack of memory. To hide its identity the attacker can spoof his source IP address or use botnets and include random *From* URIs in the sent requests.

Incomplete transactions   With brute force attacks, a system's memory is depleted by creating a larger number of transactions than the system was designed for. With the incomplete transactions attack, the attacker prolongs the lifetime of transactions and can thus deplete the memory of a server in a similar manner to the brute force attacks but using fewer requests. This is achieved by sending requests over UDP to destinations that exist but never reply. In this case the SIP proxy assumes that the requests were lost and will retransmit them in for the duration of TimerB, which has a default value of 32 seconds (Rosenberg *et al.* 2002b). After the expiry of TimerB, the proxy sends back a *408 Request Timeout response*. As the attacker will most likely not bother to send an ACK request back, the proxy will retransmit the 408 response for the duration of TimerH, which is by default also 32 seconds. This means that, for each transaction created by the attacker, the proxy will keep the transaction state for the duration of TimerB + TimerH, which taking the default values in (Rosenberg *et al.* 2002b), equals 64 seconds. Figure 8.4 depicts the message flow that can be observed in such an attack.

This kind of attack can be launched in several ways:

- Send requests to one or more user agents that are registered and hence theoretically reachable, but that do not reply to the requests. This will be the case if the receivers have no connection to the Internet, have been manipulated or the registered IP addresses are not allocated to any host.
- Another approach is for the attacker to include an IP address of a host in the *Request-URI* (name@IP address) instead of using the more usual presentation of (name@domain name). If there is no SIP implementation at this IP address, then no SIP responses will be generated.
- If a request included a *Route* header then a proxy forwards such a request based on the content of the *Route* header and not the *Request-URI*. Hence, the attacker can indicate an IP address in the *Route* header request that belongs to a host that will not reply to the request or even does not exist at all.

Similar to brute force attacks, the attacker can hide its identity by spoofing his source IP address or using botnets and including random From URIs in its requests.

Another form of this attack is realized by instructing the SIP proxy to establish a TCP connection to the destination. That is, a caller that has indicated her wish to use TCP as the transport protocol can force the SIP proxy to initiate a TCP connection to a certain destination. If the contacted callee does not respond, the resources for the transaction as well as for setting up the TCP connection, namely socket descriptor, would be blocked until a failure timeout expired.

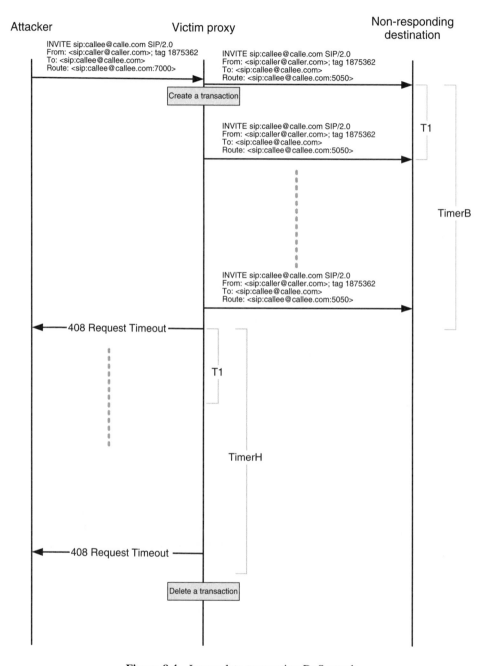

Figure 8.4 Incomplete transaction DoS attack

These forms of of attacks can be protected against using a number of measures:

- If an IP address is not reachable or does not have a SIP process, then sending a SIP request to such an IP address usually results in the generation of an ICMP error message. This reply can indicate to the proxy that it should avoid sending further requests to the destination. However, ICMP messages are often filtered by firewalls and routers and might not reach the proxy.
- Reject forwarding requests that include IP addresses in the *Request-URI*. As the task of a SIP proxy is to resolve SIP names into routes to user agents, receiving a SIP request with an already resolved name could be considered as malicious. Instead of rejecting such messages directly, the proxy could check whether the user name part of the *Request-URI* is registered and whether the registered IP address equals the one in the *Request-URI* and only reject the request if not. This requires, however, additional database queries that can increase the processing overhead at the proxy. Further, while checking the registration entries is possible for SIP proxies that act as the home proxy for the receiver, i.e. a proxy that has access to the receiver's registration status, this would not work with outbound proxies (Jennings and Mahy 2008). An outbound proxy is a proxy that is located topologically between the client and the home proxy server of the receiver. Such proxies are often used to help overcome limitations caused by firewalls or NATs. In this case, limiting the scope of IP addresses that the proxy might forward traffic to could be of some help. That is, if it is known which hosts and proxies messages can be forwarded to, then all requests that include an IP address that is not a part of this list will be rejected.
- Black list destinations that fail to generate a response to a SIP request. That is, once the SIP instance decides that a certain IP address or host name is unreachable, it is included in a black list for a certain period of time. Any other requests destined to this address are then rejected.

Where the proxy is enforcing a policy to reject requests that include IP addresses in the *Request-URI* the attacker would need to bind a SIP URI to the destination's address, e.g. the IP address or name of the destination's host. This could be achieved by sending a REGISTER request with the SIP URI and the destination's address in the *Contact* header. However, the attacker will most likely have a different IP address than the destination's host. This kind of attack can be prevented by disallowing a UA to bind a SIP URI to an address different than the source IP address of the REGISTER messages. This is unfortunately not trivial as one needs to accommodate aspects of NAT traversal in which the source IP address and the registered address are not the same. Further, as the attacker can spoof its source address to that of the destination, ingress filtering would be needed as well.

For this attack to be successful, the attacker will need to have a valid SIP account with a VoIP provider, see Figure 8.5. In many VoIP services, users can apply for a SIP address for free and without providing much personal information other than an email address. More indepth verification of the user's identity is usually only conducted when the user wishes to use a VoIP service to call into the PSTN and is, hence, expected to pay for the service. An attacker can actually prepare this attack and stay anonymous as the used email address can be a temporary one that is canceled as soon as the application for

```
Step 1:
Acquire a SIP account
        attacker@sip.com
```

```
Step 2:
Register SIP address to IP address of a destination

        REGISTER sip:sip.com SIP/2.0
        To: sip:attacker@sip.com
        From: sip:attacker@sip.com
        Contact: <sip:bob@victim.example.com>;expires=3600
        Call-ID:
        CSeq:
        Via:
        ......
```

```
Step 3:
Send requests to attacker@sip.com

        Invite sip:attacker@sip.com SIP/2.0
        To: sip:attacker@sip.com
        From: sip:fake@fake.com
        Contact: ...
        Call-ID: ...
        CSeq: ...
        ......
```

Figure 8.5 Steps for launching a incomplete transaction DoS attack

the SIP address is successful. The attacker could naturally also just use a stolen account. However, as the attacker has to generate many accounts for the attack to be effective, the effort of preparing the attack in this way might already be prohibitive. Using the *Route* header for directing a request toward a nonresponding address therefore seems to be the most efficient way of launching this kind of attacks.

Besides depleting the proxy's memory, the broken session attack can also be seen as an amplification attack. When using UDP as the transport protocol, for each request sent by the attacker, the proxy generates a number of retransmissions before terminating the transaction. The exact number of these retransmissions depends on the values of the retransmission timers and can be calculated as in Equation (8.2) with C set to 0. If all requests are directed to the same victim, then by sending requests at a rate of r, the victim will receive requests at a rate of

$$r + r \times \left(\left\lfloor \frac{\ln\left(\frac{TimerB - T1}{T1} + 1\right)}{\ln 2} \right\rfloor \right)$$

Incomplete transactions with cooperating hosts By sending provisional responses, e.g. 1xx replies, to INVITE requests, a receiving user agent can prolong the lifetime of a transaction to several minutes. In this attack scenario, the attacker sends INVITE requests to destinations that reply with 1xx messages but do not send final responses. With this approach the attacker needs to send even fewer requests to deplete the memory of the proxy.

In the incomplete transaction scenario it was sufficient for the attacker to send requests to an IP address which was not deploying any SIP software. For further prolonging the lifetime of the transactions, the used destinations have to cooperate and reply with provisional responses to the received requests. So while the attacker would be able to spoof the IP addresses of the requests and use random *From* URIs, it will not be possible to use the address of existing but uncooperating destinations. To achieve this the attacker has to either manipulate hosts or actively generate some SIP accounts and bind hosts to these accounts that reply to all incoming INVITE requests with provisional responses only. Obviously these preparation steps increase the complexity of launching the attack and, hence, reduce its attractiveness.

8.5.2.2 Memory Depletion Attacks Using SIP Authentication

For user authentication, SIP uses the HTTP digest authentication which is based on a challenge–reply concept, see section 3.9.1.

Using digest authentication, the authenticating server generates a nonce and adds it to a *401 Unauthorized* or *407 Proxy Authentication Required* response in a *WWW-Authenticate* or *Proxy-Authenticate* header. The user agent then abandons the first transaction and generates a new request with his credentials calculated based on the shared secret with the provider and the nonce. In between generating the *407* response and receiving the new request from the user, the server usually maintains a copy of the nonce. This procedure can be misused in a similar way to the incomplete transactions attack. As depicted in Figure 8.6 the attacker generates a request and includes in the *From* header the identity of a subscriber of the attacked proxy or registrar server. Once asked to verify his identity the attacker just ignores the challenge and starts a new request with another session identity. The attacked server keeps the allocated memory for a certain period of time. The memory of the attacked server is, hence, consumed by the saved nonce and the transaction data. This comes in addition to the wasted CPU resources for calculating the nonces.

One approach for avoiding the depletion of memory is to use stateless authentication. The goal of stateless authentication is to construct the authentication process in a way that does not need to store transactions or challenges at servers. This requires putting all verification information in a self-contained manner in SIP messages. An operational constraint here is that the authentication process should be conducted by any server in a server farm. That is, the request containing the credentials might be processed by a different server than the one which generated the nonce.

One solution for this is to use predictive nonces that allow for stateless authentication. Predictive nonces are calculated in such a way that they are valid only for validated messages within a certain time window. This is achieved by calculating the nonces as a cryptographic hash of some request header fields, a timestamp and a secret shared among servers in a server farm. When a request including an *Authorization* header arrives at a server, a nonce is generated again using the same criteria, namely, request headers, timestamp and secret, and is compared with the one in the request. This way the server can verify whether the nonce was created by itself or by another a server in the same

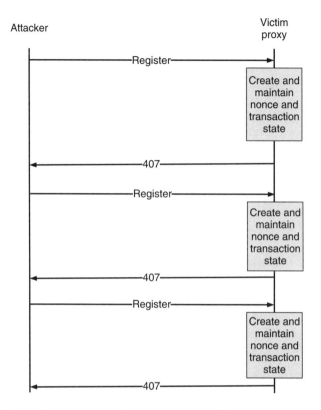

Attacker   Victim
                                                            proxy

Figure 8.6   Memory exhaustion attack on SIP registrar servers

server farm. If that is not the case, the request is rejected. Predictive nonces also provide
a simple form of message integrity.

This method works without any changes to the protocol. The only requirement put on
interoperability is not to change the header fields that were used to calculate the nonce
in resubmitted requests. There is unfortunately a pitfall in generating stateless replies to
INVITE request. If a stateless server sends a negative response to such a request, an ACK
request will confirm the receipt of the response. The server is then required to consume
the ACK request and not to forward it anywhere. To implement it statelessly, the ACK
request must include a piece of information saying "this is an ACK request to this server's
reply, do not forward". This in-message knowledge allows a server to decide whether to
consume or relay the ACK request. The only such a place in the SIP specification is the *To*
tag. The solution is then as follows: a stateless server puts its signature in the *To* tag in the
replies to the INVITE request. If present in an ACK request, the ACK request is dropped.
Unfortunately, this approach fails with re-INVITEs request as they already include *To*
tags that are valid for the duration of the SIP dialog and are generated by the callee. If
the server wants to authenticate a re-INVITE request it cannot put its own *To* tag in the
replies. As the subsequent ACK requests are not recognized as belonging to the rejected
INVITE request, cannot be consumed and are mistakenly forwarded. These ACK requests
will then be dropped by downstream servers and some unnecessary traffic is forwarded

in the network. A straightforward fix would be to modify the SIP specifications so that user agents do not send an ACK request in reply to non-200 or to have another place in requests to mirror server's signature. None of these fixes is, unfortunately, possible without harm to interoperability.

The authentication process could also be used to launch another form of the incomplete transaction attack described previously. If a proxy server wants to authenticate all session establishment requests, the proxy will reject the first INVITE with a *407* response. If no ACK request was received, the proxy will retransmit the *407* response for the duration of TimerH and the proxy will keep the transaction state for the duration of TimerH. Unlike the incomplete transaction attacks described previously, the attacker does not need to create SIP accounts or use nonresponding user agents. It is sufficient for the attacker to use valid subscriber names to trigger an authentication process. Valid names can be discoved by either searching the web for addresses including the attacked server domain name or by generating names randomly and testing all the generated names.

8.5.2.3 Memory Depletion Attacks on SIP User Agents

SIP user agents whether hard phones, adaptation devices or gateways to the PSTN can in general be expected to have less memory resources than a proxy as they are expected to deal with only a smaller number of parallel calls. If more requests are received than the user agent can handle then the excess requests are rejected. Still, an attacker can render a user agent unusable by continuously initiating calls to the user agent. Once an INVITE request is received, the user agent replies to the request and reserves the needed resources. If the attacker does not acknowledge the reception of the response, the user agent will retransmit the response until a time out expires, namely TimerB, which is usually 32 seconds. As the user agent can support only a certain number of parallel calls, say n, by sending n INVITE requests, the user agent will be blocked for the duration of TimerB. In Figure 8.7, the attacked user agent has a capacity of only one call at a time. Once attacked, all other incoming calls will be rejected with a busy signal.

As a side effect, where spoofed IP addresses are used in the INVITE messages, the responses are sent to another host and the attacker can actually launch an amplification DDoS attack on a victim in a similar manner to the TCP SYN attacks.

8.6 CPU Depletion Attacks

By overloading the CPU, an attacker can on the one hand reduce the ability of the server to process and handle legitimate requests. On the other hand, as the CPU is a shared resource between all processes running on the server, the entire server can be rendered unusable this way. A high CPU load can be incurred by either increasing the number of requests the server needs to process or by increasing the complexity of the processing.

By launching a UDP or ICMP flooding attack, an attacker depletes not only the server's bandwidth but also its CPU. As the server needs to process many more requests than usual, the CPU load goes up. If the CPU cycles required for processing the received requests exceed the CPU available to the server, the server starts dropping the excess traffic and the CPU will be occupied to 100 %.

By causing the server to conduct complex processing steps, the attacker can achieve the same results with much less traffic. Such processing steps might include complex

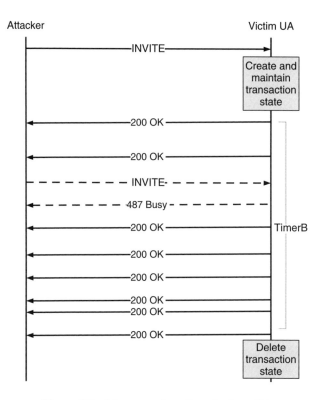

Figure 8.7 Memory exhaustion attack on UA

security checks or the launching of certain applications. Such attacks require the attacker to have good knowledge of the services and applications provided by the server and to craft the data packets used for the attack to specially target certain features of these services and applications.

When a SIP component receives a request, it needs to parse the request and react to it, e.g. send a response or forward it. Further, infrastructure components might authenticate the user, query a database for user's privileges, profile and registration information, write logging and billing information into a database and possibly execute certain services, e.g. user-specific configuration options such as certain call routing preferences. As all of these processing steps require various amounts of CPU resources they can become the target of CPU depletion attacks.

8.6.1 Message parsing

In order to figure out how to handle an incoming message, SIP components have to parse at least a part of the message. By flooding a SIP component with a large number of SIP messages, the CPU resources of the receiver can be rapidly exhausted. An attacker can increase the CPU resources required for the parsing by using more complicated message structures. This can be achieved in many different ways, such as separating the content of some headers, like Via headers, in different nonadjacent headers or adding then to the end of the message.

Because of their limited processing resources, user devices are especially vulnerable to these attacks. SIP proxies and other infrastructure components should be deployed in a scalable manner so as to be able to withstand a sudden increase in the received number of requests. This can be further optimized by parsing only those parts that are needed for correct functioning.

8.6.2   Security checks

When using the HTTP digest mechanism, see section 3.9.1, to verify the identity of a user, a SIP server needs to generate a nonce and then check the credentials of the user. This processing overhead could naturally also be used as the target for CPU depletion attacks. However, the checking uses hashing schemes such as MD5, which require a relatively low calculation overhead. A server exhibiting signs of overload due to security checks is a good indication of a bad implementation or under-designed hardware.

A more effective attack on the security mechanisms of SIP is achieved by misusing the strong identity mechanism described in section 6.4. The attacker sends requests with the *From* header set to the address of a valid user and a link to the certificate of the service provider of the user in the *Identity-Info* header. The verification instance at the targeted service provider tries to retrieve the certificate indicated in the *Identity-Info* and tries to use it to verify the identity of the caller. While the verification will fail as the attacker does not have access to the private key of the assumed caller's service provider, the verification service will waste time retrieving the certificate and CPU resources for verifying the identity.

8.6.3   Application execution

A SIP server might need to execute a certain application or script after receiving a request. The amount of CPU resources used depends on the application type and its complexity. Where the application server is located on the same hardware platform as the SIP server, the CPU resources used for the execution of the applications are no longer available for processing SIP messages. Where the application server is located on a different hardware platform, then some form of inter-process communication between the SIP server and the application server is needed which also requires some resources. Attacks targeting the application execution could be launched by sending requests to users for which the attacked SIP application server executes some applications whenever receiving a request. To make sure that such users exist, the attacker might register himself as a legal user and ask for the execution of a complicated application whenever a request is addressed to this registered user. The attacker can now send requests to the registered user and overload the server in this way. Therefore, the provider should offer users only simple and secure tools for service creation that do not allow the user to specify an application that causes infinite loops (Lennox *et al.* 2004). Note however that when, as in this case, the requests are sent to one or a few receivers, detecting such a misuse is rather straightforward and such an attack is, hence, not very beneficial for the attacker.

8.7 Misuse Attacks

Flooding attacks require the attackers to generate a large amount of traffic that depletes the victim's resources. In contrast, misuse attacks require only very few data packets that target certain features of the protocol or implementation weaknesses of the victim's operating system and applications. Here, we can mainly distinguish between attacks targeting the operating system implementation including the networking stack, e.g. the TCP/IP implementation, and memory management in the victim's operating system or applications, and attacks targeting the SIP protocol implementation itself.

The TCP/IP implementations of most widely used operating systems have been rigorously used, tested and improved over recent years. Hence, most of the attacks listed in this section will most likely only apply to old and nonmaintained systems. However, the IP stacks used on SIP devices are often implemented especially for the device's hardware and might not have gone through the same degree of extensive testing. Hence, it would not be surprising for some SIP devices become the victims of attacks that no longer present a threat for other IP devices.

8.7.1 TCP/IP Protocol Deviation Attacks

Protocol deviation attacks usually exploit vulnerabilities of some operating system (OS) implementation concerning the network part of the OS, i.e. the TCP/IP stack. These attacks can lead to system crashes by emitting only a few or even one "killer" packet. These attacks are sometimes referred to as "nuke attacks". In the following some examples of such attacks are provided. Books and resources such as (Maiwald 2003) or (McClure et al. 2003b) offer more details and other types of attacks.

Ping of Death   One popular example of such an attack is the so-called Ping of Death, where an oversized ICMP echo request packet is sent to the victim host. ICMP messages are sent within IP datagrams and they are de-multiplexed by IP for further handling by ICMP. Regular echo request packets only have a length of 64 octets, but the size is not limited by any means. Thus, an ICMP message can be transported in two or more IP packets using fragmentation and will be reassembled before getting processed. In this way, it is possible to send an echo request whose content comprises more than the maximum allowed size of 65,507 octets. Depending on the OS implementation, this may result in a system crash or reboot. It should be noticed that the attack with oversized packets is not limited to ICMP but can also be realized with other protocols.

Land attack   Another example of a protocol deviation attack is the "Land attack" (McClure et al. 2003b). In this case, the attacker sends a TCP SYN packet with the source and destination addresses set to the victim's address. This causes various systems to crash.

Teardrop attack   Another fragmentation attack is the "teardrop attack" (McClure et al. 2003b), which uses the vulnerability of some operating systems to overlapping IP fragments. In this kind of attack the attacker uses erroneous packet header information indicating overlapping fragments of packets. To reassemble such packets, data in some packets must overwrite data in others. If the software is not prepared to handle erroneous packet header information, attempts to re-assemble these packets can cause the host to crash.

8.7.2 Buffer Overflow Attacks

Buffer overflow attacks are some of the most common attack scenarios on all kinds of software. A buffer is a continuous block of memory with a finite size that serves as a temporary data storage area. Because of poor programming methods, e.g. with no proper error handling and checking, an attacker can cause an application or the operating system to write more data into a buffer than the size of that buffer. This causes the extra data to overwrite other information adjacent to the buffer in the memory, such as a procedure return address. This can cause the computer to return from a procedure call to malicious code included in the data that overwrites the buffer. This malicious code can be used to start a program of the attacker's choosing or provide access to the victim's computer so that the attacker can install malicious code.

A wide variety of misuse attacks exist and are known. Actually with every new application or new version of an existing implementation one can expect that some sort of implementation bugs that might allow for a buffer overflow attack might be introduced as well. The only way to protect against this misuse scenario is to keep the system up to date by applying the latest bug fixes and patches.

8.7.3 SIP Protocol Misuse Attacks

Besides overloading the infrastructure components or reducing their availability, an attacker can also aim at making a user unreachable or prevent a user from successfully using the service. Unlike the attacks described in the previous sections, these attacks are more targeted at users and user agents.

The SIP protocol misuse attacks described in this section aim at reducing the availability of a certain user or even preventing her from using the SIP-based services. While such attacks are less of a threat to the service provider, they are annoying to the affected users. This kind of attack requires the attacker to be able to eavesdrop on a user's requests and generate appropriate replies or requests that lead to an interruption of the user's calls. That is, the attacker needs to learn all the session parameters and include these in his messages in the appropriate way, e.g. use the same session identity, increment the sequence number and preferably use the same contact information and IP addresses as the involved parties in the session so as to overcome any firewalls and NATs that might be located between the attacker and the victim.

The efficiency of these kinds of attacks depends very much on the location of the attacker and his ability to eavesdrop on the users' communication. An attacker that is located on the same link or in the same network as the victim would only be able to interrupt services of users in this network. An attacker located close to a proxy server would be able to eavesdrop on the communication of all the users subscribed to this proxy and would hence be able to interrupt the service of all of these users.

The basic scheme for these attacks is for the attacker to generate a message that causes one of the parties involved in the communication session to believe that the communication session is interrupted.

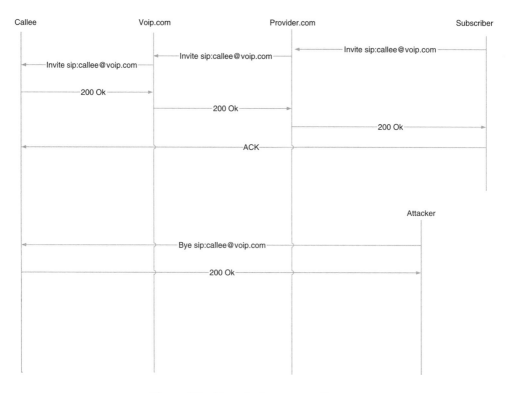

Figure 8.8 Bye attack sequence diagram

8.7.3.1 BYE Attacks

In a BYE attack the attacker tries to terminate the communication session of the victim prematurely. By eavesdropping on the user's communication, once a session is established the attacker can issue a BYE request to one of the involved parties in the session, see Figure 8.8. The BYE request includes the same session identity and an incremented call sequence (*CSeq*) number.

To protect against this kind of attack, the BYE messages should be authenticated or the communication between the communicating parties needs to be cryptographically protected using TLS. Unfortunately, authenticating BYE requests is not a viable solution if both the caller and callee belong to the same service provider. If the caller and callee belong to two different providers then without deploying some form of authenticated identity as described in Chapter 6, an attacker can assume the identity of the caller or callee and terminate the call.

8.7.3.2 CANCEL Attacks

With the CANCEL attack, the attacker can prevent a caller from establishing a communication session. After eavesdropping on a user's INVITE request, the attacker sends a CANCEL request to the callee directly afterwards. The CANCEL request includes all the main information included in the INVITE request, e.g. same *To*, *From*, and the same

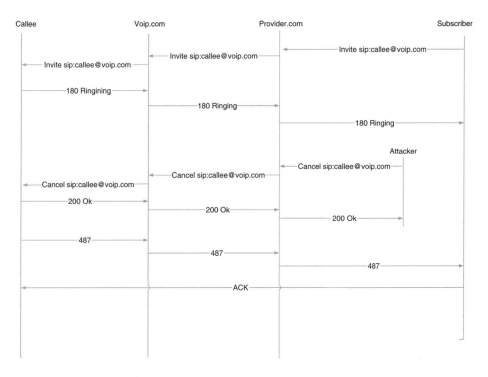

Figure 8.9   CANCEL attack sequence diagram

CSeq value. This leads the callee to believe that the caller has canceled the call and to terminate the INVITE transaction, see Figure 8.9. Note that, for this attack to be successful, the CANCEL request has to arrive at the callee before he has issued a final response. Otherwise, the CANCEL to the request will be ignored.

Unlike BYE requests, CANCEL requests cannot be authenticated as they use the same sequence number as the original INVITE. When authenticating a request the user agent must resend the request including the credentials and an incremented sequence number. If CANCEL requests are to be authenticated, the resent CANCEL requests can no longer be associated with any INVITE requests. Hence, the only possible protection in this case is to secure the communication between the caller and callee to prevent eavesdropping or use authenticated identities as described in Chapter 6.

8.7.3.3   Failure Response Attacks

Instead of sending requests to terminate a session, an attacker who manages to capture a request could simply reply with a failure response indicating that a session cannot be established using a 4xx, 5xx or 6xx response, as shown in Figure 8.10. The attacker could also ask the caller to retry another destination by sending a 3xx response. By directing the user to another destination with a 3xx response, the attacker can force the user to communicate with the attacker himself or some other malicious entity. This could have serious consequences. If the session was established between a user and the phone service

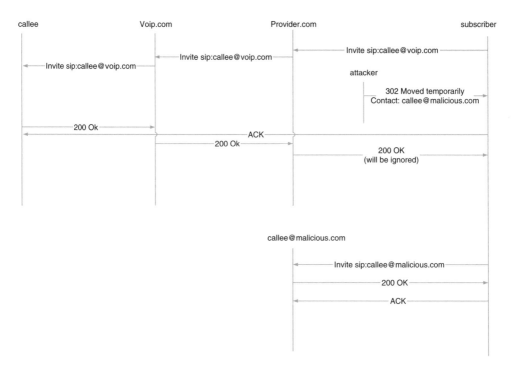

Figure 8.10   Negative response attack sequence diagram

of a bank, the attacker could redirect the call to a malicious IVR system that asks the user to reveal his bank access information.

As responses are not authenticated in SIP, the only possible protection against this attack is to use secure communication links. Note, however, that for this type of attack to be successful, the response issued by the attacker will have to reach the caller before the responses generated by the callee. Otherwise, the responses of the attacker will be ignored. To increase the chances of success of such attacks, the attacker might launch a denial of service attack on the callee so as to slow down the callee or cause the INVITE request to be lost.

8.8   Distributed Denial-of-service Attacks

In order to launch an effective memory or CPU depletion attack, the attacker requires a large amount of computing resources, high bandwidth access and low chances of being detected. Launching such attacks from a single host incurs high costs on the attacker in terms of the needed computing power and bandwidth. Actually, with access links to the servers of high-profile companies, the typical target candidates for DoS attacks, reaching a speed of multiple Gb/s, it is nearly impossible to overload such servers from a single attacking host. Further, to avoid being detected, the attacker should spoof the sender IP address in the packets it is generating using some randomly chosen IP address. With the increased use of ingress filtering (Baker and Savola 2004), network providers can block traffic that seems to carry spoofed IP addresses, e.g. addresses that do not belong to their address space. This makes hiding the attack source rather difficult and increases the possibility of being detected. To escape these limitations, attackers usually deploy distributed denial of service attacks. By launching an attack from multiple hosts insteads of a single one, the attacker can use the computing resources and access links of these hosts and hence reduce his own costs. Further, as each one of these hosts is contributing only a small percentage to the attack, intrusion detection tools that measure, for example, how much traffic is generated by these hosts might not be able to detect anomalous behavior. Finally, by having the source of the attack distributed among a large number of hosts, the originator of the attack is more difficult to detect.

In general one can distinguish between amplification attacks and botnets. Amplification attacks misuse certain protocol or service features to amplify the amount of data sent. This is achieved without having to manipulate any hosts or servers. With botnet attacks, the attacker manipulates a number of hosts and uses these hosts to launch the attacks on its behalf.

8.8.1 DDoS Attacks with Botnets

The term Bot is derived from the word Robot. Robot comes from the Czech word for "worker". In the computer world "bot" is a generic term used to describe an automated process. Botnets started appearing in the year 2000 and have since developed enormously in terms of size and complexity (Canavan 2005).

A Botnet consists of thousands if not millions of bots (Rajab *et al.* 2007) that are controlled by an attacker. In general, a botnet consists of the attacker himself, a number of master servers and a large number of slave hosts, so called zombies or bots, see Figure 8.11. Master servers are usually some Internet servers that are used to deal with a lot of traffic and to communicate with a lot of other hosts, e.g. web or mail servers. Using some form of a misuse attack, e.g. a buffer overflow, as described in section 8.7, the attacker installs some software packages on the master servers. These software packages allow the attacker to indirectly communicate with the slaves and among others to schedule an attack, configure the destination of the attack and identify the number of slaves. The slaves are usually residential hosts that have been turned into bots by luring their owners into installing some software packages, Trojans, that seem to be useful or attractive. While some of these malicious software packages actually provide the promised benefit, they also establish a communication channel with the masters. Over this channel the slave sends information about the infected host such as the access information of the user to some services or the email address of the user. The master sends the slave commands about which targets to attack or updates of the attack software so that it gets more difficult to detect. As an attacker usually deploys thousands of such bots to launch an attack, the resources used by each slave in terms of CPU and bandwidth are usually minimal. With their systems not behaving out of the ordinary, the owners of such manipulated hosts have no clear indication that their computers have been manipulated.

The exchange of information in botnets is very often realized using the Internet Relay Chat (IRC) network (Oikarinen and Reed 1993) for communicating with the slaves. IRC is a popular multiuser online chatting system that allows users to create two-party or multiparty interconnections and exchange messages in real time with each other. The IRC network architecture consists of IRC servers that are located throughout the Internet.

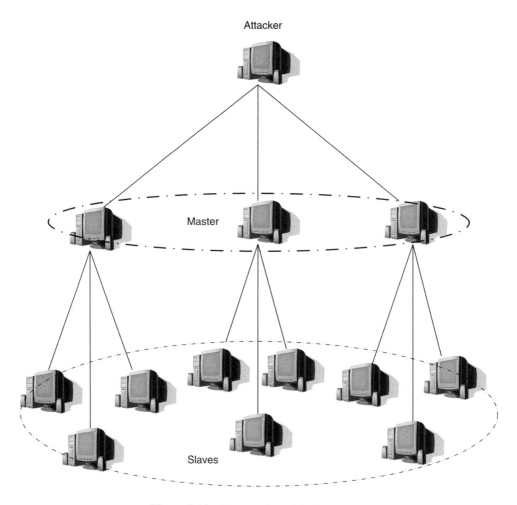

Figure 8.11 Master–slave DDoS structure

IRC chat networks allow their users to create public, private and secret channels. Public channels are channels where multiple users can chat and share messages and files. Public channels allow users of the channel to see all the IRC names and messages of users in the channel. Private and secret channels are set up by users to communicate with only designated users. Both private and secret channels protect the names and messages of users that are logged on from users who do not have access to the channel. By using the IRC network the attacker can communicate with the slaves using legitimate IRC ports. This reduces the possibility of getting blocked by firewalls and disguises the communication as part of the IRC communication. Further, the attacker can use the IRC mechanisms for discovering which slaves are available and can be used for launching an attack.

More recent botnets are increasingly using peer-to-peer technology (Grizzard *et al.* 2007). With the master–slave type of botnets the usual approach for detecting a botnet and disabling it is to detect the master servers and block the communication to them.

With peer-to-peer technology master servers no longer have to be centralized, making the detection and disabling of botnets more difficult.

Botnets can be used to launch any attacks that could be launched as single source attacks. For example a large TCP SYN attack can be launched by having multiple bots each sending only a few TCP SYN requests to the victim and not acknowledging the ACK-SYN message by the victim. Similarly, a SIP flooding attack can be launched by having multiple bots sending INVITE requests to the victim. An example of an IRC-based bot has already been released by researchers with the goal of demonstrating the feasibility and effects of such attacks.

8.8.2   IP-based Amplification Attacks

With DDoS amplification attacks, an attacker sends traffic to an IP broadcast address with the address of the victim as the source address, see Figure 8.12. When receiving a data packet sent to the broadcast address rather than to a specific address, the routers within the network forward the packet to all the hosts in that network.

A major advantage of this kind of attack, is that the attacker can use the systems within the broadcast network as zombies without needing to infiltrate and manipulate them. Two prominent examples for such attacks are Smurf and Fraggle attacks. To better hide his

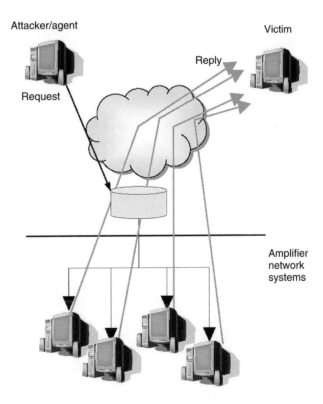

Figure 8.12   Amplification attacks

traces, the attacker can manipulate other hosts, called the agents, to send the broadcast message on its behalf.

Smurf attacks In a DDoS Smurf attack, the attacker sends ICMP packets (Postel 1981a) to the network broadcast address. The source address of the ICMP packets is set to the victim's IP address. Attackers typically use ICMP echo requests, which are packets that ask the receiver to send back an ICMP echo response packet. The ICMP packets are sent to the broadcast address and are received by all of the hosts within the network. Each of these hosts returns then an ICMP echo response to the victim. Thereby, with this type of attack the original packet is amplified tens or hundreds of times.

Fraggle attacks A DDoS Fraggle attack is similar to a Smurf attack in that the attacker sends packets to a broadcast address. Fraggle is different from Smurf in that it uses UDP packets sent to the UDP Echo service on port 7 instead of the ICMP Echo service. There is a variation of the Fraggle attack where the UDP packets are sent to the chargen service. The chargen service replies to any UDP packet with a message containing a random set of characters. The return address of the UDP packet sent is set by the attacker to the victims' UDP Echo service, creating an infinite loop as the victim's system will resend an echo packet back to the character generator, and the process repeats. This attack generates even more bad traffic and can create even more damaging effects than just a Smurf attack.

In the late 1990s a lot of hosts and networks were subject to Smurf or Fraggle amplification attacks. This has changed in recent years with the increased introduction of ingress filtering. Further, while in the late 1990s routers were expected to forward packets destined to broadcast addresses by default, this has changed and routers are now usually configured to drop such traffic (Senie 1999). Additionally, routers and hosts are now generally configured so as not to reply to requests that are destined for the broadcast address.

As an example of an immature implementation of the IP stack with regard to handling messages arriving over broadcast addresses in SIP devices, we observed during some practical tests with SIP phones from different vendors that sending a SIP INVITE message on the broadcast address caused all phones of a certain manufacturer that were connected to the same link to start ringing.

8.8.3 DNS-based Amplification Attacks

In a DNS amplification attack, the attacker initiates DNS requests to multiple DNS servers with the sender address spoofed to the address of the victim. The DNS servers then send a response to the victim. By requesting especially large DNS records, the attacker can cause the DNS servers to generate rather large responses, i.e. much larger than the request itself. To ensure that the replies to the requests are large, the attacker can manipulate an authoritative DNS server and add a DNS record with especially large entries. To avoid overloading the authoritative server, the attacker sends the DNS requests first to recursive DNS servers. These servers forward incoming requests that they cannot serve to the authoritative server and then cache the response so that future requests can be served directly. Investigations described in (Kaminsky 2006) suggest that a large proportion of the DNS servers on the Internet forward a request regardless of whether this request is arriving from a host that is supposed to be served by this DNS server or not. The attacker can misuse these recursive servers to amplify the attack. The amplification effect is reached here by having the servers reply to DNS requests that have a size of only a few tens of bytes with responses multiple kilobytes in size.

8.8.4 Loop-based Amplification Attacks on SIP Services

In the loop-amplification scenario, the attacker needs to convince a proxy server to rewrite a request to a location which resolves to the server itself. This will result in the server overloading. As an example for such an attack a user might register at one domain, e.g. victim.com, as user@victim.com with a contact address of user@victim.com. When a request is sent to user@victim.com, the proxy checks the user location database, rewrites the *Request-URI* of the request to the contact address indicated in the user's registration and forwards the request to victim.com, i.e. to itself. The same procedure is started again, see Figure 8.13. To counter such an attack the provider might check the contact addresses indicated in the registration messages and reject registration messages in which the contact equals the user's identity.

A more sophisticated variant of this kind of attack can be realized by registering two users at the same provider or two different providers with the contact address of each user

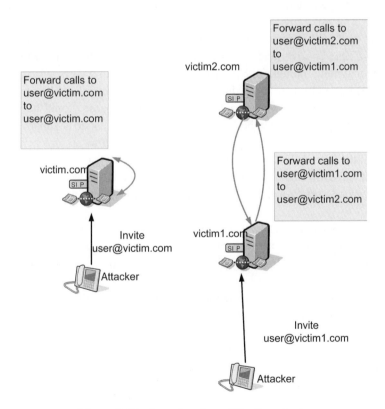

Figure 8.13 Loop-based amplification attacks

indicating the address of the other. That is, user@victim1.com registers user@victim2.com as his contact address and user@victim2.com registers user@victim1.com as his contact address. This causes all messages sent to either user to be looped through the two providers, see Figure 8.13.

The SIP specification provides two means to prevent infinite loops; namely, loop detection and the *Max-Forwards* header. To detect loops a proxy adds a *branch* parameter to its *Via* header before forwarding the request. The *branch* parameter consists of two parts: the first part must be unique and the second part consists of a hash over various routing-related information. A common way to create this value is to compute a cryptographic hash of the *To* tag, *From* tag, *Call-ID* header field, the *Request-URI* of the request received, the topmost *Via* header and the sequence number from the *CSeq* header field, in addition to any *Route*, *Proxy-Require* and *Proxy-Authorization* header fields that may be present. When receiving a request the proxy checks the *Via* list for any entries that it might have added previously. If such a *Via* header is found then the proxy calculates the second part of a branch and compares it with the values included in the found *Via* header. A match between the two values indicates that the request was looped. Because parsing the *Via* list and determining for each entry whether it was added previously by the proxy can be resource-consuming, the SIP specifications (Rosenberg *et al.* 2002b) only recommend the SIP proxies to do loop detection and leave it up to the implementors to decide whether to implement it or not.

A simpler method for detecting loops is that of the *Max-forwards* header, which is similar to IP's *Time-To-Live* field. When a request is generated, the *Max-Forwards* header indicates the maximum number of SIP proxies this request can traverse. At each traversed proxy this header is checked and reduced by one before forwarding the request further. If the value of the *Max-Forward* reaches zero then the request should be dropped and a 483 (too many hops) response be generated. Note that a sender is supposed to set the initial *Max-Forwards* value, leaving the length of a loop up to an attacker. Hence it is advisable that the proxies limit the value of Max-Forwards. That is, the proxy might set a maximum limit for the value Max-Forwards that it might accept. If a request is received with Max-Forwards set to a higher value, the proxy can reset this value to its defined limit.

The effects of such attacks is further amplified in case requests timeout. When a request is forwarded, a transaction stateful proxy expects to receive a response for this request before a time, T1, expires. Without a loop detection mechanism, the request will loop from one proxy to the other until *Max-Forward* reaches 0. With a recommended value of 70 for *Max-Forward* and 0.5 seconds for T1, the probability that the request times out before receiving a response is very high. Once a request times out, the proxy retransmits the request, which gets looped as well and further increases the load on the proxy.

Similar to the discussion of memory exhaustion attacks, amplification attacks require the attacker to register himself with the provider beforehand. This allows the provider to discover the identity of the attacker once the attack has been recognized. However, as it is often possible to subscribe to a VoIP service without providing any personal information except for an email address, this aspect will probably not deter attackers.

8.8.5   Forking-based Amplification Attacks on SIP Services

To enable services that might require several destinations to be contacted in parallel or sequentially, SIP supports *call forking*. When a SIP proxy receives a request that is destined for a user that has registered more than one contact address, the request is forked to all of the different contact addresses. An attacker can misuse this feature by registering multiple contact addresses with the same address, resulting in a much higher overhead for every request sent to the address.

Figure 8.14 depicts a possible usage of the forking feature to further amplify the loop attack. The attacker registers $N + 1$ accounts at $N + 1$ providers. At each provider the attacker registers N contact addresses pointing to the other accounts. With such a scenario one request causes the generation of M requests:

$$M = \sum_{n=0}^{n=F-1} N^n \qquad (8.5)$$

with F as the initial value of the Max-Forward header.

In addition to the overload due to message processing using forking to amplify an attack, such attacks are especially attractive, as forking proxies need to be transaction stateful.

Similar to the loop attacks it is advisable here to set a maximum limit to the value the *Max-Forwards* header can take. Because of the higher danger of forked requests, the proxy might have a value for normal requests and a lower one for forked ones. This reduces the risk of extended loops by using large Max-Forwards values.

8.8.6   Reflection-based Amplification Attacks on SIP Services

Similar to the reflection attacks described in section 8.8.2, reflection attacks can be replicated at the SIP level by forcing a proxy to forward messages to some victim. Here one can distinguish two possible attack methods:

- **Response forwarding**–an attacker can force a SIP proxy to act as a reflector by including the victim's address in the top-most *Via* header of requests sent to different reflectors. The reflectors may be any SIP server including registrars, proxy servers and SIP phones. The reflectors will send back a negative response (e.g. not found, authentication challenge, call does not exist, etc.) to the victim.
- **Request forwarding**–similarly, innocent proxy servers may be misused to route requests to a victim. By adding the address of the victim in a *Via* or *Record-Route* header, an attacker can force a proxy server to route a request to the victim. The request reflection attack has higher harm potential as request processing logic is typically more complex than response processing logic: it may be necessary to consult an SQL server to determine user location, resolve an unexisting DNS destination, verify MD5 credentials and other CPU-consuming operations. Even worse, techniques such forking in attack reflectors may be used to amplify the attack strength.

Note, however, that unless requests are forked each of the sent requests will only cause the generation of one response of similar size. Hence, the amplification effects of such

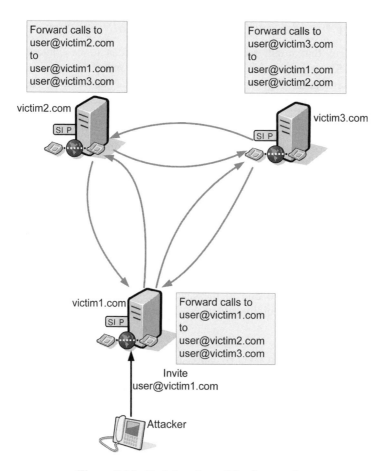

Figure 8.14 Fork-based amplification attacks

attacks is rather limited. One reason for using this kind of attacks is to overcome possible
firewall and NAT components that might only pass traffic arriving from known proxies
and drop the traffic otherwise.

8.9 Unintentional Attacks

Unlike the attacks described in the previous sections, unintentional attacks are not
launched on purpose by malicious attackers. Such attacks can in general be explained by
bad design of a system or implementation or configuration mistakes.

8.9.1 Flash Crowds

Flash crowds indicate cases when a lot of users request the same service at nearly the
same time. This is often observed in traditional telephony networks when TV viewers are
asked to vote on a certain event, for example.

The first such attack on the Internet was related to the distribution of the "hosts.txt" file. Until 1984 a centrally managed file named "hosts.txt" was used for mapping host names to IP addresses. Hence, whenever a new host was introduced to the Internet, this file was updated. For the other hosts to be able to contact this new host, they updated their local version of the hosts.txt file. This meant that regularly, i.e. whenever a new version of the hosts.txt file was announced, an increasing number of hosts was downloading this file. With each addition of a new host the number of hosts downloading this file was increasing and hence increasing the load on the server managing this file. The introduction of the Domain Name System (DNS) (Mockapetris 1987) successfully prevented this kind of attack and allowed the Internet to scale up to its current size of millions of hosts.

Another form of flash crowds is today known as the Slashdot effect. The Slashdot effect is the term given to the phenomenon of a popular web site, e.g. slashdot, linking to a smaller site, causing a huge influx of web traffic to be directed to that site. This sudden increase in traffic causes the smaller site to slow down or even temporarily close.

8.9.2   Implementation and Configuration Mistakes

Mis-interpretation of the standards or plain misconfiguration devices might result in a high volume of requests overloading servers. In (Clayton 2006) a number of such examples are listed, in which some applications or access devices generated hundreds of thousands of Network Time Protocol (NTP) (Mills 1992) requests in a matter of seconds. This caused the NTP servers to be overloaded, leading to a high delay in their reaction to legitimate requests and an overall reduction of the performance of all devices and hosts that depended on these servers. Similar scenarios were observed with DNS as well (Clayton 2006).

Combating unintentional attacks is rather difficult as the operator of the attacked server can only see that the load on the server has increased but not why. The generated traffic is usually legitimate and needed, albeit not at the expected rate. Faulty devices and applications can be identified as those that generate traffic at a higher rate than they are supposed to. Identifying the source IP addresses of these devices and applications and dropping all traffic from them reduces the load on the affected servers. However, this means that legitimate users that are using these devices and applications are blocked as well. If the fault is common to a popular device or application that is used by many users, then simple blocking would affect a large part of a provider's subscriber base. This would most likely lead to an increase in the number of calls to the support centers and bad publicity to the provider.

8.10   Address Resolution-related Attacks

The Domain Name System (Mockapetris 1989) is the basis for most currently available Internet services, including web and email. It is a globally distributed and managed database, providing an essential service for Internet applications and users, namely name resolution, which is the mapping from human readable textual host or service names (e.g. www.iptel.org) to a numerical IP address (e.g. 213.192.59.68).

Figure 8.15 depicts the basic operating mode of DNS. When an end system wants to resolve a domain name, it contacts a DNS server. The IP address of this server can be made known to the end system either by manual configuration or as part of the interaction

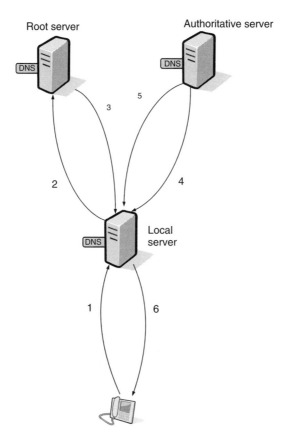

Figure 8.15   Name resolution with DNS

with a DHCP server during which the end system is allocated an IP address. If the queried
name is known to the DNS server, then it returns the IP address to the end system. This
is the case when, for example, the server is responsible for the queried domain name, i.e.
it is the authoritative server, or where the server has resolved this address previously and
has kept the result in its cache. If the DNS server cannot answer the query, the server will
issue a recursive request to other name servers that might be able to provide an answer.
The server will eventually receive a response, either containing the valid mapping or an
error message that no mapping is possible. In the former case, the mapping will be stored
in the server's internal cache for a limited period of time. The name server also sets a time
limit for the query. If no answer is received within this limit, the address is considered
to be irresolvable.

The Domain Name System plays a key role in every SIP network in the three following
ways (Rosenberg and Schulzrinne 2002c).

- Many of the header fields in a SIP message contain Fully Qualified Domain Names
  (FQDN) that need to be resolved, using the DNS, to the IP address of the next-hop
  SIP entity. These headers include the *Contact*, *Via* and *Route* headers.

- To interconnect the Public Switched Telephone Network with a SIP network, ENUM telephone number mapping is used (Peterson *et al.* 2004). ENUM allows the mapping of a PSTN telephone number (+493034637170) to a valid SIP URI. This address can then be mapped to an IP address.
- SIP can utilize different transport protocols (UDP, TCP, SCTP or TLS). To determine which protocol to use, a SIP entity might issue a DNS NAPTR request for the domain of destination SIP URI (Mealling and Daniel 2000). The response contains then one or more service names that point to the hosts that provide the SIP service and indicate which transport protocols to use.

In short, the DNS database may be queried a number of times before a message can be processed and forwarded. Attacks on the DNS servers themselves or the communication between the SIP entities and the DNS server can therefore have severe effects on the performance of SIP servers and clients.

8.10.1 DNS Servers Security Threats

There are several classes of threats that undermine the DNS service and consequently have negative effects on the operation of SIP servers. The most important are the following (Atkins and Austein 2004):

- Packet interception–within DNS, queries and responses are usually sent in single unsigned and unencrypted UDP packets. This makes attacks like man in the middle and eavesdropping combined with bogus data injection quite easy to launch. The only prerequisite for the interceptor is the ability to capture packets sent by the victim and to reply to them before the actual DNS server. This way, the attacker can direct the victim to a malicious SIP server.
- Transaction identity guessing and query prediction–the message headers of DNS queries and responses contain a 16-bit identification field which is used for matching queries with responses. With some knowledge of the records a client might try to resolve, an attacker can construct replies that seem to the client as coming from an authentic DNS server. If the attacker manages to guess the transaction identity used in the queries, the client will believe that the reply has come from the authentic server. As the transaction identity is only 16 bits long the chance of a right guess is 1 to 2^{16}. As the attacker can send multiple replies before the reply from the DNS server reaches the client, the chances of a right guess are actually higher.
- Cache poisoning–with a cache poisoning attack the attacker tries to add false information to the cache of a DNS server (Kaminsky 2008). When a DNS server sends out a query for a certain domain (e.g. example.com) the attacker uses one of the above-mentioned interception and guessing attacks and replies with false information (e.g. 209.195.132.165 instead of 208.77.188.166). All requests to example.com from users served by the attacked DNS server will then end up at another domain, e.g. hacker.com.
- Betrayal by trusted servers–supposedly trusted DNS servers can by accident or by intent become untrustworthy. Besides accidental betrayal, e.g. server bugs, break-ins, etc., the server may be intentionally configured to give answers that are not expected so that users are directed to services related to the provider of the DNS server.

8.10.2 Effects of DNS Attacks

Successful attacks on the DNS service can have devastating results. The users would be directed to malicious SIP proxies which would route their session establishment requests to the attacker. Users' web requests would be routed to malicious servers that can disguise themselves as a trustworthy sites, collect personal information that the users might reveal and use this information for fraudulent actions later.

By manipulating the DNS information, SIP traffic gets forwarded toward malicious hosts that can misuse the traffic in various ways:

- Black holing–by assuming the role of the SIP proxy serving a certain domain, the malicious server can black hole traffic by either dropping or rejecting SIP requests sent to this domain. This way the attacker can prevent calls to subscribers of this domain. Another form of this attack is call hijacking. In this case, the attacker does not drop or reject incoming requests but forwards or redirects them to another destination.
- Traffic manipulation–by placing itself on the signaling path, the attacker can act as a man in the middle and receive traffic intended for a legitimate SIP receiver, manipulate it and then forward it to the receiver. The manipulation of the SIP traffic can have different goals:
  - Bid down attacks–some of the security protocols, such as TLS (Dierks and Rescorla 2008) or DTLS (Rescorla and Modadugu 2006), negotiate the security mechanisms to use as part of the SIP signaling messages. By acting as a man in the middle, the attacker can manipulate or remove the security-related parts in the SIP messages. While this still allows for a successful establishment of the voice call, the call will not use encryption or only a weak encryption mechanism, allowing the attacker to eavesdrop on the communication.
  - Codec manipulation–the SIP user agents add the necessary information for negotiating the media format and types as part of the signaling messages. By manipulating this information, the attacker can force the user agents to use a codec of lower quality or to restrict the communication to certain formats, e.g. only audio even though both end systems can also support video.
  - Signaling manipulation–the attacker can add or remove certain headers and information and hence manipulate the routes taken by the message, any billing-related information or the identity of the caller.

8.10.3 Countermeasures and General Protection Mechanisms for DNS Services

One approach for securing the communication between the clients and servers would be to use some encryption mechanisms such as IPsec or TLS. This approach, however, imposes a fairly high processing cost per DNS message, as well as a high cost associated with establishing and maintaining bilateral trust relationships between all the parties that might be involved in resolving any particular query. Further, for heavily used name servers such as root zone, this cost would be prohibitively high. Finally, the underlying trust model only provides a hop-by-hop integrity check on DNS messages. An attacker can still overcome this security model by manipulating an authoritative DNS server.

DNSsec (Eastlake 1997) provides a more lightweight solution that supports end-to-end protection. However, the introduction and deployment of DNSsec has proved to be cumbersome, especially with regard to the configuration and replacement of the needed encryption keys. Further, usage of DNSsec significantly increases the size of DNS response packets making, among others, the servers which implement DNSsec more efficient as DoS amplifiers. Finally, the need to validate the cryptographic signatures of the DNS messages increases the processing load at the clients.

8.10.4 DNS-related Attacks on SIP Services

Whenever a SIP server encounters a fully qualified name in a header of a SIP message that is necessary for routing (e.g. *Via* or *Route* field), it issues a query to the local domain name server. On average it takes 1.3 DNS queries to receive an answer with the mean resolution latency less than 100 milliseconds (Jung *et al.* 2001). The resolution latency is considerably increased in the following cases:

- Irresolvable names–a DNS name can be irresolvable if there is an authoritative name server responsible for it, but this server happens to be unreachable or is not responding to DNS queries. Where no reply is returned for a DNS query, the query will be retransmitted after a timeout. The default configuration of the widely used Berkeley Internet Name Domain Server (BIND) (Terry *et al.* 1984) uses a timeout of 5 seconds and the query is sent twice, which means that a SIP server will need 10 seconds before deciding that a DNS name is irresolvable.
- Congested networks and overloaded servers–if the authoritative server for a certain name is overloaded or connected to the Internet over a heavily congested link then the resolution latency will most likely have a higher value than the average 100 milliseconds.

With DNS-based attacks (Zhang *et al.* 2007), the attacker aims at misusing and increasing the processing resources needed for resolving domain names. This can be achieved by causing the SIP server to resolve domain names that are either irresolvable or are served by overloaded DNS servers.

This kind of attack can be mounted by sending SIP requests to the SIP server with an irresolvable domain name included in a header that is used by the SIP server for routing the messages, e.g. *Via* or *Route* headers or in the *Request-URI*. Such requests are otherwise well-formatted SIP requests that comply with the SIP standard in every respect.

An attacker can ensure that a domain name is irresolvable by launching a denial of server attack on the authoritative server of this domain. Another approach is to actually register a number of domain names and set the addresses of the authoritative servers of these domains to hosts that do not reply to DNS queries or do not exist at all. To register a domain name the attacker is supposed to provide her name, address and payment information for a domain name registration company. However, as the name and address information are usually not verified and stolen credit cards can be used for payment, the attacker can falsify this information and hide her identity.

While all kinds of SIP servers are affected by DNS attacks to some degree, such attacks are especially fatal for components that interact directly with the user agents and are expected to forward traffic to various destinations, e.g. Proxy Call Session Control

Functions in IMS networks and outbound proxies. In the context of IMS networks, a P-CSCF is responsible for receiving traffic from roaming users and forwarding it to the home domain of the users. Outbound proxies provide a similar functionality in VoIP service, especially in conjunction with NAT traversal.

The actual effects of a DNS-based SIP attack depend very much on implementation of the SIP server itself. In general it is possible to distinguish between two types of implementations, namely, synchronous and asynchronous. Servers implemented in a synchronous manner issue a DNS query and block until they receive a reply. Asynchronously implemented servers issue a DNS query, maintain state information describing the SIP request, the status of its processing and the issued DNS query and can continue processing other SIP requests. Once the DNS reply is received, then the processing of the SIP request is continued.

Effects of DNS-based DoS attacks on synchronously implemented SIP servers A SIP server implemented in synchronous manner sends a DNS query and waits for an answer. While waiting for the reply, the SIP server is blocked and unable to process new requests. To enable the processing of multiple requests in parallel, the implementation architecture consists of a dispatcher and multiple processes or threads. When a request arrives at the proxy, it is dispatched by the dispatcher to a free process. Each process is then responsible for parsing the message, initiating any DNS requests or requesting the execution of an application and finally forwarding the message. State information can be shared among the processes using some form of shared memory.

Under an attack, the parallel processes or threads get blocked. That is, if the server consists of N processes, then by sending N queries with unresolvable DNS names, an attacker can block all processes of the proxy until a negative response is received from the DNS server. During this time, the proxy would not be able to process any new requests. Hence, if it takes on average t seconds to discover that a destination is unresolvable, then the attacker can block the proxy by sending N/t requests per second. Taking the above-mentioned average value of 100 milliseconds for a DNS query and a SIP server consisting of 100 processes, then sending 1000 requests per second with irresolvable names is sufficient to block the SIP server. When using irresolvable names for the attack, it takes each process 10 seconds before deciding that a query has failed. In this case the attacker can manage to block the SIP server by generating only 10 requests per second.

Effects of DNS-based DoS attacks on asynchronously implemented SIP servers With an asynchronous DNS implementation, whenever the routing of a SIP request requires the resolution of a DNS name, the SIP server interrupts the processing of the SIP request, issues a DNS query, keeps state information about the status of the processing of the SIP request and is then free to process new SIP requests. Once the response to the DNS request arrives or a timeout expires, the SIP server will be notified and the processing of the request that caused the DNS query can continue.

With this approach, a DNS attack does not cause the SIP server to block. However, since the server needs to keep state information about the state of the SIP requests and DNS queries, implementation complexity and memory requirements increase considerably. With a sufficiently large number of DNS queries this attack scenario can be considered as a memory depletion attack.

8.10.5   Protecting SIP Proxies from DNS-based Attacks

Reducing the number of pending DNS request can help to reduce the effects of DNS attacks. The *Via* list in SIP requests indicates the path taken by the request so far. That is, the caller adds its address as the first entry in the *Via* list. Each proxy that receives this request adds its address to the list as well. The final receiver of the request adds the *Via* list to its response and then sends the response to the top-most *Via* entry. Each proxy receiving the response removes the *Via* entry indicating its address and forwards the response to the new top-most entry. For the case of multihomed proxies or user agents or for the case of traversing a network address translator, the address indicated in the *Via* entry might differ from the IP address of the entity that forwarded the request to a proxy. For this proxy to exactly identify the IP address to which to forward the replies to, a proxy can add a *receive* parameter to the top-most *Via* entry in the received request. When receiving the response to this request, the proxy does not forward the response to the address indicated in the *Via* entry but to the address it had previously added in the *receive* parameter. To avoid the need for resolving a DNS name included in a *Via* entry of a response, one can always add a *receive* parameter to the *Via* entry of the request with the IP address of the sending entity.

An effective approach for reducing the need to interact with DNS servers is to locally cache both positive and negative (Andrews 1998) results of the DNS queries at the SIP server directly. Whenever the SIP entity needs to resolve a SIP URI it first checks the local cache and only contacts a DNS server if the cache did not include an appropriate result.

8.11   Attacking the VoIP Subscriber Database

VoIP providers usually offer their subscribers two kinds of interfaces, illustrated in Figure 8.16. The signaling interface is based on SIP and enables the subscribers to use the VoIP and multimedia services offered by the provider. The provisioning interface allows the subscribers to customize their personal information. That is, subscribers to VoIP services usually have profiles that indicate their preferences, e.g. whether calls should be forwarded to some other destination, personal information such as identity and password and various other information such as call logs and billing records. This information is usually maintained in a database and users can access parts of this information through web interfaces.

The user profiles are used by the SIP infrastructure components for authenticating and authorizing the users and taking the appropriate decisions on where to route requests originating from and destined to the users. The SIP components update some of the information maintained in the databases such as the call logs and call detail records.

By getting access to the subscriber database, an attacker can cause severe damage by:

- deleting subscriber profiles and preventing legitimate users from accessing the service—this means loss of revenue for the provider and loss of the phone service for the subscribers;
- deleting or manipulating CDRs or the access rights of the subscribers;
- getting access to the user's call lists, i.e. lists indicating who the user called or was called by, thus intruding on the user's privacy;

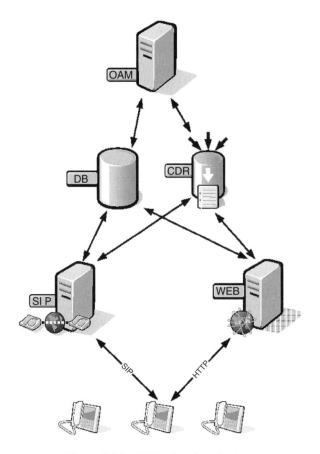

Figure 8.16   VoIP subscriber database

- accesing user profiles that usually include the subscriber's passwords in a form that allows the attacker to assume the identity of the subscriber and correctly authenticate himself.

This structure allows for two kinds of attack possibilities, namely web-based attacks and SIP-based attacks. In both kinds of attacks, the attacker tries to get access to the subscriber database with the goal of manipulating its content or stealing the passwords of subscribers.

8.11.1   Web-based Attacks on the Subscriber Database

Attacks on web sites are among the most common security threats faced today on the Internet (Richardson 2007). In general, these attacks do not cause the web server to become unavailable or have negative effects on its performance. A successful attack could, however, allow the attacker to get access to the subscribers' personal information or manipulate the subscriber database itself (McClure *et al.* 2002).

8.11.1.1   Cross-site Scripting

Cross-site Scripting (XSS) (Fogie *et al.* 2007) is an attack technique that allows the attacker to inject some code into the attacked application in order to get access to private information of the users of this application. The code itself is usually written in HTML/JavaScript, but may also extend to VBScript, ActiveX, Java, Flash or any other browser-supported technology.

Web sites use scripting languages such as PHP (Lerdorf *et al.* 2002) for displaying dynamic content. For example a web site, www.example.com, that allows its users to search through its database would use a PHP script *search.php*. A user looking for articles about *DoS* sends an HTTP request with DoS as the input parameter for the PHP script in the form of: `http://www.example.com/search.php?string=DoS` The result of this search would be a web page with the related articles as well as heading including something like "10 results for DoS".

A web domain that is vulnerable to XSS attacks displays the user's input on the resulting web page without checking whether the content was malicious or not first. An HTTP request in the form of: `http://www.example.com/search.php?text=<img` `src="http://attacker.com/image.jpg"` causes the vulnerable web site to display the image of the attacker.

The code injected by the attacker runs within the security context (or zone) of the attacked domain. With this level of privilege, the code has the ability to read, modify and transmit any sensitive data accessible by the domain. This allows the attacker to steal personal information of the user such as the cookies that maintain the user's access information to the web domain, to redirect the user's browser to another location or possibly to show fraudulent content delivered by the web site they are visiting. Cross-site scripting attacks essentially compromise the trust relationship between a user and the web site.

There are two types of cross-site scripting attacks, non-persistent and persistent. Non-persistent attacks require a user to visit a specially crafted link with the malicious code included as part of the URL. This can be achieved by sending the user an email disguised as an offer from the user's VoIP service provider with an attractive offer and asking the user to click on the link included in the email. Upon clicking the link, the script included in the URL is activated and could for example read the cookie which includes the user's authentication information toward the service provider and send it to the attacker or redirect the user to a web site looking like that of the service provider and ask the user to enter his login name and password. This information can then be used for fraud purposes. Persistent attacks occur when the malicious code is submitted to a web site where it is stored for a period of time. Such sites could be message board posts, web mail messages and web chat sites. This code can then be activated simply by viewing the web page containing the code.

XSS attacks are protected against in one of two ways: careful programming style and data inspection. Already when developing the web site, the programmers need to consider possible XSS attacks. A careful programming style implies restricting the possible input to the different fields of a web page, filtering suspicious parameters and inspecting and quoting parameters included in the URLs to prevent them from getting executed.

Because of the complexity of web sites and the fact that the content of a web site is often developed by different parties with varying degrees of proficiency and understanding of the security risks, web sites should also be protected with firewalls that inspect the traffic addressed to the web site and filter HTTP requests that include scripts or any suspicious looking data.

8.11.1.2 SQL Injection

SQL (Structured Query Language) (Groff and Weinberg 1999) is a programming language used for the retrieval and management of data in relational database management systems, database schema creation and modification, and database object access control management.

Based on data added by a user in the input fields of a web form, a web server constructs SQL commands and executes these commands. These commands can include expressions for retrieving data from the database, searching for certain strings or deleting or replacing certain fields of the database. In an SQL injection attack, the attacker includes SQL expressions in the input fields of a web form. The web server then constructs a SQL command that includes the SQL expression and this expression, hence, gets executed.

For example, a web page that authenticates the user can include fields for the user name and password. Once the user writes the name and password the web server constructs an SQL command for looking into the database and checking whether the provided password and name correspond to each other. Such a command could be in the form of:

```
SELECT Name From subscriber WHERE User_ID='<username>' AND
                  User_Password='<password>';
```

An attacker could use as the password something like

```
a';DROP TABLE users; SELECT * From data;
```

which would result in a SQL command in the form of:

```
    SELECT Name From subscriber WHERE User_ID='a' AND
User_Password='a';DROP TABLE users; SELECT * From data;'
```

This statement causes the deletion of the "users" table as well as the selection and display of all data from the "data" table, c.g. in essence revealing the information of every user.

For SQL injection attacks to be successful, the attacker has to know the names of the used tables. While this information is not trivial to guess, it is still possible and actually rather simple if the application is open source.

This kind of attack can be protected against in a similar manner to the XSS attacks, namely by careful programming and inspection. During the development phase the developers need to restrict the possible input and quote the included parameters. Similar to the case of XSS attacks, traffic inspection components can be used to filter requests to the web server that include SQL statements. By restricting the access rights of the applications accessing the database, some higly damaging manipulation attempts such as deleting a table or displaying it in its entirety can also be prevented.

8.11.1.3 HTML Tampering

The interaction between a user and a web server, for example for subscribing to a new service or changing some of the user's preferences, is often realized using HTML forms. These forms allow the user to change certain fields or write some information. As this is often conducted over several steps, some information might have to be saved across the different steps in order to avoid forcing the user to repeatedly specify the same information. Because of the stateless nature of web applications, for applications that lead a user through a series of input forms, the application must temporarily store field data entered on previous pages. This is often done by adding this information as hidden fields in the HTML forms.

However, HTML forms have one major weakness: users can save the form to a file, edit it, then use the edited version to submit data back to the server. By manipulating the data stored in the HTML form, an attacker can get access to the data of other users. To prevent this kind of attack, the web application should not only store the state information in some hidden fields in the HTML form but also generate a fingerprint of the stored data. The fingerprint is generated by running a hashing scheme, e.g. MD5, over this data plus some secret information known only to the web server and is then stored in the form. When the form is returned to the web server, the correctness of the data can be verified by comparing the included fingerprint and the result of running the hashing scheme over the stored data plus the server's secret. Without knowing the server's secret, an attacker would not be able to manipulate any data or the fingerprint without the manipulation being detected.

8.11.2   SIP-based Attacks on the Subscriber Database

As part of serving a SIP request, a SIP proxy usually needs to query a database and retrieve the user's profile, subscription and authorization information as well as the user's preferences. This information is needed for authenticating the request, checking whether the user is allowed to use the requested service and for correctly forwarding the request. As input to the database queries, the server often uses information included in the *To* and *From* headers.

Similar to the case of web services, SQL databases are also widely used as part of VoIP services as well for saving and managing subscriber data. Hence, when a request from user "jku" needs to be authenticated, a SIP proxy using a SQL database checks a table, e.g. "subscriber", using a statement such as the following:

```
SELECT password From subscriber WHERE username='jku' AND
                realm='provider.com'
```

Similar to attacks on the web server, a malicious user can launch a SQL injection attack on such SIP proxies by inserting malicious SQL code in the headers of his SIP requests.

Figure 8.17 depicts such an attack in which the attacker adds the malicious code in the authorization header. This message can be any SIP message requiring authentication by

```
INVITE sip:callee@voip.com SIP/2.0
Via: SIP/2.0/UDP client.provider.com:5060;branch=z9hG4bK74b03
Max-Forwards: 70
From: subscriber <sip:subscriber@provider.com>;tag=9fxced76s I
To: callee <sip:callee@voip.com>
Call-ID: 3xRb9vxSit55XU8o9@provider.com
CSeq: 1 INVITE
Contact: <sip:subscriber@provider.com>
Proxy-Authorization: Digest username="jku;
    Update subscriber set first_name='malicious' where
username='jku'; ",
        realm="provider.com",
        nonce="wf84f1ceczx41ae6cbe5aea9c8e88d359", opaque="",
        uri="sip:callee@voip.com",
        response="42ce3cef44b22f50c6a6071bc8"
Content-Type: application/sdp
Content-Length: 151

v=0
o=subscriber 2890844526 2890844526 IN IP4 provider.com
s=-
c=IN IP4 191.0.3.102
t=0 0
m=audio 48181 RTP/AVP 0
a=rtpmap:0 PCMU/8000
```

Figure 8.17   SQL injection with SIP messages

a SIP server. The code can be included either in the *username* or in the realm fields in the *Authorization* header.

As soon as the proxy receives this SIP message with the malicious *Authorization* header, it generates and executes the following SQL statement:

```
SELECT password From subscriber WHERE username= 'jku'; UPDATE
    subscribe SET first_name='malicious' WHERE username='jku';
```

While the message authentication fails in this case, the second command, if executed, results in changing the first_name of user "jku" to "malicious". It is also possible for a malicious user to attempt to delete complete tables or change the passwords of other users. This way the information in the database effectively becomes useless and can lead to situations in which a proxy accepts all incoming traffic regardless whether it can be authenticated or not or rejects all traffic.

This kind of attacks can be prevented using similar measures as described in section 8.11.1.2, namely by filtering the input, restricting the access and avoiding the usage of data strings included in the SIP messages as direct input for the SQL queries.

8.12   Denial-of-service Attacks in IMS Networks

On the one hand IMS networks are based on the SIP specifications. On the other they also impose more stringent access control mechanisms. Hence, the ability of subscribers to IMS networks to launch an attack and hide their traces is much reduced. However, the complexity of the IMS architecture can also introduce additional security threats that are not considered an issue in VoIP networks.

One can expect that to become a subscriber to an IMS service, one would have to go through the same due diligence procedures as in current PSTN services. That is the existence, the address and the monetary situation of the subscriber are investigated before accepting someone as a subscriber. Compared with VoIP services in which it is often sufficient to provide a valid email address to become a subscriber, these checks usually deter anyone from launching an attack. Further, when deployed over UMTS networks, before being allowed to use the IMS service, the subscriber needs to authenticate himself to have access to the physical network resources as well as to the multimedia services. Hence, spoofing IP addresses or using fake SIP names is also not possible.

While these measures surely reduce the possibility of denial of service attacks, they are unfortunately insufficient. An insistent attacker can get legal access to an IMS network by stealing the identity of other people, e.g. by using stolen credit cards, mobile phones or other stolen identification information. Further, getting access using a prepaid service often does not require any verification. Once having passed the authentication hurdle, the attacker can launch most of the attacks presented in the previous section with even more devastating effects.

Further, IMS networks will most likely be introduced as islands that are connected to the Internet, e.g. to other VoIP providers, over gateways. Figure 8.18 depicts such a deployment scenario in which IMS networks are connected over secured links with each other. To enable calls to other non-IMS VoIP providers IMS networks deploy gateways to strip out network specific information and forward and receive SIP requests from non-IMS providers. Attackers could also flood these gateways and launch DoS attacks over these open connections, bypassing the IMS authentication and authorization steps.

In the following we investigate the applicability of various DoS attacks that were discussed in the previous sections to IMS networks as well as some more IMS-specific attacks. The general conclusion here is that IMS networks are just as vulnerable to DoS attacks as VoIP services. The major difference lies in the hurdles imposed on the attacker in actually launching such attacks. As an attacker cannot spoof her identity, attacks on an IMS network can only be launched using legitimate accounts. If an attacker does not want to be detected, then such accounts can only be acquired using a stolen identity or by manipulating IMS clients.

8.12.1 Bandwidth Depletion Attacks

Components that connect an IMS network to the Internet can be a target of bandwidth depletion attacks, just like any other components on the Internet. A misbehaving IMS client can just as easily become the source of a flooding attack. Further, as IMS clients are basically IP devices, they could be turned into bots just like any other IP device and used to launch DDoS attacks on components inside the IMS network as well in other networks.

8.12.2 Memory Depletion Attacks

Similar to the case of bandwidth depletion attacks, IMS components that interface the Internet can be the target of memory depletion attacks such as TCP SYN or IP packet

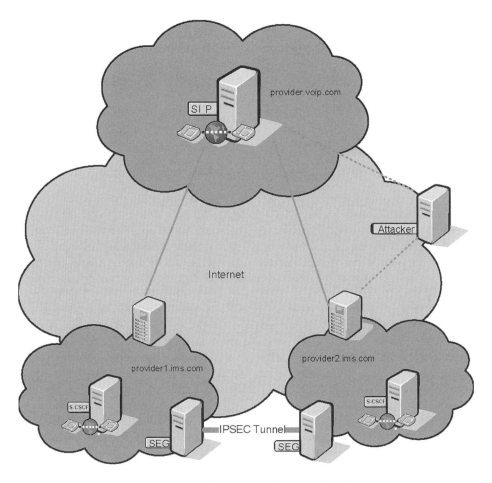

Figure 8.18   Deployment of IMS as islands

fragmentation. Misbehaving IMS clients could also launch such attacks on components inside the IMS network.

Launching SIP-based memory depletion attacks using the incomplete transactions scenario as described in section 8.5.2 requires having an IMS subscriber that is registered but does not reply. For the case of VoIP services, setting up such an account often does not require more than a valid email address. In the case of IMS it is expected that the users will be more thoroughly checked and, hence, the ability of a provider to trace back a malicious account to a legal person is greater than in VoIP networks. Hence, for an attacker to successfully launch such attacks in the IMS environment without facing the danger of being detected would require using either false identities or manipulating IMS end devices.

Launching SIP-based memory attacks from an IMS network into the Internet, e.g. depleting the memory of a VoIP server, would require the manipulations of IMS clients. For a forking attack the attacker has to set up some call routing preferences. As an attacker would probably not set up his own account for fear of being detected, such attacks can only be launched using stolen identities or manipulated clients.

Nonintentional attacks due to configuration errors can lead to trouble in IMS as well. A user nonintentionally misconfiguring his forwarding rules on his different accounts can also lead to a memory depletion attack.

8.12.3   CPU Depletion Attacks

The need to use stolen identities or get access to and manipulate an IMS client increases the hurdles for an attacker successfully launching memory-depletion attacks compared with the more open VoIP services. However, the higher complexity of the IMS technology means, that once an attacker manages to get access to a valid account more damage can also be caused in the network. This includes misuse of certain CPU intensive technologies such as compression and encryption. Similarly, the negative effects of misconfiguration or implementation errors are higher as well.

8.12.3.1   Misusing Signaling Compression (SigComp)

To reduce the bandwidth requirements of transporting SIP messages over wireless links, in IMS compression of SIP messages is suggested. While the bandwidth savings can be significant, using compression adds a considerable processing overhead at the P-CSCFs.

Flooding attacks are especially annoying when SigComp is enabled. That is, an attacker can send useless messages to the P-CSCF. As the messages are compressed, the P-CSCF will detect that these messages are useless only after having decompressed them. Considering that the decompression task is a highly CPU-consuming task, by sending a large number of such messages the P-CSCF can be overloaded rather quickly.

Further, a malicious client could also increase the processing load by using complex decompression schemes that require a large number of operations. The effect of such an attack can be reduced by limiting the maximum number of operations that can be used for decompression.

Before exchanging SIP messages using SigComp, IMS client and the P-CSCF need to exchange some state information that is then maintained at both ends for some time. This can be done only once when the user registers to the IMS network or each time a new sessions is created. An attacker could misuse this to launch a memory depletion attack by requesting the creation of SigComp session state several times at the P-CSCF and not using it.

While such attacks usually require the attacker to manipulate an IMS client, due to the complexity of the technology such attacks could also occur unintentionally due to implementation or configuration errors.

8.12.3.2   IPsec-based Attacks

IMS uses IPsec for securing the communication between the IMS components of one provider as well as between the IMS components of different providers. An attacker could misuse this by sending useless traffic on ports over which an IMS component expects encrypted traffic. In this case, the IMS component first needs to decrypt the traffic before being able to classify it as useless.

8.12.4 Protocol Misuse Attacks

As IMS components are primarily IP-based devices, an attacker misusing a known implementation mistake at an IMS component can cause the component to crash similarly to VoIP components. SIP misuse attacks are, however, more difficult to launch. IMS traffic is supposed to be encrypted and hence an attacker should not be able to eavesdrop on the traffic in an IMS network, which is a prerequisite for launching SIP misuse attacks.

Where a SIP session is established without encrypting the traffic on all links of the path, e.g. when establishing a communication session between an IMS client and a user agent in a non-IMS environment, these attacks can still be launched.

8.12.5 Web-based Attacks

Similar to any VoIP service, an IMS network includes web servers, databases and provisioning servers. Hence, all attacks described in section 8.11 are applicable in the IMS context as well. IMS clients will often have email and web browsing capabilities. This makes them vulnerable to web-based attacks just like any other IP device.

Further, IMS SIP messages can be used to carry SQL commands and, hence, can be used for SQL attacks as well.

8.13 DoS Detection and Protection Mechanisms

Successful DoS attacks on popular services usually get wide press coverage. This can then easily lead to high losses of revenue, reputation damage to the service provider and a reduction of the users' trust in the service provider. To avoid these consequences and reduce and possibly prevent attacks, academia and industry have provided over recent years a wide range of anti-DoS mechanisms and solutions (Mirkovic *et al.* 2004). These mechanisms can be roughly grouped into three classes, namely, detection, prevention and reaction, as shown in Figure 8.19. The basis for any security system is knowledge of possible security threats. While some of these threats can be specified theoretically, knowledge gained by observing the actual traffic in the network and monitoring the traffic pattern is of utmost importance for the proper design and functioning of any security system. Further, by monitoring and analyzing the traffic, it is possible to detect attacks and launch the proper reaction mechanisms.

By detecting an attack in a timely manner, an administrator can aim at reducing the negative effects of the attack and keep the attacked services running, albeit in a degraded manner. To reduce the possibilities of being attacked in the first place and reduce the effects of an attack, various protective measures need to be deployed. As a VoIP infrastructure is a collection of SIP, DNS and web servers as well as databases, the service provider must not only use general security mechanisms but also deploy various protective measures specifically designed for each type of server used.

8.14 Detection of DoS Attacks

The basis for protecting a system against denial of service attacks is to understand these attacks and have the ability to identify and detect them. A prerequisite for Intrusion Detection Systems (IDS) is the ability to collect and analyze network traffic and classify it

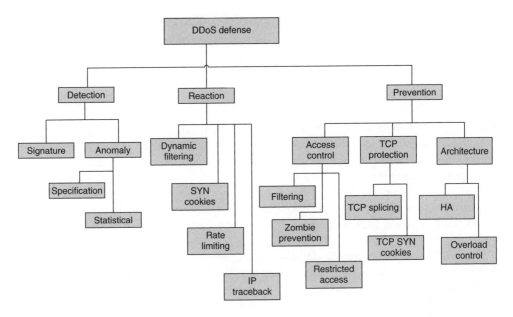

Figure 8.19 Classification of defense measurements against DoS attacks

into normal and abnormal traffic behavior. This decision can be based on either preknown attack signatures or by testing for anomalous traffic patters.

8.14.1 Signature-based DoS Detection

Several DoS attacks show certain predictable traffic patterns. For example, the packets of a "Land attack" as described in section 8.7.1 always have the same source and destination addresses. Signature-based detection tools deploy a list of predefined signatures covering known attacks. Traffic arriving at the system to be protected is analyzed and the resemblance to any of the known signatures is investigated.

A major advantage of this kind of detection is that, once the signature of an attack is well known, detecting such attacks is usually rather efficient in terms of the required processing resources and detection speed.

However, intrusion detection systems based solely on signatures cannot detect new attacks, so-called "zero-day attacks". Further, as attacks might have many variations, the list of signatures that such systems need to maintain and keep up to date will grow with time and can become rather long. This in turn might have negative effects on the performance of such systems.

8.14.2 Anomaly-based DDoS Detection

In the context of DoS detection an anomaly is defined as a deviation from the expected behavior, e.g. deviation from the specification of a protocol. A major advantage of anomaly-based detection compared with signature-based detection is that previously unknown attacks can be detected as well. That is, traffic that shows noticeable deviation from the expected behavior can be readily categorized as malicious. However, it is not always possible to define an exact model for the expected traffic. The less accurate the model is and the higher the variance of the traffic, the more likely it is that anomaly-based detection systems will lead to false alarms.

8.14.2.1 Specification-based Anomaly Detection

For all protocols used in the Internet, there are specifications that clearly describe what the protocol messages should look like, how the involved systems should react to these messages and what to do if something goes wrong. By checking whether traffic conforms to the protocol specifications, an intrusion detection system can detect attacks by observing packets that arrive unexpectedly. This might include, for example, observing ICMP reply messages when no ICMP requests have been issued or a SYN-ACK TCP message when no SYN message has been sent by the system that is protected by the intrusion detection system.

To detect such anomalies, the intrusion detection system maintains state information describing which requests have been sent or received by the hosts that are to be protected by the intrusion detection system. This has the disadvantage that a considerable amount of memory is needed by the intrusion detection system. With the increased number of sessions, searching the list of all available sessions for each incoming packet will also become a CPU-intensive task. Further, in many attack scenarios the sent messages actually conform to the standards and will, hence not be discovered by this approach.

Anomalous session setups can be detected using an IDS running a SIP state machine that acts as a SIP client and server. By monitoring the exchanged SIP messages and acting as a SIP server when observing a request and as a SIP client when observing a response, such IDS systems can detect whether session initiation requests were are not properly terminated or BYE requests were received for sessions that were never created (Sengar *et al.* 2006).

A major advantage of standards-based anomaly detection is that the chance of false positives is rather small as all legitimate traffic must conform to the specifications.

8.14.2.2 Statistics-based Anomaly Detection

The normal behavior of a network or a system is the expected behavior of the network or system when no attacks are launched against them. The normal behavior is determined by observing certain parameters such as the rate of incoming messages, inter-arrival rate, type of messages or distribution of source and destination addresses. Which parameters are observed depends very much on the protocols and applications used, the attacks to be detected and the available resources for the intrusion detection system.

Besides defining the normal behavior of protocol specific parameters, e.g. relation of SYN and FIN messages in TCP, INVITE to BYE in SIP or the frequency of certain headers or fields in the messages, it is very common to use general measurements indicating the expected number of messages per second, average inter-arrival rate and packet length. The measurement values are then analyzed through various mechanisms in order to decide whether an attack is being launched and when the attack started. While the analysis algorithms vary in their complexity, accuracy and efficiency, they usually compare the analysis results and some thresholds that define normal behavior. When the measured values exceed the used thresholds by some value, it is assumed that an attack has been detected.

Statistics-based detection schemes are especially useful for detecting flooding, memory depletion and loop-based attacks on SIP servers. In these scenarios the amount of traffic exchanged between a limited number of user addresses is much higher than between other users. This difference can be used as an indication of an attack.

Sequential change-point detection   One popular approach for detecting anomalies is based on using the Sequential Change-Point Detection algorithm (Basseville and Niki-forov 1993). This algorithm can help detect the point at which an attack starts. To use change-point detection, the information needs to be organized as a time series. If the series shows a change at time T, then this can contribute to a denial of service attack starting at time T. The cumulative sum (CUSUM) algorithm (Wald 1947) is a class of change-point detection algorithm that operates on continuously sampled data and requires only small amounts of memory and computational resources.

To identify and localize a DoS attack, the CUSUM algorithm identifies deviations in the actual vs. expected local average in the traffic time series. An attack is highly probable once the difference exceeds some upper bound by some threshold. Through the settings of the threshold and upper bound, the CUSUM algorithm can trade off detection delay and false-alarm rates.

Wavelet analysis   Wavelets provide for concurrent time and frequency description, and can thus determine the time at which certain frequency components are present (Carl *et al.* 2006). For detection applications, wavelets separate out time-localized anomalous signals from background noise; the input signal contains both. Ideally, the signal and noise components will dominate in separate spectral windows. Analyzing each spectral window's energy determines the presence of anomalies. Wavelets were used in (Barford *et al.* 2002) to determine anomalies such as network failures or misconfigurations, DoS attacks and flash crowd events. This was achieved by decomposing traffic data into distinct time series of average IP/HTTP packet sizes per second, flows per second and bytes per second. Applying wavelet analysis to each time series resulted in time-localized high- and mid-band spectral energies. The low-frequency content was considered to describe the daily or weekly activity. Whether an attack was observed was determined based on the values of the high- and middle-spectral energies.

Backscatter analysis   In order to hide their traces, attackers often use spoofed source addresses. The basic assumption of backscatter analysis (Moore *et al.* 2001) is that the used spoofed addresses are chosen with a uniformly distributed probability. Hence, as described in the examples of TCP SYN attacks or the ICMP attacks, an attacked system replies to the messages sent by the attacker with messages sent to a uniformly distributed address range. In general, systems communicate only with a rather small and restricted number of other hosts. Hence, observing that some host is sending messages to a uniformly distributed address range is a very good indication that this host is under attack. Other approaches use the ratio of observed SYN and FIN TCP packets (Wang *et al.* 2002). In the normal case, this ratio should be close to 1. In case of an attack there will be many more SYN packets than FIN packets as the connections will not be properly established and terminated.

The use of statistical models for detecting anomalies can be efficient and simple for detecting sudden and large changes in the traffic due to DoS attacks. However, providing accurate models and using the right parameters can be rather difficult. The deviation between the monitored traffic and the expected behavior is usually expressed in terms of thresholds, i.e. whether the difference between some monitored parameter exceeds some predefined value. Hence, for statistical approaches to be efficient in the face of changing networking conditions, the thresholds must be updated dynamically. For example, as the Internet traffic can vary with time of day, one cannot use a single threshold for describing the maximal bandwidth usage in a network for the entire day but must use adaptive thresholds. This means, however, that an attacker that manages to increase the bandwidth usage in the network slowly also causes the adjustment of the detection thresholds and will not be discovered. Further, these algorithms observe the traffic for some period of time and then decide whether an attack is in progress. By setting the observation periods much longer, these schemes can collect more information and make a better judgment with regard to the attack, however, at the cost of a slow detection. Setting the observation periods short makes these schemes more reactive; however, this also increases the possibility of false alarms.

Further, as the analysis is done using often rather complex statistical models, the more parameters are observed, and the higher the granularity of the traffic model, the more processing resources will be needed by the intrusion detection system.

8.15   Reacting to DoS Attacks

Using the detection methods described in section 8.14, it is possible to detect that an attack is being launched and possibly identify the hosts involved in the attack and the attack signature. This information can then be used to react and mitigate the effects of the attack. Because of the different natures of the attacks, there is no one solution that can solve all problems. To be able to efficiently react to the various DoS attacks, various mechanisms have to be available and are often used together.

8.15.1   Dynamic Filtering

Once the detection mechanisms have identified the source or the signature of the attack, the firewalls used to protect the victim can be dynamically configured to filter out the malicious traffic. This is often also referred to as Access Control Lists (ACL). Note that, unless the characterization of the attack is very accurate, filtering can run the risk of accidentally denying service to legitimate traffic.

8.15.2   Rate Limiting

A flash crowd scenario or a DoS attack can cause overload not only at the targeted victims, e.g. SIP servers, but also at the routers in the network. This leads to packets being dropped at those routers and affects not only the flows that are causing the overload but also all other flows traversing the overloaded routers. In (Mahajan *et al.* 2002), the authors propose an approach in which the routers penalize aggregates of flows that are causing the overload. These aggregates can be defined as the traffic generated by some address or destined for one or more addresses, e.g. victims. It is naturally also possible to use more complex aggregate definitions. However, as routers need to deal with a large number of packets, increasing the complexity of the aggregation policies would have negative impact on the performance of the routers. Once one or more aggregates are identified as the cause of the overload, the router starts dropping the packets belonging to these aggregates with a higher probability than other packets. This means that traffic not belonging to the DoS attack is prioritized and has a higher chance of reaching its destination. As an optimization, this approach can be extended with a push-back mechanism. After identifying the misbehaving aggregates, the routers inform their neighboring routers, through which these aggregates were received, to throttle these aggregates as well. This reduces the load closer to the source of the attack and the overload on the entire network. However, this approach requires cooperation between the routers and the administrators of these routers, which might be possible as long as the routers belong to same administrative domains but is not supported across domains currently.

Most modern routers allow the administrator to limit the rate of certain traffic, e.g. ICMP replies, TCP SYN packets, traffic from certain sources or to certain destinations. Once an attack is detected, the administrator of the routers can restrict the traffic that was identified as contributing to the attack and, hence, reduce the effects of the attack on the downstream links.

The same concept could be deployed with SIP servers. Once it was detected that the amount of signaling traffic exchanged to or from certain addresses exceeds some limits, the SIP servers could limit the rate to those addresses.

Note that, while limiting the rate of certain traffic type can reduce the effects of the DoS attack on the network and servers, it will also lead to the drop of legitimate traffic. Also, some malicious traffic will still reach its destination.

8.15.3 IP Traceback

The optimal approach for defending against a denial of service attack is to detect the source of the attack and block the malicious traffic before it enters the network. A theoretically interesting approach for achieving this is the IP traceback approach. IP traceback was first suggested by (Savage *et al.* 2001). With this approach, routers include with some low probability their IP addresses into the packets that traverse them. Once sufficient packets have been received by the victim, it can construct the path that was traversed by the attacking flow. This approach requires the victim to receive a large number of packets from the same attack flow, i.e. packets that have traversed the same path. To reduce the number of needed packets and make the detection faster, various modifications for this approach have been proposed that add more information to the IP packets and hash this information so that it fits in the IP packets (Ma 2006; Song and Perring 2001).

Another approach for IP traceback is for the routers to collect information about the packets they are forwarding. Once an attack is detected at some router, the data collected by neighboring routers can be queried and the source of the attack can be detected step by step. As collecting a copy of each packet requires a huge amount of memory, in (Snoeren 2001) a modified approach is presented in which only a fingerprint of the packets is collected, e.g. the packet headers and a small fraction of the data are collected and saved as a hash. Once a router classifies a packet as suspicious, it can query the fingerprint of this packet at the neighboring routers. While this approach is much more efficient in terms of required memory, the hash space has to be chosen as rather large to avoid collisions of hash values. This naturally reduces the savings effects. On the other hand, if only a small hash space was used then collisions would be unavoidable and two different packets might mistakenly be classified as equal, resulting in the determination of a false path.

Regardless of the deployed approach for IP traceback, all the mechanisms require cooperation either between the routers, e.g. use similar marking schemes, or between the administrators of the routers, e.g. to allow access to collected information at the routers. Currently both are unrealistic, at least across administrative domains. With the diversity of the Internet it not possible to achieve collaboration between distant ISPs without a legal framework that forces the ISPs to cooperate more closely.

8.16   Preventing DoS Attacks

The best defense against denial of service attacks is to prevent them from reaching the victims in the first place. With the current networking technologies this is unfortunately still not achievable. However, by deploying preventive measures, the ease of launching and the effects of a denial of service attack can be considerably reduced.

8.16.1   Access Control

The most effective measure for reducing security threats is by preventing the attacker from getting access to the system components or sending traffic to it. In general we can distinguish between three aspects here, namely access to the network, to the service components and to the service itself.

8.16.1.1   Controlling Access to the Network

In order to avoid being detected, an attacker would in general aim to hide its location, e.g. IP address. This is usually achieved by either spoofing the source addresses in the sent packets or by utilizing a botnet. Therefore, increasing the complexity of hiding the attackers' location, and hence increasing the possibility of them being detected, will deter attacks. Another aspect of controlling the access to the network is by filtering malicious traffic before it actually reaches the servers.

Ingress filtering   By spoofing their IP addresses, attackers can hide their own identity and utilize amplification attacks to target the attack to one or more victims. To prevent attackers from using spoofed addresses, ISPs are increasingly using so-called "ingress filtering" (Ferguson and Senie 1998). Ingress filtering describes an approach in which routers only carry traffic that arrives from a valid source address. Depending on the location of the router and the number of hosts or networks connected to this router, the valid address might be the address allocated to a host by the network or any one of the addresses managed by the provider to whom the router belongs. The stricter the address range, the less useful spoofing becomes to the attacker. Hence, this approach is most effective in access networks in which the routers are responsible for a restricted range of IP addresses.

Static filtering   As already mentioned in section 8.14.1, the signatures of various attack scenarios are well known. By implementing these signatures in firewalls, it is possible to filter out malicious traffic before it reaches the victims. Many firewalls already include such signatures and proactively drop suspicious traffic.

Zombie prevention   Botnets and zombies are now being considered as the highest risks on the Internet. As an attack can be launched by thousands of zombies, the reactive approaches for DDoS protection are often inefficient. Using thousands of filters at a firewall incurs high processing requirements and might cause the firewall itself to become the bottleneck. Also, as the single zombie contributes only few packets to the attack, rate-limiting is also not efficient as using too strict rate-limiting prevents the legitimate applications on the zombie from communicating. Using too loose a limit is insufficient for preventing the malicious traffic. Traceback mechanisms are in this case inefficient as well. As the zombies might be distributed around the world, tracing all zombies back is inefficient and requires a high level of cooperation between ISPs, which is not the general case today. Therefore, the best approach for combating zombies is to prevent hosts from becoming zombies in the first place. This requires more security measures at the hosts and higher security awareness of the users of those hosts.

8.16.1.2   Controlling Access to the Service Components

A VoIP service resembles to a great extent other Internet services such as mail or web service. It consists of components that need to be accessible by the subscribers to this service, relies heavily on other services such as DNS and DHCP and uses some form of a database for maintaining subscriber profiles and access information. Hence, securing a VoIP infrastructure should follow the same guidelines for securing other Internet services.

A basic security rule indicates that all ports and protocols that are not needed by the users of a system should not be accessible by the users. Hence, the SIP infrastructure needs to be protected by firewalls that block all traffic that is not needed for enabling the SIP service, such as telnet connections from the user to the SIP proxy. Further, SIP traffic is usually received at a predefined port, namely 5060. Hence, traffic that is sent on other ports should be rejected.

While it is rather straightforward to block traffic belonging to known protocols and to restrict the usage of signaling messages, restricting the usage of ports and connections used by the media traffic is more complex. The port numbers and IP addresses used for exchanging media traffic in SIP services are negotiated dynamically during session setup and are not known in advance.

To be able to restrict the use of ports and addresses that might be used by media traffic, Application-Level Gateways (ALGs) are often used. ALGs are usually implemented as firewalls that observe the progress of a session establishment process. Once the two communicating parties have exchanged the addresses and ports to be used for the media exchange, the ALG opens the needed ports to allow the media traffic to pass through. When a BYE request is sent to terminate the session, the ALG closes the ports that were opened for this session. While the deployment of such ALGs can restrict the number of open ports at a firewall, it increases the complexity of the firewall and introduces additional processing delays during the session establishment process.

ALGs are often used in the enterprise environment with the goal of controlling the incoming and outgoing VoIP calls as well as preventing any undesired traffic from entering the enterprise network. Another approach that is more often used by service providers is the deployment of session border controllers. These components, see section 3, terminate all incoming calls and re-initiate them toward the SIP servers of the provider. Similar to the ALGs, the SBCs can impose various restrictions on the incoming traffic and reject certain types of traffic. Further, SBCs often manage not only signaling traffic but also media traffic. SBCs can also impose various restrictions on the media traffic and only allow media packets that belong to previously established sessions to enter the network.

Beside the public interfaces toward the subscribers, SIP components usually also have interfaces for administration and configuration purposes. These interfaces should be protected through secure login mechanisms and reachable only through secure communication links and not from the public Internet.

The subscriber databases can be seen as the crown jewels of any VoIP infrastructure. By getting access to the subscriber database, an attacker can not only steal the service but also manipulate the accounts of other users and cause a lot of damage to the provider both financially and to its reputation. To avoid this, the databases should be located in a secured network segment that is not accessible from the Internet.

As depicted in Figure 8.20, a SIP proxy should be part of at least two network segments:

- A network segment connecting the proxy with the subscribers. This segment can be further protected by a firewall or hidden completely through a session border controller.
- A network segment connecting the proxy with the databases and used for administration and configuration purposes. Access to this segment should only be highly protected and restricted to system administrator and trusted hosts.

To terminate VoIP calls in the PSTN or receive calls from the PSTN, the VoIP provider can either interact with PSTN gateways belonging to other providers or manage its own gateways. Similar to the other components in the VoIP infrastructure, the access to administration and configuration interfaces of the gateways must be secured as well. Further, the communication with the gateways must be secured so as to ensure that only authenticated and authorized traffic gets forwarded to the PSTN. Therefore, gateways should only accept signaling traffic that is arriving over secured links from trusted proxies. Such secure links can be established using TLS or IPsec. Any other signaling traffic should be rejected.

8.16.1.3   Controlling Access to the Service

Most of the attacks on the SIP service require the attacker to have a VoIP account. Hence, the first step to reducing the possibility of fraud is to use the authentication and authorization mechanisms supported by SIP.

Further, attacks such as incomplete transactions, loop or fork-based amplification, see section 8.5.2.1, require the attacker to gain access to a number of different accounts. Currently it is usually sufficient to provide a valid email address to get such an account. By increasing the hurdles for getting a VoIP account, e.g. by requesting some verifiable personal information, the chances of detecting the true identity of an attacker will increase, which should deter some if not all attackers from launching such attacks.

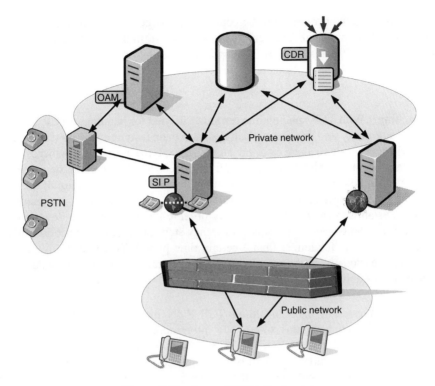

Figure 8.20   Secure VoIP infrastructure

8.16.2   Memory Protection

Memory depletion attacks are to some extent more malicious than flooding or CPU exhaustion attacks. In comparison, the attacker needs far fewer signaling packets to achieve the same results. Therefore, in the case of SIP servers, one has to protect against both TCP and SIP memory depletion attacks.

8.16.2.1   TCP Memory Protection

TCP SYN attacks constitute the major security threat for TCP servers. In general, once a TCP server has acknowledged a TCP SYN request and created some state information for it, it can suffer from a DoS attack. Hence, the protection measures (Eddy 2007) aim at either reducing the state memory used during the connection establishment or preventing malicious TCP SYN packets from arriving at the server.

TCP splicing   In order to distribute the load of a server between multiple instances, proxies are often used to mediate between the clients and the servers. Besides load distribution, these proxies can also protect the servers from TCP SYN attacks by deploying so-called "TCP splicing" (Maltz and Bhagwat 1998). With TCP splicing, the proxy intercepts TCP SYN messages sent by the clients and establishes a second TCP connection to one of the servers. The protection effect is achieved by delaying the establishment of the

TCP connections to the servers until the connection between the proxy and the client has been successfully established. The servers are shielded from the attacks and receive only legitimate requests. Proxies offering TCP splicing must be optimized to support a much higher number of TCP connections than the servers, as otherwise they would become the victims.

TCP watching   With TCP splicing, the resources needed for handling a TCP SYN attack are located at a proxy and the servers can dedicate their resources for serving legitimate requests. However, where there was no attack, use of the proxy is not necessary and increases the session establishment delay and introduces another node in the network that might also fail and contribute to an increase in the overall failure possibility of the service. To overcome these negative side effects, some routers and firewalls do not intercept the TCP session establishment messages but only observe them. Once these devices observe a high number of half-open TCP connections, they start terminating connections that stay in the half-open mode for longer than a certain time period. This is done by sending an RST message to the servers that causes the servers to release the resources dedicated to the connection establishment.

TCP SYN cookies   For each incoming TCP SYN message, a TCP server sends a response message and maintains some state information until the response message is acknowledged by the client. When using TCP SYN cookies, the TCP server encrypts the state information and includes it as part of the sequence number included in the response.

The sequence number included in the response is chosen such that the difference between this number and the sequence number included in the TCP SYN message includes the following information:

- Time–a value indicating the time of the generation of the response. This value is encoded in 5 bits as T modulo 32 with T being a 32 bit long counter that is incremented every 64 seconds.
- Maximum segment size (MSS)–the MSS used by the TCP server. This value is encoded in 3 bits.
- Connection details–the server includes a hash of the client and server IP addresses and port number as well the the time value in 24 bits.

Once an acknowledgment to the server's response is received, the server compares the acknowledged sequence number with the current time, and MSS and IP and port numbers used. This information is then used for establishing the TCP session.

As the TCP server does not need to maintain any state information between the initial TCP SYN message and the final acknowledgment, there is no risk of memory depletion due to a flooding attack.

8.16.2.2   SIP Memory Protection

For SIP, we can identify two possible memory depletion attack scenarios, namely attacking the authentication process and transaction processing.

Stateless authentication The authentication process is best protected using stateless authentication, as described in section 8.5.2.2. This is to some extent similar to the TCP cookies approach described above and relieves the server from having to keep any state information when authenticating a request.

Transaction memory protection Protecting the transaction processing is rather difficult as a SIP server cannot distinguish between a malicious request and a good one in advance. Besides the approaches listed above in terms of rate-limiting, filtering and securing the access to the service, a number of further general guidelines for reducing the possibilities of such attacks can be used:

- Dynamic black listing–once a server detects that the number of requests sent to or received from certain addresses exceeds certain thresholds, these addresses can be put on a black list. Requests destined to black listed addresses are rejected. Further, destinations that do not reply should also be added to the black list. This way requests destined to a malfunctioning device are rejected directly and do not cause the generation of state information on the SIP server and the sending of retransmissions.
- Reduction of the maximum value of the *Max-Forwards* header–by default a value of 70 is suggested; however, where a server suspects that it is the target of a looping attack, this value should be reduced.
- Loop detection and reduction of the breadth of forking–as described above, a forking attack can result in a large consumption of bandwidth, memory and CPU resources. To reduce the possibilities and effects of such attacks, SIP servers should deploy loop detection to identify looped packets and drop them. Further, to reduce the effects of a forking attack, the maximum number of destinations to which a request can be forked should be kept minimal (Sparks *et al.* 2007).
- Stateless processing–the best approach for protecting against memory depletion attacks is to not use memory at all. While this is not always possible, SIP servers that can process SIP messages without having to maintain transaction states should do so. This applies especially to SIP load balancers that distribute traffic between SIP servers that provide the actual service.

DNS caching Caching the results of DNS resolution queries locally at the SIP proxy can considerably reduce the time needed for resolving DNS names and speed up the processing of SIP messages.

8.16.3 Architectural Consideration

Regardless of the deployed security measures, it will not be possible to entirely deter attackers from launching DoS attacks and to entirely protect a service from the negative results of such attacks. Hence, it is necessary to design the service in such a manner that it can withstand possible DoS attacks and continue functioning, albeit in a degraded manner. This requires the service architecture to take into account the effects of sudden increases in load as well as the possibility of some components failing.

8.16.3.1 Highly Reliable Service Architecture

Similar to any other Internet service, a VoIP service must be provided in a highly reliable manner. This not only includes ensuring the high availability of the SIP components themselves but also all other components required by the SIP service, including DNS servers, DHCP servers, firewalls, NATs and databases.

With a DHCP server down, IP phones and other components in the network cannot acquire an IP address and initiate or receive calls. This can be an especially big risk in enterprise and residential environments. As the number of servers in a VoIP service provider infrastructure is relatively small, the different servers can be statically configured without the need for DHCP.

The failure of a DNS server prevents any VoIP component from resolving SIP URIs and from initiating and forwarding SIP requests. If the DNS server of an enterprise fails, then all users attached to this enterprise will not be able to use the VoIP service. If the DNS server of a service provider fails, then all subscribers of this provider will suffer the consequences.

The failure of either DHCP or DNS renders the VoIP service unavailable. To prevent this, the administrators of these services need to implement various measures that aim at reducing the probability of a failure and the effects of a failed server. Among these measure one can list:

- Use the newest software versions with patches against the most recently detected software bugs and security vulnerabilities.
- Provide DNS and DHCP servers in a redundant configuration with the different servers attached to independent electricity plugs and network access components. It is also an advantage if the different servers use different implementations and run on different operating systems.
- Protect the DNS and DHCP servers from non-relevant traffic using firewalls in a similar manner to the protection of the access to the SIP component, as described in section 8.16.1.2.

Besides these general guidelines the reliability of DHCP servers can be further increased by using multiple DHCP servers. Each server could allocate addresses from a separate pool of addresses. This achieves redundancy at the cost of address space usage.

The availability of the DNS service can also be enhanced using the following measures:

- Use multiple DNS services by partnering with another organization or purchasing external DNS services.
- Where the DNS name space can be divided into an external and internal view, then separate servers should be used for serving the internal and external DNS data.
- Separate machines for serving an organization's DNS data (authoritative servers) and fetching data from outside sources (caching or recursive servers) should be used.

8.16.3.2 Overload Control

A DoS attack on a SIP component usually has the aim at overloading one or more resources of the SIP components. These resources might include the CPU, memory or any other resource needed for processing incoming SIP messages. Once one or more resources are overloaded the SIP component will no longer be capable of dealing with the incoming traffic, which will lead to the dropping of the excess traffic. As described in section 8.3, this may result in an even higher traffic volume due to the retransmission behavior of SIP. Work done in (Ohta 2004) suggests that overload situations not only reduce the performance of a SIP server but can finally lead to a complete failure of the VoIP service. The most straightforward approach for handling such situations is to ensure that the available processing resources of a SIP component are sufficient for handling SIP traffic arriving at the speed of the link connecting this component to the Internet. With modern access lines reaching gigabit speeds, provisioning the VoIP infrastructure of a provider to support such an amount of traffic, which is most likely several times the normal traffic, can easily become rather expensive.

SIP does not provide much guidance on how to react to overload conditions. According to (Rosenberg *et al.* 2002b) a server that is not capable of serving new requests, e.g. because it is overloaded, could reject incoming messages by sending a *503 Service unavailable* response back to the sender of the request. This signals to the sender that it should try forwarding the rejected request to another proxy and not to use the overloaded proxy for some time. Further, the *503* response includes a *Retry-After* header indicating the period of time during which the overloaded server should not be contacted. While this reduces the load on the overloaded proxy, it results in directing the traffic that has caused the overload to another proxy, which might then get overloaded itself. Figure 8.21 depicts a scenario in which a load balancer distributes the traffic to two different proxies. In the case of a DoS attack it is most likely that all the SIP servers in a SIP cluster will be affected and will be overloaded at the same time. When the first server replies with a *503*, the load balancer will forward the traffic destined to that server to the other server. With the additional traffic this server will become overloaded as well and will issue *503* replies. Shifting the traffic from one server to another has only made the situation worse for this server. This shifting of traffic can also lead to an on–off behavior. That is, consider the case when an attacker is generating traffic that is causing both servers to run at 100 % of their capacity. When one of them issues a *503* response, the traffic destined for it will be forwarded to the other, which will now receive traffic at 200 % of its capacity. This server will, hence, issue a *503* response. Where the *Retry-After* value of the first server expires before that of the second server, then that server will suddenly receive traffic at 200 % of its capacity and will reject the traffic with another *503*. This on–off behavior can actually even lead to a lower average throughput, making the *503* approach not optimal for the cases in which a SIP component receives SIP messages from only a small number of other SIP instances. Where a SIP server receives requests from a large number of user agents, then the *503* approach can work much more efficiently as only the user agents that receive the *503* response will try another destination. Further, the on–off behavior will not be observed in this case as spreading out the *503* among the clients has the effect of providing the overloaded SIP instance with more fine-grained controls on the amount of work it receives. Naturally, if the senders are bots that do not respect the *Retry-After* header, using a *503* will not be suficient for protecting the server from getting overloaded.

As guidelines for providing mechanisms for solving the overload control issue, in (Rosenberg 2006) the author discusses the requirements an overload control scheme should

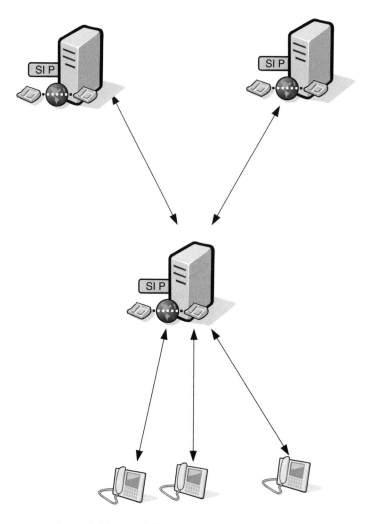

Figure 8.21   Load distribution and use of 503 replies

fulfill. Among these requirements is that an overload control scheme should enable a server to support a meaningful throughput under all circumstances, should prevent the forwarding of traffic to other servers that might be overloaded themselves and should work even if not all servers in the network support it.

The overload handling approach described in (Hilt *et al.* 2007) is based on having the overloaded proxy exchanging load information with its neighboring proxies. The neighboring proxies then adjust the amount of traffic they send to the overloaded proxy based on this information. In (Shen and Schulzrinne 2008) various overload control schemes that use buffer occupancy as an indication of overload situations are presented and compared. These schemes determine the load of the SIP server by observing the length of the incoming buffers and inform their neighbors about the load situation.

A different approach for avoiding overload by separating different types of SIP messages into different queues and allocating specific shares of the CPU to the different queues is described in (Acharya *et al.* 2005). This mechanism can help in preventing overload situations when, for example, the overload is caused by a certain type of message. However, this approach does not help against memory attacks or flooding attacks that use different kinds of SIP requests.

In (Sisalem and Floroiu 2009) the authors present a mechanism that can be used by a SIP proxy, for example for reducing overload by rejecting incoming requests as soon as certain thresholds are exceeded. This is based on the observation that rejecting traffic is much more efficient in terms of resource usage than processing and forwarding it. By reacting to overload early on, the overloaded instance can use resources that have been used for processing requests for rejecting them. Instead of using all the resources for processing incoming requests and then either dropping the excess traffic or rejecting it using *503*, a SIP server dedicates a certain portion of its resources for rejecting requests, enabling it to handle a much larger number of requests and reduce the probability of messages getting dropped.

8.16.3.3 Overprovisioning and Load Distribution

When dimensioning a server to offer a certain service, one should take care that the resources of the server are sufficient for providing the service even under worst-case scenarios. Further, by distributing the load on multiple servers, a service can still be offered even if one or more servers fail or are overloaded. It is especially important to be able to deal with flash crowd situations without becoming unavailable. If the resources available for providing the service are higher than those available to the attacker, then the service will very likely withstand all kinds of attacks. However, with the attackers relying more on distributed attacks launched by thousands of zombies, providing enough resources to withstand such attacks becomes expensive and impractical.

8.17 DDoS Signature Specification

To be able to protect a system from DDoS attacks, it is important to understand the attack patterns and determine the signature of these attacks. While there are different approaches for achieving this, the most common approach is unfortunately based on forensics. That is, after a successful attack, the attacked systems are analyzed using collected logging and network traffic information. This requires a constant and detailed monitoring of the networks and systems and only helps to protect a system after an incident. In the following we discuss other approaches that can be used to determine attack patterns before actual harm is done.

8.17.1 Fuzzing

To test the quality of a software implementation one usually conducts conformance testing. That is, the software is tested with regard to its conformance to certain specifications. Fuzzing tests aim at testing the software's reaction to requests and messages that do not conform or only partially conform to the specifications. This is often realized by issuing requests and messages that include random input or have some control or useless characters embedded between legitimate parts or by issuing messages and requests that are too long or too small (Miller *et al.* 1990). Another form of fuzzing that is especially used in testing of networking protocols is the so-called "stateful fuzzing" (Banks *et al.* 2006). That is, instead of just sending a corrupted message to the software, the fuzzing tool emulates the state machine of the protocol with small variations, e.g. jump over some states or not start from the specified initial state.

Fuzzing tests are not adequate substitutions for conformance tests. However, they are very useful in detecting vulnerabilities in implementations that might cause the software to crash or to go into infinite loops. Once such vulnerability is detected, the implementation of the software should be updated. To protect the system until the software is patched, the signature of this vulnerability can be added to static filters used for protecting the system.

8.17.2 Honeypots

Honeypots (Honeynet Project 2004) are systems and networks whose only goal is to lure attackers into trying to access them. As these systems do not offer any services to legitimate users, any traffic observed on them can only be generated by attackers. The collected traffic in the honeypots can then be used to identify attack patterns and attack targets without endangering systems used by legitimate users. Honeypots can either emulate some services or consist of complete networks with various hosts running all kinds of services. The traffic or actions observed at the honeypots are logged. Further, while honeypots accept all incoming traffic from the attackers, outgoing traffic must be carefully controlled so as to prevent the hosts of a honeypot from becoming the attacker's basis for attacking further systems.

As all traffic and logs collected in honeypots are generated by attackers, the amount of information that needs to be analyzed for detecting attack patterns or new vulnerabilities is much lower and has a higher information density than would be the case at a live system. As honeypots can only be used for analyzing attacks that have been launched at the honeypots, they are only of use if they are actually attacked and have, hence, only a limited vision on possible attacks.

Honeypots can also be used to slow down attacks. By involving the attacker in long exchanges, for example by using a very small TCP window size, the attacker's time and resources are wasted and give the administrators time to protect the live systems.

9

SPAM over IP Telephony

9.1 Introduction

Thinking they had discovered a revolutionary marketing tool and an easy path to wealth, in April 1994 the lawyers Laurence Canter and Martha Siegel sent out usenet posts offering their services in filling out the documents needed for participating in the Green Card Lottery. These emails were sent to at least 6000 Usenet discussion groups and hence reached a large portion of the Internet users those days. The strong reaction of the users in the form of emails to the ISP of Canter and Siegel as well as letters to the American Immigration Lawyers Association finally led to the termination of the Internet access of the couple and the disbarment of Canter from practicing law. These sanctions did not stop the couple from launching a company dedicated to this kind of Internet marketing and writing books about using the Internet as a marketing tool. While it is unclear if the couple made a significant fortune with their marketing activities, they did pave the way for one of the most annoying aspects of the Internet today, namely spam. Since this first incident, the level of spam on the Internet has increased significantly. Exact numbers are difficult to obtain and vary depending on the measurement data and measurement approach as well as the source of the data. However, statistics collected by various antispam organizations and companies suggest that the portion of spam emails compared with the overall number of exchanged emails has risen from less than 10 % in 2001 up to more than 80 % today.[1]

In general, spam describes information, often of dubious nature, sent to a large number of recipients without their prior consent. While spam is often used to describe emails about hot stocks, revolutionary medicine or with links to sites with adult content, it can be generalized to all kinds of communication such as SMS messages, telemarketing calls or bulk mail and faxes.

Besides containing information of dubious nature, spam is often used to distribute viruses and worms as well as for conducting fraud. In Phishing, the spam messages try to lure the receiver into revealing personal information such as the bank account access code, social security number or the credit card PIN number.

---

[1] http://www.spam-o-meter.com/stats/

SIP Security   Dorgham Sisalem, John Floroiu, Jiri Kuthan, Ulrich Abend and Henning Schulzrinne
© 2009 John Wiley & Sons, Ltd

One of the defining characteristics of spam is that the user receiving the spam has not asked for it, did not give her consent to receive it and did not solicit it. Hence, spam is also often referred to as unsolicited communication.

While most evident in the Internet, unsolicited communication is also popular in the traditional PSTN environment as well. Telemarketing calls with offers for new types of insurance or other financial services and faxes with commercial offers are widely used marketing tools. However, with telemarketing, mail or faxes, the spammers have to pay for the costs of the phone minutes, stamps or faxes and are usually identifiable. This has limited the scope and amount of such marketing tools. With email the costs for the spam are shifted from the spammer to the receiver and the ISPs. Receivers need to spend time determining that an email is spam and deleting it. The ISPs need to transport and store the spam emails even though they will be deleted at the end. This increases the memory and bandwidth usage at the ISPs without any financial return.

With SIP-based VoIP services ever gaining in popularity, it is only reasonable to expect that spammers will use the SIP technology to conduct their business. In this chapter we draw parallels between unsolicited communication using SIP and other technologies, describe the different types of spam that can be realized by SIP and the features that make SIP attractive to spammers. The major part of this chapter will investigate some of the approaches used for fighting spam. In this context general approaches used for fighting email spam will be discussed and their applicability to SIP will be investigated. Further, some approaches specific to SIP and VoIP will be presented as well.

9.2   Spam Over SIP: Types and Applicability

SIP supports various types of communication including voice, video and text. While this allows service providers to offer a wide range of services using the same infrastructure, it also allows spammers to send different kinds of spam (Rosenberg and Jennings 2008).

Spam over IP Telephony (SPIT)   This type of spam is rather similar to current telemarketing activities. The spammer initiates VoIP calls to a large number of addresses and, once a call is answered, the spammer either conveys his information by playing a prerecorded audio message or engages the callee in talk and tries to convince the callee to buy or do something.

Spam over Instant Messaging (SPIM)   This kind of spam can be defined as the bulk sending of unsolicited instant messages (Campbell *et al.* 2002). The information to be distributed by the spammer is included in textual or graphical form in the messages. In this context, SPIM is similar to email spam. While the MESSAGE method of SIP is the most obvious candidate for this kind of spam, any other request that causes content to appear automatically on the user's display could be used as well. That is, if for example the user's phone displays the *Subject* header of an INVITE request, then the spammer can include the information in this header.

Spam over Presence Protocol (SPPP)   This type of spam is defined as the bulk sending of presence requests (Rosenberg 2004a), e.g. SUBSCRIBE requests (Roach 2002). By accepting a subscription request, the user would usually put the spammer automatically into his buddy list. This would allow the spammer to send IM messages to the user or even call him without being blocked by possible antispam measures.

9.2.1 General Types of Spam

Broadly speaking, the goal of sending spam is to generate some revenue for the spammer. While in general one would think that this is achieved by selling things like health products or software or even PhD attests, spammers actually follow different strategies of various levels of sophistication.

Direct passive marketing   This is the most straightforward approach in which the spammer offers certain products for sale. Products offered using email spam are often of questionable value, such as health products, software, electronic devices or learn-free PhDs. The spammer makes money in this case by scamming the buyers by not delivering the promised goods at all, by delivering broken or nonfunctional goods or by making some profit on each delivered item. SPIT as well as SPIM could be used for this type of marketing as well. A prerecorded voice or voice and video message would be used for praising the product and streamed once a call has been successfully established. While preparing a convincing audio/visual presentation of the product is likely to be more costly than a simple text message, it will also more likely be more convincing than a simple mail.

Direct interactive marketing   Telemarketing calls in which the salesperson tries to convince the callee to buy some insurance or financial product require interaction between the two. Using VoIP and SIP for this kind of spam would enable the spammer to replace the PSTN technology with a less costly one.

Web clicks   Instead of actually convincing someone to buy a product or subscribe to some service, the spammer can make money by luring people into visiting her web site. This is usually achieved by offering some kind of service or content for free or on a trial basis and often with adult content. The spammer gets advertising revenue from banner ads displayed on those pages each time someone visits this web site. This kind of spam can be conducted using IM messages as well. Using voice calls to lure a user to visit some web site is probably less efficient than a text message with a direct link to that site.

Stock promotion   By promoting some lightly traded stocks and luring people into buying them, a spammer can manage to significantly increase the value of this stock. The spammer makes his gain here by buying the stock in advance and selling after having increased its value. Similar to the case of passive marketing, both IM and VoIP calls can be used to advertise the offer.

Call-back   An often-seen fraud method in mobile networks is the call-back fraud. In this scenario the fraudster calls a mobile phone number but hangs up before the callee has the chance to answer. Out of curiosity the callee calls the fraudster back, not knowing that he is calling a premium phone number belonging to the fraudster. Such a scenario can be realized using SIP rather easily as the fraudster is only replacing the used PSTN technology with VoIP equipment.

Selling spam services   Denial of service attacks and spam campaigns are usually launched using botnets consisting of thousands of bots. Such networks are usually not owned by the attackers or spammers themselves but rented from other miscreants. These miscreants manage to establish a botnet by sending mails that either contain the spyware, e.g. the software that would turn the host into a bot, disguised as useful software or interesting content or lure the receiver to visit some web site and download spyware, again by disguising it as interesting content or useful software. While such types of spam could be conducted using IM messages, using VoIP calls is more complex. VoIP devices including phones, adapters and servers often use proprietary operating systems and hardware. Hence, even if it was possible to transport spyware using SIP messages and make the receiver execute the received code, the miscreant would need to port the spyware to a large number of proprietary operating systems and to send to each victim the appropriate version. This would naturally make the distribution process more complex and less attractive.

9.3   Why is SIP Good for Spam?

The business model behind spam is rather simple. The spammer distributes his message to a large number of receivers and hopes that the message will actually convince some of those receivers to buy the advertised product. For a successful campaign the spammer hence needs to reduce her own costs, reach a large number of receivers and avoid getting blocked. SIP can help with all of these factors.

Cost reduction   As a simplistic comparison we can consider the following scenario. A flat rate telephony service with PSTN costs in Germany around $US30 per month and allows the spammer to call any national number for any number of minutes. If a call takes on the average 2 minutes, the spammer will be able to conduct 30 calls per hour for around 0.14 cents per call. A VoIP-based flat rate telephony service can also cost around US$30 per month. In addition to the VoIP service, the spammer needs to subscribe to a broadband Internet access. Such a service cost another US$30 per month, which considering the German and US market seems to be a rather conservative assumption. With a 2 Mbit/s access to the Internet and a bandwidth usage of 64 kbit/s pe call, the spammer can actually generate more than 30 parallel calls and hence 900 calls per hour for US$60 per month. This means that, using SIP, the spammer can reduce her costs by a factor of 15. This calculation does not take into account that spam calls using VoIP can be generated using an off-the-shelf PC whereas in the case of PSTN some special hardware is needed. Note also that a flat rate telephony service is only needed if the spammer wants to also reach users with only a PSTN line. VoIP calls that originate and terminate in the Internet are usually free of charge and would not require a flat-rate telephony service. Thereby, with the increased deployment of VoIP, the spammers will be able to reduce their costs further toward the costs of the Internet access.

Increasing scope   While the cost savings on a national basis are already significant, the spammer can achieve even greater savings when going international. International calls are usually not covered by basic flat-rate service offers in PSTN. Hence, launching a spam campaign on an international scale would be rather costly. With VoIP, a spammer located in Germany can subscribe to a flat-rate VoIP service in the USA and launch a spam campaign in the USA at a cost similar to one in Germany. Further, with the increased deployment of VoIP and the increased number of users that can be reached over VoIP directly without traversing a PSTN link, a spammer would be able to launch spam campaigns in a similar manner to email. Thereby, under the assumption that VoIP calls are free of charge, as is currently the case, a spammer sitting in the USA can initiate calls to receivers all around the world.

Identity hiding Usually spammers try to hide their identities, especially when distributing spam with fraudulent content. With email the spammer can forge the sender information in the emails, send the emails over open email relays or directly to the recipients so as to avoid authentication, by using stolen email accounts and spoofing the sending IP address. Open email relays are mail servers that are either manipulated or relay emails without authenticating the sender first. Therefore, from the point of view of the receivers, the spam emails are coming from the relays and the receivers cannot determine the exact address of the sender. The high bandwidth usage when sending millions of messages over the same access to the Internet would probably alert the ISP to possible misuse. To avoid arousing suspicion, the spammer can distribute the load using a botnet consisting of thousands of bots. With spam over PSTN this is more difficult. While the spammer can annonymize his calls, the origin of the call can still be traced by the network and the high number of calls placed by the spammer could make the network suspect that the phone line is being used for spam and block it. In this regard, SIP-based services are more like email services. The spammer can use botnets for distributing the load, use false sender information and send the SIP requests directly to the recipients without passing an authenticating proxy first.

High hit rate In order to reach a large number of possible users, the spammer needs to have a large list of valid email addresses. Such lists are constructed as a combination of different methods. Web crawlers analyze web sites for any strings that might be an email address, e.g. contain an "@" character. With a directory-based approach, the spammer assembles either automatically or by guessing a large number of user and domain names and uses these as possible email addresses. Finally, there have been reports of stolen lists of user names from various ISPs and large organizations. SIP end points are addressed either using email-like addresses or E.164 phone numbers (ITU-T Rec. E.164 2005). For the email-like addresses the same methods as are used for harvesting email addresses can be used. E.164 numbers are much more predictable in their structure and values than email addresses, which could theoretically consist of all possible letter combinations. Hence, using SIP and phone numbers it is possible to get much more accurate address lists and the spammer can reduce the number of calls destined for nonexistent users compared with email spam.

Regulation The number of telemarketing calls that can be classified as spam is still relatively small compared with email spam. Because of the high costs of international calls, telemarketers usually restrict their activities to their domestic markets. However, in most countries there are some regulations that outlaw unsolicited marketing calls and spammers risk heavy penalties. By reducing the costs of conducting spam across international borders, VoIP will enable spammers to conduct their business across borders and hence will no longer fall under the jurisdiction of such regulations.

9.4   Legal Side of Unsolicited Communication

From the point of view of the receiver, spam emails are annoying. The user has to check the content and delete them, wasting some seconds. SPIT calls will most likely be even more annoying as the user will be alerted by the phone ringing, possibly late at night, and will have to listen to at least some of the transmitted message before classifying the call as SPIT. However, although annoying, SPIT and spam do not cause significant financial harm to the receiver. Looking at spam from a more global view, though, the financial harm caused by spam is actually significant. Employees spend several minutes a day in deleting useless emails, companies need to buy antispam products and the administrators need to acquire the knowledge to fight spam and to properly operate antispam software. The ISPs need to carry spam emails and often save them on their email servers until they are retrieved by their subscribers. This requires additional processing power and memory and disk space as well as bandwidth resources. While it is difficult to accurately estimate the total costs of spam, various resources estimate overall costs at anywhere between US\$2 and US\$10 billion.[2]

Besides the financial harm, spam and other kinds of unsolicited communication intrude on the user's privacy. This happens in the form of collecting and selling personal information of the user such as the user's mail address and phone number without the user's permission. Further, the privacy of the user is abused by sending him information and content which he did not explicitly ask for.

In order to combat unsolicited communications, legislators all around the world have issued laws and rules that define and prohibit different kinds of unsolicited communication (Hladjk 2005a, b). In their attempts to protect the people's privacy and property, legislators need also to take into consideration aspects of free speech and legal marketing activities. Both activities often require sending some kind of information to a large number of recipients, sometimes even unsolicited. With the different interpretation of what constitutes free speech and legal marketing activities, different nations have set up different laws and regulatory frameworks that vary in their scope and strength. Still, these different regulations share the goal of protecting the privacy and property of their citizens.

9.4.1   Protection of Personal Privacy

In the context of unsolicited communication, the privacy of a person can be intruded upon in one of two ways, namely misuse of personal information and intrusion on the privacy of the person by contacting him without his prior consent.

The E-Privacy Directive (European Parliament 2002) of the European Parliament prohibits the exchange of unsolicited electronic communication unless the explicit consent of the receiver has been obtained in advance. The directive defines communication as

> any information exchanged between a finite number of parties by a publicly available electronic service

The phrase

> a finite number of parties

---

[2] See http://www.messagelabs.co.uk/

aims at indicating that, for this kind of communication, an address is required both for the receiver and the sender and hence distinguishes this kind of communication from broadcast services. More precisely, the Directive covers different means of electronic communication, namely automatic calling machines, fax and electronic mail. In this context electronic mail covers

> any text, voice, sound or image message sent over a public communication network which can be stored in the network or on the recipient's equipment until it is collected by the recipient

This includes not only email but also Short Message Service and Multimedia Messaging Service.

To still enable direct marketing and advertising of certain products and goods, the Directive indicates that the receiver's consent can be obtained implicitly in some cases. This can be the case if a customer relationship already exits between the advertiser and the recipient. The recipient must, however, be given a free and simple means of refusing future messages. Also the messages must be labeled clearly as advertisements.

In contrast to the general *opt-in* approach of Europe, the US CAN-spam (Federal Trade Commission 2004) favors an *opt-out* approach. Under the US CAN-spam Act unsolicited emails are not unlawful as long as the recipient has not explicitly asked not to receive the email. To comply with the Act, unsolicited messages must further include a valid sender address which the recipient can use to opt-out and the advertising content of the message must be displayed clearly.

Compared with the opt-in approach, the opt-out approach is actually of little use. While law-abiding advertisers will in general respect the recipient's wish to not receive future emails, others will actually use the opt-out mails to verify the recipient's address and hence improve the quality of their lists of addresses used for sending spam. Because of the danger of abuse, recipients might not opt-out and use other means for filtering out unsolicited messages. Thereby, with the opt-out approach many recipients receive unsolicited communication legally because they did not explicitly object. Finally, with the number of businesses and email accounts that do not want to receive spam, maintaining an opt-out list that is up-to-date and is easily accessible to anyone who wants to send an email becomes a technological challenge.

With regard to the use of personal information such as email address, name or phone number, both the European and American regulations prohibit the collection of data without the prior consent of the person or if the person, web site or online service has indicated that this is not allowed. Further, collected information cannot be sold or transferred to third parties without the explicit consent of the owner of the information. Additionally, it is not permitted to obtain email addresses by automatic means that generate possible addresses by combining names, letters or numbers into numerous permutations.

9.4.2 Protection of Property

Besides stealing our time, various kinds of unsolicited communication can also cause financial harm that could build a basis for legally pursuing the spammers. Actions based upon so-called *unjust enrichment* can be granted when the defendant saves costs or increases his income by making use of other people's property without their consent.

This is the basis upon which unsolicited faxes are prohibited. The sender of unsolicited faxes shifts his costs to the recipients, who have to pay for paper and ink when receiving the faxes. For email, this argument is a little more difficult to justify. If the recipient is using a flat-rate access, then receiving the unsolicited email does not incur any costs. If, however, the recipient pays for his Internet access per minute, then downloading spam mails incurs some costs. In this case the unjust enrichment argument might still apply.

Not being the target of the unsolicited communication, the argument of unjust enrichment is not easily applicable by an ISP acting against a spammer. In this case, *Trespass to chattels* has been alleged by service providers in some cases of unsolicited communications (Kabel 2003). Trespass to chattels is committed when a person uses or interferes with another's personal property without authorization. In the case of a service provider this could mean the excessive use of bandwidth, memory and processing resources needed for handling the large number of spam emails. This could also mean the unauthorized use of email servers as relays.

Another approach for a service provider to protect itself against spammers is to indicate in their terms of usage contracts with their subscribers that any form of spam is prohibited and that the subscribers need to abide by *netiquette* rules. While not being an explicit law or regulation, it is well established that sending unsolicited emails to public discussion groups is contrary to the customs of the Internet. Based on this, a service provider can have a legal basis for terminating the subscription of a spammer.

9.4.3 Legal Aspects of Prohibition of Unsolicited Communication by Service Providers

In order to protect their subscribers from unsolicited communication on the one side and ensure a reputation as a spam-free provider on the other, service providers aim to identify unsolicited messages and block them. While this sounds like a logical approach, it actually intrudes on the fundamental rights of the subscribers.

Various conventions on human rights and freedom indicate that the interception, opening, reading or delayed reception of communications or impeded sending of a message is considered as an intrusion into the right of correspondence. By blocking a message, the freedom of speech of the sender is limited. From the point of view of the recipient, the blocking of the message is a kind of censorship. Further, in order to determine that a message is some sort of unsolicited communication, the provider often needs to scan and interpret the content of the message, which intrudes on the privacy of communication. Thereby, without a valid reason, such as the public safety, national security or the economic well-being of the country, such actions will be illegal.

For a service provider to be able to protect its infrastructure and reputation as well its customers and still avoid intruding on the privacy of its subscribers, it needs to implement various measures that put the subscribers in charge of the antispam measures. The service provider needs to inform its subscribers about the measures taken for detecting and blocking unsolicited communication and the customers of the service provider need to agree to these measures. The provider should also offer its subscribers a cost-free way of opting out of such protective measures. Further, the subscriber must have a cost-free way of viewing messages classified by the provider as spam so as to allow the subscriber to take the final decision on whether a message is to be actually deleted or not. In this way the possibility of a false positive, e.g. a message that is wrongly classified as spam, can be reduced. Finally, the provider must clearly describe its policy for how collected data, such as samples of spam messages, addresses of receivers and senders, is dealt with and under what terms and for what purposes such data can be exchanged with other providers.

Even though such measures improve the legal position of the providers, they do not safeguard the providers against legal actions. While antispam measures are justifiable for spam with fraudulent content, the senders of messages with political or religious content could claim censorship and limitation of freedom of speech. Also, it is difficult for an ISP to judge whether the sent content might actually be desired by the recipients, e.g. the recipient actually opted in to receive advertisements about watch replicas.

9.4.4 Effectiveness of Legal Action

The various antispam laws and regulations impose high penalties on spammers that can range from high financial penalties to imprisonment. Still, the effectiveness of these laws has been very limited. While these regulations have managed to keep unsolicited calls and faxes at a low level, they have failed completely in the case of email.

Unsolicited communication can take different forms and with different content ranging from fraud, cyber crime or data protection infringement. It is often not clear which authority needs to be involved in enforcing the regulations. Further, the regulations often only have a national scope, so enforcing them when the miscreant is located in a different country is even more complicated.

The success of regulations in the case of unsolicited calls and faxes stems from the difficulty of hiding the origin of the callers and the high costs of conducting calls and sending faxes across national borders. Spammers conducting spam on a national level fall under the jurisdiction of such regulations and risk heavy penalties. This is most evident in Europe, in which the opt-in approach is used.

In the case of email and VoIP, the cost barriers for sending bulk messages or generating thousands of calls no longer exist and the scope of the spammers' actions is no longer geographically restricted. Further, while the open nature of the Internet has supported the introduction of novel services, it unfortunately also provides a much wider range of possibilities for the spammers to hide their traces. Even though some of the legal frameworks are intended to have an international scope, without being able to locate the miscreants, it is not possible to take legal actions against them.

9.5 Fighting Unsolicited Communication

The proliferation of spam has led to the development of various antispam technologies. While most of the technologies developed help to reduce the amount of spam seen by the end users, none of these technologies has proved to be sufficient. Hence, in general antispam products deploy multiple technologies at the same time. This section will give an overview of the antispam solutions currently deployed. Further, the applicability of these solutions to VoIP and SIP is discussed.

In general antispam solutions can be classified into different categories, as illustrated in Figure 9.1. These mechanisms aim to detect whether a call or an email is to be considered as spam based on the identity of the sender, the content of the messages or the behavior of the sender and whether his sending behavior conforms to certain profiles. Once a message is classified as spam, it can be dropped or labeled as spam and put in a special

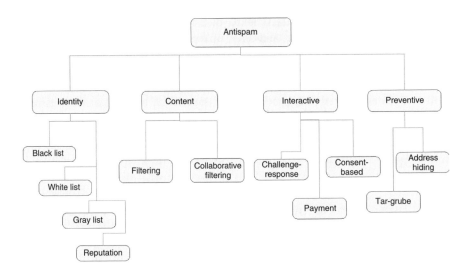

Figure 9.1 General classification of antispam technologies

folder where the receivers can check them again if required. Preventive approaches aim at making life more difficult for spammers by hiding the addresses of users or slowing down the rate at which a spammer can send emails or initiate calls.

9.5.1 Antispam Measures Based on Identity

Based on the identity of the sender, e.g. the email or SIP address in the messages or the IP address, the email or call can be classified as spam, not spam or suspected spam. As the mail server only needs to check the identity information of the messages, the processing load caused by such solutions is relatively low. This approach has been rather successful in instant messaging services. In such environments both the senders and receivers are authenticated by the same entity and hence identity forging is not trivial. In a more open environment like inter-domain VoIP calls, in which the caller and callee belong to different service providers, a caller can easily forge the identity of some other user. Hence, for identity-based antispam measured to be efficient, some form of strong identity is required, see Chapter 6.

9.5.1.1 Black Lists

Black lists are also often called block lists to avoid the negative associations with the political term used in the McCarthy period in the 1950s in the USA. Black lists are in general lists of addresses of suspected senders of spam. These lists are maintained by certain organizations and can be queried by email servers using DNS, hence such lists are often called DNS blacklists (DNSBL). Once an email server receives an email, it can check whether the sender of this mail is a spammer by issuing a DNS request in one of two forms:

- ip4r—in this approach the IP address of the sender is reversed and attached to the zone of the provider of the DNSBL. That is, if the sender has the IP address 192.212.212.2, the mail server issues a DNS request querying 2.212.212.192.example.dnsbl.com. If this IP address is included in the black list then the email server will get a reply indicating some IP address, otherwise it will get back a negative reply.

- RHSBL–with a Right-Hand-Side Black List (RHSBL) the mail server takes the domain name in the "FROM" header of the email and checks if it is black listed.

There are currently a few hundred black lists that differ in their scope, listing policies and business models. While some black lists are intended for certain user groups, e.g. subscribers to some mail lists or residents of certain countries, some are open for all. Some of the widely used lists can be used free of charge, and some require some fees from frequent users.

Providers of black lists can collect addresses of suspected spammers in various ways. Spamtraps are email clients that have email addresses that are not published anywhere. Hence, any email arriving at these clients will be originated by spam tools that generate email addresses in an automatic manner. Black lists include also addresses of open mail relays, that is, mail servers that forward any incoming traffic without authentication. Further, black lists can include addresses reported by users or lists of well-known spammers.

Another approach for detecting malicious users is based on observing the sending behavior of the users. Users who are sending too many emails or contact the destination email servers directly are possible spammers. To avoid wrongly black listing a user, one has to take into consideration the use of mailing lists and marketing campaigns that respect the legal restrictions of sending commercial mails.

Black lists could be used in VoIP in the same way as in email. That is, SIP proxies could query a black list using the same DNS queries as is done by email servers. The black lists can include the addresses of SIP proxies that do not authenticate users and addresses of end hosts that are known to generate SPIT calls. Further, SPIT traps could be used as well. That is, SIP user agents can be deployed that have valid addresses that are not published anywhere. Hence, if a call reaches such a trap then it must have been generated by a spammer. Users could also report spammers to their providers.

Spammers could also be detected by observing their calling behavior. Spammers will be initiating more calls than average users, they will initiate multiple calls in parallel and, as they will often call automatically generated addresses, the failure ratio of the call generation will also be higher than with normal users. By observing such characteristics it is also possible to identify spammers and black list them. However, one must be cautious here not to black list call centers or telemarketers that adhere to the legal restrictions.

The main advantage of DNS black lists is that an email or VoIP server can detect in a simple manner whether a message is spam and drop it before it reaches the destination. Unfortunately, they introduce also a number of problems and have their deficiencies as well:

- The black list concept is based on the assumption that spam emails or calls include some information revealing the identity of the spammer such as the IP address of the host used by the spammer for generating the spam or even the emial or SIP address of the spammer. However, in order to achieve high volumes of spam and to hide the identity of the spammer, a large portion of spam today is generated through botnets.

By blacklisting the IP addresses of the bots, the owners of these manipulated hosts will be blocked as well. This might actually be useful as the owners of these hosts will notice that something is wrong and will need to fix their machines. This will, however, often lead the owner of the machine to call his service provider and complain, which is something the service provider naturally wants to avoid. Further, IP addresses are often assigned to a host only on a temporary basis. That is, a host will use the address for a while and then the address will be allocated to another host. If the IP address is added to a black list, then any host using this address will be prevented from sending mails or issuing VoIP calls, even though the spammer is no longer using this address. While all providers of black lists offer some way of unlisting a black listed address, this is usually a complex and time-consuming procedure.

- Black lists sometimes include the addresses of open relay servers, web servers or even complete ISPs that are suspected to be the source of spam. While this stops the spam from these sources, it also affects the legal users of these servers.
- Unless some caching mechanism is used, an email or VoIP server will need to issue one DNS request for each incoming message. Actually, as different list maintainers have different policies about which addresses to include in their lists, it is advisable to query more than one list at the same time so as to reduce the chances of blocking a user that might have been mistakenly added to one list. This means that a number of DNS queries are issued for each incoming request, which introduces some processing delay at the servers.

Instead of black listing IP addresses, it is also possible to black list the names, e.g. email or SIP addresses, of the senders. However, this is unlikely to be very effective. Email and SIP addresses can be obtained at no cost and, once an address is blacklisted, the spammer can start using a different one. Also, spammers can even forge their names and then either use open relays that do not authenticate the requests or send the requests directly to the destinations, in which case the identity of sender is not authenticated as well.

9.5.1.2 White Lists

As the name already suggests, white lists are the opposite of black lists. That is, a server deploying a white list will reject all calls and messages arriving from sources not listed in the white list. While conceptually rather simple, white lists have a number of problems, namely the introduction problem, identity theft and anonymous callers. A caller that is not on the white list of a callee needs to be introduced to the callee before being added to the callee's white list. In cases when the identity of a sender is not authenticated using some secure authentication mechanism, a spammer can just use identities that are white listed and avoid being blocked. Finally, calls from legitimate users that might be using someone else's phone or a public phone or have annonymized their identity would also be rejected.

White lists are used very successfully in instant messaging systems. In such systems the buddy list of a user is used as the white list. To be added to the buddy list of another user a consent-based mechanism is used, see section 9.5.4.1, in which one user asks the other if they can communicate with each other before the first exchange of messages. Further, IM systems are usually closed systems in which the operator of the system can authenticate all users. Hence, stealing the identity of another user is not trivial and would require stealing the password of a user or some other kind of manipulation. Hence, while white lists can be used for messages and presence solutions based on SIP, for such an approach to be successfully used in VoIP or email communication, further solutions for solving the introduction problem and identity theft must be provided as well.

9.5.1.3 Gray Listing

Gray listing is a complementary mechanism to white and black lists (Wolfe *et al.* 2004). When a message is received by an email server from a sender that is neither listed on a white nor on a black list, then the message is rejected temporarily. For the case of email, senders who implement the Simple Mail Transfer Protocol specifications (Klensin 2001) correctly will retry sending the message later. The retransmitted message will then be accepted by the server and forwarded to the client. Gray listing is based on the assumption that spam software is rather simple and is optimized to send a lot of messages but does not care about retransmissions. In this way, messages from legal users are never dropped unnecessarily and are always forwarded to the receivers, albeit slightly delayed. Messages from spammers are dropped with very little effort. Extending spamming software so as to consider retransmissions would incur higher costs for the spammer as the spam software would need to remember which messages need to be retransmitted, which increases the complexity and amount of memory used. Further, the introduced delay would give anti-spam solutions more time to identify possible spam campaigns and the characteristics of the spam emails, and introduce new filters for dropping messages that comply with these characteristics or come from an identified spammer.

Another approach for gray listing is called no-listing. In this approach, an email service is provided by two servers, a primary and a secondary server. The address of the primary server is announced through DNS as the contact address of the email service. This address is not reachable, though. An standards compliant sender would first try the primary one and, if this failed, would try the secondary one. A spammer would probably give up after the failure message from the primary server. The efficiency of this approach is, however, questionable. Spammers often use the secondary mail server directly as these servers often have less stringent antispam policies and are less overloaded.

Gray listing can also be used with SIP. In this case, an INVITE request arriving from a source that has not been seen before, e.g. on neither a white list nor a black list, would be ejected by the proxy that is serving the destination with a "480 Temporarily Unavailable" response. In such a response the proxy would include a "Retry-After" header indicating when the call should be retried. Another approach for realizing gray listing that can be used when SIP messages are transported over UDP is to simply drop the first request. An RFC compliant SIP sender would then retransmit the request after a timeout. However, the efficiency of gray listing in the case of VoIP is rather questionable. Unlike the sending of emails, establishing a call requires the sender to wait for a response from the destination and maintain some session information anyway. Hence, incorporating retransmissions into the SPIT software does not substantially increase the complexity and resource usage. Therefore, the basic assumption upon which gray listing is based does not apply. Further, the delay introduced by gray listing will annoy legitimate users, who will have to wait longer until their calls are established. Gray listing could, however, be useful for protecting a SIP service from presence and instant messaging spam as well as denial of service attacks, as in these cases the goal is to send as many messages as possible and the spammer does not need to wait for a response from the destination. In this case, the attacker might actually prefer using very simple software that simply sends a SIP request and not bother with processing the responses. This would save the attacker a substantial amount of memory. In such a case gray listing can be very efficient.

Besides the questionable applicability of gray listing to VoIP, it has some general disadvantages too. Senders who do not comply with the RFCs, e.g. do not retry the sending of the messages, would be blocked. Also, a retransmitted request might arrive from a different IP address, which would be the case if a message was sent from a cluster of servers. In this case the retransmission would be blocked as well. Finally, the sever deploying gray listing would need to maintain some state information describing which requests were rejected so as not to falsely reject the retransmissions. Under a denial of service attack in which a server is receiving a large number of different requests, the amount of memory consumed in building the gray list can overload the server and hence itself become the target of the DoS attack.

9.5.1.4 Reputation Systems

An identity listed in a white or black list indicates what the owner of the list believes in the trustworthiness of the identity. With a reputation system, black and white lists can be established based on what other people believe about the trustworthiness of others, i.e. the reputation of an identity. Reputation systems (Resnick *et al.* 2000) are widely used in different Internet communities and services. With some online-auction systems, for example, the decision to buy from a certain user is often influenced by what others say about this user. In such systems the users of the system rate each other after having interacted with each other, e.g. bought something. The trustworthiness of a user in such a community is then the sum of the positive and negative ratings he has received from others.

In principle, using a reputation system for building up white and black lists is rather simple. Users who have a high number of positive ratings will be added to the white list and those having negative ratings will be black listed.

Using a reputation system for building a white list would mean that a user new to the system might find his calls and messages listed automatically as spam because he still does not have a sufficient number of positive ratings. When used to establish a black list, a spammer could use a new identity each time each time the old identity gets black listed and would hence have a fresh start. Further, in order not to get black listed, a spammer might blackmail others and threaten them that if they give him a negative rating he will do the same to them. This fear of retaliation reduces the efficiency of such a system.

Another aspect of using reputation systems is their vulnerability to sybil attacks (Douceur 2002). In systems where new identities, i.e. sybils, can be obtained cheaply, an attacker can collect a large number of identities and use them to increase his reputation or decrease the reputation of others.

Reputation systems used in online-auction sites or hotel reservation systems require explicit user feedback. For users to provide negative feedback they have to be annoyed enough to actually provide the feedback. Some incentives are also needed to encourage users to provide positive feedback, such as getting positive feedback from the other side as well. Another approach for a reputation-based system that does not require such feedback is to infer the reputation implicitly using the black and white lists. People listed on a white list have themselves white lists. Hence, if someone is on the white list of somebody that is on my white list then the probability that this person is trustworthy is rather high. Similarly, if someone is on the black list of somebody that is on my black list then that someone is probably not so trustworthy. In this way, a circle of trust can be established. The reputation of a person can be weighed by how strong the relation to that person is in this circle. A person that is one hop way, e.g. on the white list of someone that is on my white list, is to be trusted more than someone that is 10 hops away. While this approach for building a reputation avoids the sybil attacks, it requires that the white lists of different people can be linked together. Further, such a circle will be broken if a spammer manages to get on the white list of someone in this circle.

Reputation systems have another two disadvantages that make their usability questionable. First, the need to collect the information in one central place requires different providers of VoIP services to exchange the collected ratings. This is often not desired by the providers and is often prohibited by privacy laws that prevent the providers from distributing user-related information to outsiders. Secondly, for reputations system to work, similar to black and white lists, they require strong identities, otherwise a spammer can just use the identity of someone with a high positive rating. Recent studies (Jagatic *et al.* 2007) suggest that, without strong identities, reputation systems can even have negative effects as people tend to easily trust other people who are listed in their white lists.

9.5.1.5 Statistical Analysis of User Behavior

One would expect the usage behavior of the vast majority of Internet users in terms of number of sent and received emails or phone calls to be very different from that of a spammer. For one thing, normal users can be expected to receive more emails or calls than they send, mainly because they receive a lot of spam. Spammers will in general receive only very few emails or calls and will send a lot. Further, emails and calls generated by normal users will only rarely result in an error. In contrast, spammers often use addresses that are automatically generated or are no longer valid. This leads to a relatively high ratio of transport errors. Finally, whereas normal users would be expected to generate tens or a few hundred emails and calls a day, a spammer will be sending thousands if not millions.

With these obvious differences between spammers and normal users, it would be tempting to build an antispam solution based on monitoring the activity of the users and statistically analyzing their behavior. Users with an anomalous behavior would then be blocked. Unfortunately, though, such a solution would not be very effective. Already the monitoring part is rather problematic. Calls and emails can be classified based on either the sender identity as indicated in the emails or SIP messages or the IP addresses of the senders. Unless some form of strong identity is used, it is possible to falsify the sender information in emails as well as SIP messages, making it ineffective to classify the mails or calls based on this information. On the other hand, the source IP addresses of the messages are only available to the ISP of the spammer as once the SIP messages traverse a proxy the original IP address will no longer be available. Unless the ISP is also the VoIP or email provider of the spammer, he does not have the incentive to monitor the traffic and do the analysis. Doing the monitoring requires the ISP to check all the traffic of all of its users, whether this is email or SIP traffic, and do the analysis. Besides the possible legal implications of an ISP eavesdropping on the traffic of its users, detecting spammers and blocking their traffic would be more of a benefit to users of VoIP and email providers and not necessarily the ISP. Further, even if it was possible to monitor the traffic, detecting spammers would still be a nontrivial problem. While the usage behavior of different users might have similar characteristics, there are still large differences. An office worker will probably exchange many more emails than a blue collar worker. Employees of a marketing department are very likely to be sending more emails and making more calls than they receive. Further, spammers are very likely to use botnets for generating their emails and calls. On the one hand, in such a case the increase in the amount of sent emails and made calls by the manipulated hosts might be only minor and might not be detected. On the other hand, detecting such a change might also not be beneficial. Blocking all mails and calls from the manipulated hosts will not only block the spam emails and calls but also the legitimate calls and emails made by that host. The owner of the host that is no longer capable of communicating with others will most likely complain to his provider, which will increase the load on the call centers and help desks of the provider. In the worst case the owner of the manipulated host might be very unhappy with this blocking and decide to change provider. Naturally there are other possibilities for solving this issue, such as informing the owner that his host is being used as a botnet and asking the owner to clean up the host. In any case such an approach would incur additional costs to the provider. Thereby deploying such a solution at an ISP would not only incur additional costs for the traffic monitoring and analysis, but might also cause the customers of the ISP to become unsatisfied with the service, thereby further reducing the incentives for the ISP to use this approach.

9.5.2 Content Analysis

Content analysis techniques try to identify the content of a message and detect whether it contains content that is probably not wanted by the receiver. When using content analysis, an antispam software analyzes the content of a message and tests for certain phrases that are often seen in spam messages. The efficiency of this analysis is increased by using probabilistic classifiers such as Bayesian filters (Sahami *et al.* 1998). This enables the antispam software to adapt to changes in the content without having to generate new rules.

Content analysis is one of the most widely used approaches for detecting email spam. As the SIP signaling messages themselves carry little content, the equivalent in the VoIP area would be the analysis of the audio or video content sent by a spammer.

Content analysis has proved to be a valuable tool for detecting spam emails. However, its applicability to VoIP is rather questionable. On the one hand, to be able to analyze a call, the call has to be established first. That is, the callee has to have already been annoyed by the ringing of the SPIT call before the analysis can start, which is obviously too late. On the other hand, analyzing speech or video messages is more complex than the analysis of text and is more resource-consuming in terms of memory and CPU needed.

While ineffective for preventing SPIT calls, content analysis can be used for fighting spam over IM or presence. With both services, the spammer sends his content as text messages and hence the same tools and mechanisms as used for email can be used here as well. Further, content analysis can be used to detect SPIT calls that have been routed to a voicemail box. By analyzing the content of calls recorded in a voicemail box, it is possible to detect SPIT calls and save them in a certain folder or alert the user that these calls are very likely to be SPIT. This would help the users to reduce the time wasted in listening to SPIT calls.

9.5.3   Collaborative Filtering

Collaborative filtering is based on the assumption that the same or similar messages are sent to a large number of receivers. Once one receiver decides that a message is actually spam, he can provide a central entity with a summary, e.g. hash of this message. The central entity then shares the hash of the message with the antispam solutions deployed at email servers. When these email servers receive an email they calculate the hash of the email and compare the hash with the list of hashes they have received from the central entity. As the same hash can only be produced by the same email messages, if the hashes match then the email can be classified as spam.

As a spammer can avoid detection by including small variations in their messages, e.g. changing some letters or adding some random text in the messages, systems deploying collaborative filtering use so-called "fuzzy hashes" (Kornblum 2006). Using fuzzy hashes, it is possible to determine that two pieces of content are similar despite minor differences. There are naturally limitations on the effectiveness of fuzzy hashes. To be able to detect higher levels of variance, one either has to use larger hashes or accept a higher level of false positives. Using larger hashes increases the overhead incurred by this approach in terms of the bandwidth used for transporting the hashes between the collaborating entities, the memory needed for saving all the hashes, the CPU needed for generating the hashes and the delay needed for comparing the message hashes with the saved hashes of identified spam messages.

The hashes of the spam messages can be collected by spamtraps or be submitted by email servers or users that are able to identify a message as spam. Without strict control on who is allowed to submit a hash of spam messages, such a system can be open to abuse when hashes of legitimate mailing list messages get submitted to the central entity.

While perfectly valid for spam over IM or presence, in the case of VoIP the use of collaborative filtering is limited to voicemail services. That is, once a voicemail server or a SPIT trap has recorded a SPIT call, a hash of this call can be generated. This hash can then be used by other voicemail servers to identify possible recorded SPIT calls.

9.5.4   Interactive Antispam Solutions

One of the main reasons for the rapid increase in the number of spam messages on the Internet is the simplicity and low cost associated with generating and sending such messages. Interactive antispam solutions aim at deterring spammers from sending millions of spam messages by increasing the complexity and costs of sending such messages. This is achieved by requiring the senders to answer questions that require some form of human interaction, to pay some money or to spend some computational resources in solving a certain puzzle.

Interactive solutions can be used in conjunction with white and black lists as a solution for the introduction problem. That is, if a message is received from a sender who is not listed on either the white or black lists of the receiver, the sender is asked to perform some task and, if this is successful, then the user is added to the white list and otherwise to the black list.

Such systems have not found wide usage for combating spam and it is questionable if they are useful for combating VoIP and IM spam. Besides the technical limitations and deployment issues of the various solutions, such solutions require the interaction of the caller or add some delay to the call setup time, which can be annoying not only to spammers but also to legitimate callers. The need for interaction limits the possibility of using automated legitimate services such as emails received from email lists or prerecorded calls from a school, for example, to all students informing then that the school is closed due to a snow storm. If receiving emails or calls from such services is desired, then the sources of these services will have to be white listed manually, which incurs additional overheads on the callee in keeping the white list up-to-date.

Further, such solutions require the extension of SIP with additional headers and information. Hence, for the introduction of such solutions all SIP user agents would need to be extended to support these extensions. However adding new features and protocol extensions to existing hardware and software products is never an easy task and there will always be some components that do not support the latest standards. Simply rejecting calls from user agents that do not support interactive solutions would mean that possibly a large proportion of legitimate users would be denied access to SIP-based services.

Finally, where a call is forked to multiple destinations, the caller would have to reply to different challenges which would further increase the complexity of the system and reduce the willingness of legitimate users to use such solutions.

9.5.4.1   Consent-based Communication

To reduce the number of spam messages received by a user, instant messaging systems ask the user to give her consent before forwarding a message from an unknown sender. That is, if a sender is neither on the white or black lists of the receiver, the messages from this sender will not be forwarded to the receiver unless the receiver explicitly agrees. This is usually achieved by displaying a message to the receiver indicating that some sender would like to communicate with her.

Such an approach has been rather successful in instant messaging systems in which both the sender and receiver are using the same provider and are authenticated by that provider. In a more open system in which the senders and receivers belong to different operators, the sender might assume any identity that might be trusted by the receiver. Even if strong identities that cannot be forged are used, the spammer might use a SIP address that looks like a trustworthy address, e.g. john@bonk.com instead of john@bank.com. This might fool the receiver in to believing that the incoming message is from his bank contact and adding the spammer to his white list. Further, the spammer could actually formulate the essence of his message as part of the name that gets displayed to the receiver such as "low.cost.drugs.for.healthier.life@drugs.com". Thereby, while the spammer would not be able to send a complete message to the receiver, the receiver would still get to see the major points. Finally, having to reject requests for communication manually, which usually means clicking on some pop-up screen, will itself be rather annoying to the receiver if it happens too often.

For VoIP this approach could be realized by having the operator first generate a call to the receiver informing her that some user would like to talk to her. This approach is obviously of limited use as the call from the operator to the receiver will be just as annoying to the receiver as the SPIT call itself. Another approach is to forward all calls from unknown users to a voicemail box. The receiver can listen to the calls either while being recorded or afterwards and decide whether the caller is a spammer, in which case the caller will be black listed, or legitimate, in which case the callee will pick up the call or call back and whilte list the caller. This will, however, incur a long delay between conducting a call and actually receiving the call. Further, it requires a high level of interaction of the callee, which actually contradicts the goal of interactive approaches, which aim at increasing the costs incurred on the spammers for sending spam and not shifting the costs to the receivers.

9.5.4.2 Completely Automated Public Turing Tests to Tell Computers and Humans Apart (CAPTCHA)

The Turing test (Turing 1950) was proposed so as to enable a human to determine whether he is communicating with a human or machine. CAPTCHAs (von Ahn *et al.* 2004) are a kind of reverse Turing test in that they enable a machine to detect whether it is communicating with a machine or human. CAPTCHAs are often used by web sites to ensure that the requester of some service is actually a human being and not a machine. This is done by asking the requester to solve a problem that is very simple to solve for a human being but rather difficult for a machine, like recognizing letters and numbers in a distorted image or clicking on part of an image.

CAPTCHAs are often used for hiding email addresses. That is, an email address can only be viewed after solving a CAPTCHA. They can also be used for preventing spam generated by automated processes. Once an email is received that could be spam, the receiver sends back an email with a CPATCHA asking the sender of the original email to solve it. Only if the sender of the CAPTCHA receives an email with the right solution will the original email be considered as legitimate. Similar procedures could be used with SIP-based services. Once an IM was received, an IM including a CAPTCHA could be sent back. The sending of the CAPTCH and the processing of the result could be done automatically.

CAPTCHAs have been mostly successful in protecting web sites from being misused by automated processes, such as bots generating thousands of new email accounts which could be later used for sending spam. However, when used for preventing spam or SPIT they have major deficiencies. When two users deploy CAPTCHAs to protect themselves from spam or SPIT, they will not be able to communicate at all. So, for example, consider that users A and B use CAPTCHAs. User A sends an email to user B. User B sends in return an email with a CAPTCHA to user A and will consider the email from A as spam until a reply with the answer to the CAPTCHA is received from A. As user A also uses CAPTCHAs, A will consider the email from B including the CAPTCHA as spam and will send an email in return with his own CAPTCHA. This could be overcome with SIP by adding the CAPTCHAs into the SIP replies. This would require, however, extensions of SIP and the support of the end systems for such extensions, which would complicate widescale deployment.

Further, spammers can use cheap labor for solving CAPTCHAs or even convince Internet users to solve CAPTACHAs on their behalf. This can be achieved for example by offering a web site with supposedly attractive content and asking users that want to get access to this content to solve CAPTCHAs that were received by the spammer in return to the spam emails or calls, as illustrated in Figure 9.2.

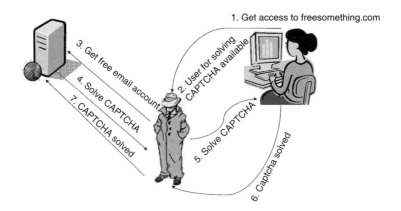

Figure 9.2   Solving CAPTCHAs using Internet users

Turing tests could also be deployed in another form for combating SPIT calls. By directing calls to an IVR (Interactive Voice Recognition) system, the caller can be engaged in a Turing test and the caller will be asked to answer a question like *does the German flag contain the color red? Press 1 for yes, 2 for no*. The answers to such questions could still be verified by low-end devices, making this technique also useful for private users. Turing tests are, however, only partially useful in blocking direct passive marketing-type SPIT calls. With *n* possible answers, even a bot generating SPIT calls has a 1/*n* chance of getting the right answer. Calls from telemarketers have a much higher success rate as the callers will be able to solve the tests or just guess the right answers. Another drawback of such an approach is that it requires legitimate callers to understand the language used for the tests. In order to reduce the usage complexity of such systems, a caller would only be offered a small number of languages to choose from for conducting the test. If the caller does not understand any of these languages, then he will not be able to conduct the tests and his call will be blocked regardless of whether it is legitimate or not.

9.5.4.3   Computational Puzzles

One approach for combating spam is to increase the costs of the spammer in terms of either the required computational resources or time needed for sending a large number of messages. With computational puzzles the sender of a message is asked to find an answer to a puzzle before accepting a message from him. While determining the answer is computationally expensive, verifying that it is correct is rather simple. Hashcash (Back 2003) is an approach that is used for reducing email spam and protecting against DoS attacks. In this approach the sender of an email is asked to find a string of characters $\mathbf{X}$ that, given an original string $\mathbf{Y}$ with $(\mathbf{X} \neq \mathbf{Y})$, would still result in $[f(\mathbf{X}) = f(\mathbf{Y})]$. In Hashcash the function f is the hashing function SHA1 (*Processing Standards Publication 180-2* 2004). Hashing functions are one-way cryptographic functions. That is, given hash(x) it is not possible to derive x. With a hash size of l bits it takes nearly $2^{l/2}$ [3] tries of

---

[3] While one would expect that 2^l tries of different $\mathbf{Y}$ values would get the same hash produced by $\mathbf{X}$, due to the so-called birthday paradox (McKinney 1966), only $2^{l/2}$ are actually needed.

different **Y** values to get a hash value equal to the one produced by **X**. To increase the complexity, hashcash requires the calculation of a so-called second pre-image. That is, instead of asking for any string **Y**, the sender is asked to determine a string **Y** that starts with the same characters as the original string **X**. This requires then 2^l tries with a brute force approach.

Hashcash distinguishes between interactive and noninteractive approaches. In the interactive approach the receiver of a message, e.g. email server, would first challenge the sender by giving him the hash value $f(\mathbf{X})$. The sender would then need to resend the message and include **Y** and $f(\mathbf{X})$ as his credentials. In the noninteractive approach, the sender would include as his credentials both **X** and **Y**. The receiver can easily verify the correctness of the solution by hashing **X** and **Y**.

The same approach may be used for SIP (Jennings 2007). When a call is received, the receiver rejects the call by issuing a response that includes the puzzle. The caller first needs to solve the puzzle and add the result in a new request. Only where the result is correct will the user be alerted.

In principle, adding computational puzzles to SIP signaling is rather straightforward. However, using computational puzzles introduces various deployment issues that make their benefits rather questionable. SIP end devices range from dedicated hardphones and software applications running over mobile phones with limited computation and memory resources to software applications running over powerful computers. This makes it is rather difficult if not impossible to find a puzzle that will not cause too much delay for users of low-end devices but is still not trivial to solve on high-end devices. Also, spammers are most likely to have large computational resources by either using powerful computers or the distributed resources of botnets. Therefore, by making the puzzles hard to solve, normal users with dedicated VoIP appliances such as hard phones or adapters will suffer from unacceptably long delays for the call setup, whereas spammers can compensate for the complexity by using greater resources.

9.5.4.4 Payment Systems

Solutions based on computational puzzles aim to deter spammers by indirectly increasing the costs of sending spam emails by requiring more computational resources. In (Abadi *et al.* 2003), the authors describe a solution that directly increases the costs of sending spam by having the sender of an email deposit a certain amount of money at a payment server and include proof of the deposit in the email. If the receiver of the email decides that the email is legitimate, the sender can reclaim the deposited money.

In (Jennings *et al.* 2007), a similar approach is proposed for SIP. In this case the callee rejects incoming calls from unknown users and requests a payment at a certain payment server first. The caller then needs to contact the payment server, make the payment and get a receipt confirming the payment. The call is then re-initiated with the receipt added to the SIP request. This technique requires two transactions. One from the caller to the callee and then the refund from the callee to the caller. Hence, for this solution to work it would require efficient and low-cost payment systems.

Similar to other interactive solutions the applicability of this solution is questionable. To avoid possible theft of the payment receipts, reusing the same payment receipt for different calls or the forging of receipts, the involved parties, e.g., caller, callee and payment system must communicate over secured links and utilize mechanisms that would prevent the misuse of the solution. This increases the complexity of deploying the system. Further, at least for some transition period, different payment servers might exist each with a different technology. This would make the communication between users not using the same payment server rather difficult.

As with computational puzzles, deciding on the appropriate level of payment is not straight forward. While asking a user to deposit a dollar or two before making a call might be acceptable by most people living in rich countries, such a requirement might be prohibitive for users in poorer countries, especially as they risk loosing the deposit if the callee classified the call as spam. However, reducing the deposit to a level that would be acceptable to users all around the world would probably result in a deposit that even spammers are willing to spend.

Automated call services could still work with this approach. However, it would require a school calling all of its students to deposit a relatively large sum of money in order to ensure that all the calls go through which would be prohibitively expensive.

Finally, this payment system might actually be used for fraud. A fraudster could for example announce an attractive service, e.g., a low cost adult chat forum, that would lure a lot of people to call in. However, instead of providing the service, the fraudster would then classify all the incoming calls as spam and keep the deposited payments.

9.5.5 Preventive Antispam Methods

The previously mentioned antispam solutions aim to discover whether some message is sent by a spammer. Preventive methods aim to prevent spam in the first place. This can be achieved by hiding the addresses of users or reducing the speed with which a spammer can send messages. This gives the administrators of the email and VoIP servers more time to detect spam attacks and increase the resources needed by an attacker for launching a large-scale spam campaign.

9.5.5.1 Address Hiding

In order to launch a spam campaign, spammers need to first collect addresses of possible receivers. This is often achieved by harvesting addresses from web sites using web crawlers. A web crawler is a software that browses through web sites and searches for text strings that might contain contact information such as email addresses or phone numbers, e.g. strings containing "@" characters or digits preceded by a "+". Hence, a straight-forward approach for reducing the chances of getting spam emails and calls is to hide contact information or to make it inaccessible to web crawlers. To increase the complexity of finding such information different strategies can be used:

- Address munging–this typically involves disguising an addresses in such a way that a human being can still read the correct address but an automatic process would interpret it incorrectly. This can be achieved by adding some usage instructions such as "user@exampleREMOVETHIS.com" or swaping certain parts such as "user at example.com". A drawback of this approach is that a person who wants to use such an address will have to reassemble the address in the correct format before using it. Further, web crawlers have become more intelligent and now identify most commonly used replacement words and do the replacement themselves.

- Address images–add the contact information in a web site as an image. This requires the spammers to do image processing, which increases the complexity of finding contact information considerably. While such an approach makes harvesting of addresses much more complex, it is also also annoying for the users, who need to type the email addresses instead of clicking on them or copying and pasting them into their applications.
- Spam poisoning–this approach aims at increasing the ratio of false addresses collected by the spammers so as to discourage them from harvesting addresses. This is achieved by polluting web sites with numerous false addresses and false links. Some solutions used for this approach dynamically generate numerous random fake addresses and links that lead to further fake addresses. These fake links are created by a web application and lead to the same application again. An address harvesting web crawler that is looking for email addresses follows these links. With each followed link new false links are created and the crawler ends up in an endless loop. Address harvesting software has become aware of such techniques and can detect addresses that do not belong to valid domains or pages that contain nothing but addresses and links. While the increased intelligence of the harvesting software makes spam poisoning less effective, it still increases the effort on the side of the spammer and the number of false addresses in the spammers lists.
- Address protection–another approach for address hiding is to only present the address after asking the requester to solve a puzzle. That is, if someone wants to get an email address from a web site, she would need to click it and, instead of getting the address itself, the user would be asked to solve a CAPTCHA first.

9.5.5.2   Temporary Addresses

Spammers harvest addresses not only from web sites but also from public mailing lists and discussion forums. Further, many web sites require the users to register and indicate a valid email address before getting access to the services of these sites. However, some studies suggest (Jacobsson and Carlsson 2003) that some of these sites actually give or sell the collected addresses to spammers and that the number of spam messages received after registering to such sites increases considerably. To still be able to join public mailing lists and register to various web services one could use different email addresses for different services and different groups of people. That is, the user would own a lot of addresses and use different addresses in different contexts. Once spam is received over a certain address then this address is invalidated.

Using different addresses requires the user to remember which address was used with which group and once an address is invalidated the user will need to create a new address and give it to those people who were using the old one. This overhead makes this approach rather complex and the user actually risks being unreachable. To reduce this management overhead some web sites offer their users temporary addresses that are only valid for a period of time ranging from a few minutes to a few months. Such addresses can then be used when registering to a web site that requires a valid address for sending back a confirmation or when joining a public discussion group. The user can then use a permanent contact address for communicating with his trusted peers and temporary addresses otherwise.

9.5.5.3 Tarpit

Spam filters usually drop emails that are identified as spam. While this has the positive effect that the number of spam mails delivered successfully is reduced, it has no other effect on the spammer. Tarpit[4] is a technology that aims not only to reduce the number of delivered spam messages but also to slow down the spammer and hence requires the spammer to use more resources. With email this is achieved by having the mail server send back a reply to the spammer that consists of multiple lines. Each line is sent very slowly and the spammer has to wait until the complete answer is received. While this approach does not prevent the spammer from opening multiple connections and sending spam to other destinations, it reduces its ability to send spam to the server using the tarpit technology and engages some of the spammer's resources in useless communication.

A similar approach could be used with SIP. A SIP end system that is located in a honeypot, for example, would reply to an incoming call by sending temporary responses. The spammer would then have to wait until a final response had been received, which would, however, only come after a long time, if ever. This would keep the spammer engaged in a useless communication and would give the administrator time to detect the spammer and take some action against him, e.g. add him to a black list.

9.6 General Antispam Framework

From Section 9.5 it is obvious that there are already many concepts and technologies that can be deployed to combat unsolicited communication. The discussion in Section 9.5 also suggests that none of the available technologies is sufficient by itself for preventing SIP or email spam. However, by combining different technologies, the risk of spam can be considerably reduced.

Antispam mechanisms used for email rely mainly either on checking the identity of the sender by using white and black lists or reputation systems, or on analyzing the content of the sent emails. While content filtering is applicable for IM spam, it is hardly usable for VoIP calls. Hence, in the case of VoIP the identity of the caller is the main basis for classifying a call as SPIT or not. In (Rosenberg and Jennings 2008) a general framework for fighting spam is outlined, consisting of the following components:

- Caller classification – callers can be classified as either callers we know directly or those who are known to other people whom we know. This knowledge is manifested in the usage of black and white lists as well as reputation systems. Using this knowledge, an anti-SPIT solutions can then make the decision on whether to accept or reject an incoming call. However, in general, only the service provider of a caller can authenticate the caller. Similar to email, a spammer can overcome this hurdle by bypassing the servers of his provider and contacting the destination directly. Thereby, the identity included in the SIP messages would be unverified and hence useless. Therefore, to be able to use black and white lists or reputation systems, some sort of strong identification as discussed in section 6.4 must be used in order to be able to classify callers correctly

---

[4] Also sometimes referenced to as Targrube or Teergrube.

- Caller introduction – where a caller is not known in advance, then some interaction is needed to introduce the caller to the callee and enable the callee to verify that the caller is not a spammer. The interactive approaches presented in section 9.5.4 can be of some help in solving this introduction problem. A widescale deployment of such interactive approaches would make SPIT more expensive and more complex and would hence reduce the activity to the spammer. Interactive approaches, however, have varying levels of usefulness for different sorts of SPIT. So while payment can be used to combat all kinds of unsolicited communication, Turing tests are more useful for combating SPIT calls with prerecorded content. Telemarketing calls cannot be prevented using such tests as the caller will be able to solve them. Another approach for dealing with unknown callers is to forward the call to a voicemail. The callee can then later classify the call as SPIT or not or the content can be analyzed using speech recognition software.

Combining strong identities with caller classification and introduction provides a powerful framework that is likely to reduce the number of successful SPIT calls, i.e. calls that actually make it to the their intended receiver. By deploying the preventive measures discussed in section 9.5.4, the total number of SPIT calls generated over the Internet can be further reduced.

However even with a widescale deployment of strong identities and interactive approaches for solving the introduction problem, SPIT will most likely not disappear completely. In general it is more acceptable to receive a spam call or email than to lose a completely legitimate one. So in order to reduce the possibility of false positives, i.e. classifying a legitimate call as SPIT, an anti-SPIT solution cannot use too strict rules such as rejecting all calls from unknown callers that do not support some introduction method. Deploying such strict rules would mean that a call from a family member calling using a phone in a hotel that does not support the introductory method used would be rejected, which is naturally completely unacceptable.

So, in summary, similar to the situation with the email, SPIT is very likely to become another widely used marketing tool. While anti-SPIT solutions will be able to combat a good many of the SPIT calls, there will very likely be some SPIT that does get through to the receivers.

Bibliography

282001 ES 2008 NGN Functional Architecture. ETSI Standard Telecommunications and Internet Converged Services and Protocols for Advanced Networking (TISPAN), ETSI.

187003 TS 2008 Security Architecture. Technical Specification Telecommunications and Internet Converged Services and Protocols for Advanced Networking (TISPAN), ETSI.

282004 ES 2008 Network Attachment Sub-System (NASS). ETSI Standard Telecommunications and Internet Converged Services and Protocols for Advanced Networking (TISPAN), ETSI.

11.11 TS 2007 Specification of the Subscriber Identity Module–Mobile Equipment (SIM-ME) Interface. Technical specification group terminals, 3rd Generation Partnership Project.

23.003 TS 2008 Numbering, addressing and identification. Technical specification group core network and terminals, 3rd Generation Partnership Project.

23.060 TS 2008 General Packet Radio Service (GPRS); Service description; Stage 2. Technical Specification Group Services and System Aspects, 3rd Generation Partnership Project.

23.101 TS 2007 General Universal Mobile Telecommunications System (UMTS) architecture. Technical specification group core network and terminals, 3rd Generation Partnership Project.

23.107 TS 2007 Quality of service (QoS) concept and architecture. Technical specification group core network and terminals, 3rd Generation Partnership Project.

24.008 TS 2008 Mobile radio interface Layer 3 specification; Core network protocols; Stage 3. Technical specification group core network and terminals, 3rd Generation Partnership Project.

24.228 TS 2006 Signalling flows for the IP multimedia call control based on session initiation protocol (SIP) and session description protocol (SDP). Technical specification group core network and terminals, 3rd Generation Partnership Project.

24.229 TS 2005 IP multimedia call control protocol based on Session Initiation Protocol (SIP) and Session Description Protocol (SDP); Stage 3 (Release 7). Technical specification group core network and terminals, 3rd Generation Partnership Project.

29.162 TS 2008 Interworking between the IM CN subsystem and IP networks. Technical specification group core network and terminals, 3rd Generation Partnership Project.

29.198 TS 2001 Open Service Architecture (OSA) Application Programming Interface (API)–Part 1. Technical specification group core network and terminals, 3rd Generation Partnership Project.

29.207 TS 2005 Policy control over Go interface. Technical specification group core network and terminals, 3rd Generation Partnership Project.

29.208 TS 2007 End-to-end Quality of Service (QoS) signalling flows. Technical specification group core network and terminals, 3rd Generation Partnership Project.

29.209 TS 2007 Policy control over Gq interface. Technical specification group core network and terminals, 3rd Generation Partnership Project.

SIP Security   Dorgham Sisalem, John Floroiu, Jiri Kuthan, Ulrich Abend and Henning Schulzrinne
© 2009 John Wiley & Sons, Ltd

29.228 TS 2008 IP Multimedia (IM) Subsystem Cx and Dx Interfaces; Signalling flows and message contents. Technical specification group core network and terminals, 3rd Generation Partnership Project.

29.278 TS 2005 Customized Applications for Mobile network Enhanced Logic (CAMEL); CAMEL Application Part (CAP) specification for IP Multimedia Subsystems (IMS). Technical specification group core network and terminals, 3rd Generation Partnership Project.

29.328 TS 2008 IP Multimedia Subsystem (IMS) Sh interface; Signalling flows and message contents. Technical specification group core network and terminals, 3rd Generation Partnership Project.

31.101 TS 2007 UICC-terminal interface; Physical and logical characteristics. Technical specification group core network and terminals, 3rd Generation Partnership Project.

31.102 TS 2008 Characteristics of the Universal Subscriber Identity Module (USIM) application. Technical specification group core network and terminals, 3rd Generation Partnership Project.

31.103 TS 2008 Characteristics of the IP Multimedia Services Identity Module (ISIM) application. Technical specification group core network and terminals, 3rd Generation Partnership Project.

33.102 TS 2002 3G security; Security architecture. Technical Specification Group Services and System Aspects, 3rd Generation Partnership Project.

33.203 TS 2007 3G Security; Access security for IP-based services. Technical Specification Group Services and System Aspects, 3rd Generation Partnership Project.

33.210 TS 2008 3G security; Network Domain Security (NDS); IP network layer security. Technical Specification Group Services and System Aspects, 3rd Generation Partnership Project.

33.310 TS 2007 Network Domain Security (NDS); Authentication Framework (AF). Technical Specification Group Services and System Aspects, 3rd Generation Partnership Project.

33.828 TR 2008 IMS media plane security. Technical Specification Group Services and System Aspects, 3rd Generation Partnership Project.

33.978 TR 2007 Security aspects of early IP Multimedia Subsystem (IMS). Technical Specification Group Services and System Aspects, 3rd Generation Partnership Project.

35.201 TS 2007 3G Security; Specification of the 3GPP Confidentiality and Integrity Algorithms; Document 1: f8 and f9 Specification. Technical Specification Group Services and System Aspects, 3rd Generation Partnership Project.

35.206 TS 2007 3G Security; Specification of the MILENAGE Algorithm Set: An example algorithm set for the 3GPP authentication and key generation functions f1, f1*, f2, f3, f4, f5 and f5*; Document 2: Algorithm Specification. Technical Specification Group Services and System Aspects, 3rd Generation Partnership Project.

Abadi M, Birrell AD, Burrows M, Dabek F and Wobber T 2003 Bankable postage for network services *ACSC: Asian Computing Science Conference, LNCS*.

Abdelnur H, Avanesor T, Rusinowitch M and State R 2008 Abusing SIP authentication. *International Symposium on Information Assurance and Security*, pp. 237–242. IEEE Computer Society, Los Alamitos, CA.

Aboba B and Beadles M 1999 The Network Access Identifier RFC 2486 (Proposed Standard). Obsoleted by RFC 4282.

Aboba B, Blunk L, Vollbrecht J, Carlson J and Levkowetz H 2004 Extensible Authentication Protocol (EAP) RFC 3748 (Proposed Standard). Updated by RFC 5247.

Aboba B, Beadles M, Arkko J and Eronen P 2005 The Network Access Identifier RFC 4282 (Proposed Standard).

Acharya A, Kandlur D and Pradhan P 2005 Differentiated handling of SIP messages for VoIP call control. United States patent 20050105464.

Adams C, Farrell S, Kause T and Mononen T 2005 Internet X.509 Public Key Infrastructure Certificate Management Protocol (CMP) RFC 4210 (Proposed Standard).

AES 2001 *FIPS 197: Advanced Encryption Standard (AES)*. Federal Information Processing Standards publication, NIST.

Al-Muhtadi J, Campbell R, Kapadia A, Mickunas MD and Yi S 2002 Routing through the mist: Privacy preserving communication in ubiquitous computing environments. *ICDCS '02: Proceedings of the 22nd International Conference on Distributed Computing Systems (ICDCS'02)*, 74. IEEE Computer Society, Washington, DCA.

Andersen S, Duric A, Astrom H, Hagen R, Kleijn W and Linden J Internet Low Bit Rate Codec (iLBC) RFC 3951 (Experimental).

Andreasen F 2008 SDP Capability Negotiation. Internet Draft draft-ietf-mmusic-sdp-capability-negotiation-09, Internet Engineering Task Force. Work in progress.

Andreasen F and Wing D 2007 Security Preconditions for Session Description Protocol (SDP) Media Streams RFC 5027 (Proposed Standard).

Andreasen F, Baugher M and Wing D 2006 Session Description Protocol (SDP) Security Descriptions for Media Streams RFC 4568 (Proposed Standard).

Andrews M 1998 Negative Caching of DNS Queries (DNS NCACHE) RFC 2308 (Proposed Standard). Updated by RFCs 4035, 4033, 4034.

Arango M, Dugan A, Elliott I, Huitema C and Pickett S 1999 Media Gateway Control Protocol (MGCP) Version 1.0 RFC 2705 (Informational). Obsoleted by RFC 3435, updated by RFC 3660.

Arkko J, Torvinen V, Camarillo G, Niemi A and Haukka T 2003 Security Mechanism Agreement for the Session Initiation Protocol (SIP) RFC 3329 (Proposed Standard).

Arkko J, Carrara E, Lindholm F, Naslund M and Norrman K 2004 MIKEY: Multimedia Internet KEYing RFC 3830 (Proposed Standard). Updated by RFC 4738.

Arkko J, Lindholm F, Naslund M, Norrman K and Carrara E 2006 Key Management Extensions for Session Description Protocol (SDP) and Real Time Streaming Protocol (RTSP) RFC 4567 (Proposed Standard).

Asokan N, Niemi V and Nyberg K 2002 Man-in-the-middle in tunnelled authentication protocols. In *11th Security Protocols Workshop*.

Atkins D and Austein R 2004 Threat Analysis of the Domain Name System (DNS) RFC 3833 (Informational).

Atkins D, Stallings W and Zimmermann P 1996 PGP Message Exchange Formats RFC 1991 (Informational). Obsoleted by RFC 4880.

Audet F 2008 The use of the SIPS URI scheme in the session initiation protocol (SIP). Internet draft, draft-ietf-sip-sips-09, Internet Engineering Task Force. Work in Progress.

Back A 2003 The Hashcash Proof-of-Work Function. www.hashcash.org/papers/draft-hashcash.txt

Baker F and Savola P 2004 Ingress Filtering for Multihomed Networks RFC 3704 (Best Current Practice).

Banks G, Cova M, Felmetsger V, Almeroth K, Kemmerer R and Vigna G 2006 SNOOZE: toward a Stateful NetwOrk prOtocol fuzZEr. *Proceedings of the Information Security Conference (ISC) LNCS*. Springer, Samos.

Barford P, Kline J, Plonka D and Ron A 2002 A signal analysis of network traffic anomalies. *IMW '02: Proceedings of the 2nd ACM SIGCOMM Workshop on Internet measurment*, 71–82. ACM Press, New York.

Baset S and Schulzrinne H 2006 An analysis of the skype peer-to-peer internel telephony protocol. *Proceedings of the INFOCOM '06*.

Basseville M and Nikiforov IV 1993 *Detection of Abrupt Changes: Theory and Application*. Prentice-Hall, Upper Saddle River, NJ.

Baugher M, McGrew D, Naslund M, Carrara E and Norrman K 2004 The Secure Real-time Transport Protocol (SRTP) RFC 3711 (Proposed Standard).

Berners-Lee T, Fielding R and Masinter L 1998 Uniform Resource Identifiers (URI): Generic Syntax RFC 2396 (Draft Standard). Obsoleted by RFC 3986, updated by RFC 2732.

Bernstein D, 2005 Cache-timing attacks on AES. Technical report, University of Illinois in Chicago.

Best K and Walsh N 2001 A URN Namespace for OASIS RFC 3121 (Informational).

Biham E, Biryukov A and Shamir A 2005 *Cryptanalysis of Skipjack Reduced to 31 Rounds using Impossible Differentials*, 291–311. Springer, London.

Biham E and Shamir A 1991 Differential cryptanalysis of DES-like cryptosystems. *CRYPTO'90 & Journal of Cryptology* **4**(1), 3–72.

Bin D, Furong W and Ke J 2006 Performance analysis of signaling using SigComp scheme in narrowband system. *3rd Consumer Communications and Networking Conference, 2006*. IEEE, New York.

Biondi P and Desclaux F 2006 Silver needle in the skype. *BlackHat Europe*.

Boer BD and Bosselaers A 1994 Collisions for the compression function of MD5. In *Advances in Cryptology, Proceedings of EUROCRYPT '93*, 293–304.

Bormann C, Burmeister C, Degermark M, Fukushima H, Hannu H, Jonsson LE, Hakenberg R, Koren T, Le K, Liu Z, Martensson A, Miyazaki A, Svanbro K, Wiebke T, Yoshimura T and Zheng H 2001 RObust Header Compression (ROHC): Framework and Four Profiles: RTP, UDP, ESP, and Uncompressed RFC 3095 (Proposed Standard). Updated by RFCs 3759, 4815.

Bormann C, Liu Z, Price R and Camarillo G 2007 Applying Signaling Compression (SigComp) to the Session Initiation Protocol (SIP) RFC 5049 (Proposed Standard).

Calhoun P, Loughney J, Guttman E, Zorn G and Arkko J 2003 Diameter Base Protocol RFC 3588 (Proposed Standard).

Calhoun P, Zorn G, Spence D and Mitton D 2005 Diameter Network Access Server Application RFC 4005 (Proposed Standard).

Camarillo G 2004 The Internet Assigned Number Authority (IANA) Uniform Resource Identifier (URI) Parameter Registry for the Session Initiation Protocol (SIP) RFC 3969 (Best Current Practice).

Camarillo G 2008 Message Body Handling in the Session Initiation Protocol (SIP). Internet Draft draft-ietf-sip-body-handling-03, Internet Engineering Task Force. Work in progress.

Camarillo G and Schulzrinne H 2004 Early Media and Ringing Tone Generation in the Session Initiation Protocol (SIP) RFC 3960 (Informational).

Camarillo G, Marshall W and Rosenberg J 2002 Integration of Resource Management and Session Initiation Protocol (SIP) RFC 3312 (Proposed Standard). Updated by RFCs 4032, 5027.

Campbell B, Rosenberg J, Schulzrinne H, Huitema C and Gurle D 2002 Session Initiation Protocol (SIP) Extension for Instant Messaging RFC 3428 (Proposed Standard).

Canavan J 2005 The evolution of IRC bots. *Proceedings of Virus Bulletin Conference*.

Carl G, Kesidis G, Brooks RR and Rai S 2006 Denial-of-service attack-detection techniques. *IEEE Internet Computing* **10**(1), 82–89.

Carpenter B 2000 Internet Transparency RFC 2775 (Informational).

Carrara E, Lehtovirta V and Norrman K 2006 The Key ID Information Type for the General Extension Payload in Multimedia Internet KEYing (MIKEY) RFC 4563 (Proposed Standard).

Casson HN 1910 *History of the Telephone*. A. C. McClurg.

Clayton R 2006 The rising tide: DDOS from defective designs and defaults. *SRUTI'06: Proceedings of the 2nd Conference on Steps to Reducing Unwanted Traffic on the Internet*, 1. USENIX Association, Berkeley, CA.

Cohen D 1977 Specifications for the Network Voice Protocol (NVP) RFC 741.

Cohen D 1997 Issues in transnet packetized voice communications. *Proceedings of the Fifth Data Communications Symposium* (September), 6–13.

Cooper D, Santesson S, Farrell S, Boeyen S, Housley R and Polk W 2008 Internet X.509 Public Key Infrastructure Certificate and Certificate Revocation List (CRL) Profile RFC 5280 (Proposed Standard).

Cooper M, Dzambasow Y, Hesse P, Joseph S and Nicholas R 2005 Internet X.509 Public Key Infrastructure: Certification Path Building RFC 4158 (Informational).

Criscuolo PJ 2000 Distributed Denial of Service Trin00, Tribe Flood Network, Tribe Flood Network 2000 and Stacheldraht. UCRL-ID-136939, rev. 1, Department of Energy Computer Incident Advisory Capability (CIAC), Lawrence Livermore National Laboratory.

Cuervo F, Greene N, Rayhan A, Huitema C, Rosen B and Segers J 2000 Megaco Protocol Version 1.0 RFC 3015 (Proposed Standard). Obsoleted by RFC 3525.

DES 1999 *FIPS 46-3: Data Encryption Standard (DES)*. Federal Information Processing Standards publication, NIST.

Dierks T and Allen C 1999 The TLS Protocol Version 1.0 RFC 2246 (Proposed Standard). Obsoleted by RFC 4346, updated by RFC 3546.

Dierks T and Rescorla E 2008 The Transport Layer Security (TLS) Protocol Version 1.2 RFC 5246 (Proposed Standard).

Dobbertin H 1996 Cryptanalysis of md5 compress. *Rump Session of Eurocrypt*.

Dotson S and Hoggan SC 2008 Proxy Mutual Authentication in SIP. Internet Draft draft-dotson-sip-mutual-auth-02, Internet Engineering Task Force. Work in progress.

Douceur JR 2002 The sybil attack. *IPTPS '01: Revised Papers from the First International Workshop on Peer-to-Peer Systems*, 251–260. Springer, London.

Droms R 1993 Dynamic Host Configuration Protocol RFC 1531 (Proposed Standard). Obsoleted by RFC 1541.

Droms R 1997 Dynamic Host Configuration Protocol RFC 2131 (Draft Standard). Updated by RFCs 3396, 4361.

Droms R, Bound J, Volz B, Lemon T, Perkins C and Carney M 2003 Dynamic Host Configuration Protocol for IPv6 (DHCPv6) RFC 3315 (Proposed Standard). Updated by RFC 4361.

Dryburgh LS 2008 Interview with Jonathan Christensen (skype). *Ecommedia Blog*.

DSS 2000 *FIPS 186-2: Digital Signature Standard (DSS)*. Federal Information Processing Standards publication, NIST.

Durham D, Boyle J, Cohen R, Herzog S, Rajan R and Sastry A 2000 The COPS (Common Open Policy Service) Protocol RFC 2748 (Proposed Standard). Updated by RFC 4261.

Dworkin M 2001 *Recommendation for Block Cipher Modes of Operation Methods and Techniques*. NIST special publication.

Dworkin M 2005 *Recommendation for Block Cipher Modes of Operation: The CMAC Mode for Authentication*. NIST special publication 800-38b draft.

E.164 ITR 2005 The international public telecommunication numbering plan. Technical report, ITU-T.

Eastlake 3rd D 1997 Secure Domain Name System Dynamic Update RFC 2137 (Proposed Standard). Obsoleted by RFC 3007.

Eddy W 2007 TCP SYN Flooding Attacks and Common Mitigations RFC 4987 (Informational).

Ehlert S, Petgang S, Magedanz T and Sisalem D 2006 Analysis and signature of skype voip session traffic. *Proceedings of the Fourth IASTED International Conference on Communications, Internet and Information Technology*.

Elwell J 2007 Connected Identity in the Session Initiation Protocol (SIP) RFC 4916 (Proposed Standard).

Euchner M 2006 HMAC-Authenticated Diffie–Hellman for Multimedia Internet KEYing (MIKEY) RFC 4650 (Proposed Standard).

European Parliament 2002 Directive 2002/58/ec of the European Parliament and of the Council of 12 July 2002 Concerning the Processing of Personal Data and the Protection of Privacy in the Electronic Communications Sector (Directive on Privacy and Electronic Communications).

Faltstrom P and Mealling M 2004 The E.164 to Uniform Resource Identifiers (URI) Dynamic Delegation Discovery System (DDDS) Application (ENUM) RFC 3761 (Proposed Standard).

Federal Trade Commission 2004 The can-spam act: Requirements for commercial emailers.

Ferguson P and Senie D 1998 Network Ingress Filtering: Defeating Denial of Service Attacks which employ IP Source Address Spoofing RFC 2267 (Informational). Obsoleted by RFC 2827.

Fluhrer S, Mantin I and Shamir A 2001 Weaknesses in the key scheduling algorithm of rc4. *Proceedings of the 4th Annual Workshop on Selected Areas of Cryptography*.

Fogie S, Grossman J, Hansen R, Rager A and Petkov PD 2007 *XSS Exploits: Cross Site Scripting Attacks and Defense*. Syngress.

Fox B and Gleeson B 1999 Virtual Private Networks Identifier RFC 2685 (Proposed Standard).

Franks J, Hallam-Baker P, Hostetler J, Lawrence S, Leach P, Luotonen A and Stewart L 1999 HTTP Authentication: Basic and Digest Access Authentication RFC 2617 (Draft Standard).

Freed N and Borenstein N 1996 Multipurpose Internet Mail Extensions (MIME) Part One: Format of Internet Message Bodies RFC 2045 (Draft Standard). Updated by RFCs 2184, 2231, 5335.

Garcia-Martin M, Henrikson E and Mills D 2003 Private Header (P-Header) Extensions to the Session Initiation Protocol (SIP) for the 3rd-Generation Partnership Project (3GPP) RFC 3455 (Informational).

Garcia-Martin M, Belinchon M, Pallares-Lopez M, Canales-Valenzuela C and Tammi K 2006 Diameter Session Initiation Protocol (SIP) Application RFC 4740 (Proposed Standard).

Grizzard JB, Sharma V, Nunnery C, Kang BB and Dagon D 2007 Peer-to-peer botnets: overview and case study. *HotBots'07: Proceedings of the First Workshop on Hot Topics in Understanding Botnets*, 1. USENIX Association, Berkeley, CA.

Groff JR and Weinberg PN 1999 *SQL: The Complete Reference*. McGraw-Hill Professional, New York.

Guha S, Daswani N and Jain R 2008 An experimental study of the skype peer-to-peer VoIP system. *Proceedings of The 5th International Workshop on Peer-to-Peer Systems, IPTP '08*, Tampa Bay.

Gutmann P 2008 Key Management through Key Continuity (KCM). Internet Draft draft-gutmann-keycont-01, Internet Engineering Task Force. Work in progress.

Handley M and Jacobson V 1998 SDP: Session Description Protocol RFC 2327 (Proposed Standard). Obsoleted by RFC 4566, updated by RFC 3266.

Handley M, Schulzrinne H, Schooler E and Rosenberg J 1999 SIP: Session Initiation Protocol RFC 2543 (Proposed Standard). Obsoleted by RFCs 3261, 3262, 3263, 3264, 3265.

Handley M, Jacobson V and Perkins C 2006 SDP: Session Description Protocol RFC 4566 (Proposed Standard).

Harkins D and Carrel D 1998 The Internet Key Exchange (IKE) RFC 2409 (Proposed Standard). Obsoleted by RFC 4306, updated by RFC 4109.

Hasebe M, Koshiko J, Suzuki Y, Yoshikawa T and Kyzivat P 2008 Example calls flows of race conditions in the Session Initiation Protocol (SIP). Internet Draft draft-ietf-sipping-race-examples-06, Internet Engineering Task Force. Work in progress.

Hautakorpi J, Ed. GC, Penfield R, Hawrylyshen A and Bhatia M 2008 Requirements from SIP (Session Initiation Protocol) Session Border Control Deployments. Internet Draft draft-ietf-sipping-sbc-funcs-07, Internet Engineering Task Force. Work in progress.

Hawkes P, Paddon M and Rose GG 2004 Musings on the wang *et al*. md5 collision Cryptology ePrint Archive, Report 2004/264.

Hilt V, Widjaja I, Malas D and Schulzrinne H 2007 Session initiation protocol (sip) overload control. Internet Draft draft-hilt-sipping-overload-05, Internet Engineering Task Force. Work in progress.

Hladjk J 2005a Effective EU and US approached to spam? Moves towards a co-ordinated technical and legal response–part I. *Communications Law* **10**(3), 71–83.

Hladjk J 2005b Effective EU and US approached to spam? Moves towards a co-ordinated technical and legal response–part II. *Communications Law* **10**(4), 111–120.

Honeynet Project 2004 *Know Your Enemy: Learning about Security Threats (2nd Edition)*. Addison-Wesley Professional, Reading, MA.

Housley R 2005 Using Advanced Encryption Standard (AES) CCM Mode with IPsec Encapsulating Security Payload (ESP) RFC 4309 (Proposed Standard).

Ignjatic D, Dondeti L, Audet F and Lin P 2006 MIKEY-RSA-R: An Additional Mode of Key Distribution in Multimedia Internet KEYing (MIKEY) RFC 4738 (Proposed Standard).

ITU-T Rec. H.248.1 2005 Gateway control protocol. ITU-T.

ITU-T Rec. E.164 2005 The international public telecommunication number plan. ITU-T.

ITU-T Rec. H.323 2006 Packet-based multimedia communicaton systems. ITU-T.

ITU-T Rec. Q.1912.5 2004. Interworking between session initiation protocol (SIP) and bearer independent call control protocol or ISDN user part. ITU-T.

ITU-T Rec. I.250 1995 ISDN user-network interface layer 3 specification for basic call control. Technical report, ITU-T.

ITU-T Rec. Q.1902.3 2003 Bearer independent call control protocol (Capability Set 2) and Signalling System No. 7 ISDN user part: Formats and codes. Technical report, ITU-T.

ITU-T Rec. Q.701 1993 Functional description of the message transfer part (MTP) of Signalling System No. 7. Technical report, ITU-T.

ITU-T Rec. Q.761 1999 Signalling System No. 7–ISDN User Part functional description. Technical report, ITU-T.

ITU-T Rec. Q.931 1998 Definition of Supplementary Services–Integrated Services Digital Network (ISDN)–General Structure and Service Capabilities. Technical report, ITU-T.

ITU-T Rec. X.680 2002 Abstract Syntax Notation One (ASN.1): Specification of basic notation. Technical report, ITU-T.

Jacobsson A and Carlsson B 2003 Privacy and spam: Empirical studies of unsolicited commercial e-mail. *IFIP Workshop on Risks and Challenges of the Network Society*, Karlstad, Sweden.

Jagatic TN, Johnson NA, Jakobsson M and Menczer F 2007 Social phishing. *Commun. ACM* **50**(10), 94–100.

Jennings C 2007 Computational puzzles for SPAM reduction in SIP. Internet Draft, draft-jennings-sip-hashcash-06, Internet Engineering Task Force. Work in progress.

Jennings C and Mahy R 2008 Managing Client Initiated Connections in the Session Initiation Protocol (SIP). Internet Draft, draft-ietf-sip-outbound-15, Internet Engineering Task Force. Work in progress.

Jennings C, Peterson J and Watson M 2002 Private Extensions to the Session Initiation Protocol (SIP) for Asserted Identity within Trusted Networks RFC 3325 (Informational).

Jennings C, Fischl J and Tschofenig H 2007 Payment for services in session initiation protocol. Internet Draft, draft-jennings-sipping-pay-06, Internet Engineering Task Force. Work in progress.

Jonsson J and Kaliski B 2003 Public-Key Cryptography Standards (PKCS) #1: RSA Cryptography Specifications Version 2.1 RFC 3447 (Informational).

Josefsson S 2006 The Base16, Base32, and Base64 Data Encodings RFC 4648 (Proposed Standard).

Jung J, Emil Sit HB and Morris R 2001 DNS performance and the effectiveness of caching *Proceedings of the ACM SIGCOMM Internet Measurement Workshop '01*, San Francisco, CA.

Kabel J 2003 Spam: A terminal threat to ISPS? the legal position of ISPS concerning their anti-spam policies in the EU after the Privacy & Telecom Directive. *Computer Law Review International*.

Kaminsky D 2006 Black OPS of TCP/IP 2005.5 *ShmooCon*, Washington, DC.

Kaminsky D 2008 It's the end of the cache as we know it. Or: 64k should be good enough for anyone. *Blackhats*, USA.

Kaplan H and Wing D 2008 The SIP Identity Baiting Attack. Internet Draft draft-kaplan-sip-baiting-attack-02, Internet Engineering Task Force. Work in progress.

Karn P and Simpson W 1999 Photuris: Session-Key Management Protocol RFC 2522 (Experimental).

Kaufman C 2005 Internet Key Exchange (IKEv2) Protocol RFC 4306 (Proposed Standard). Updated by RFC 5282.

Kazatzopoulos L, Delakouridis K and Marias G 2008 Providing anonymity services in SIP. *Workshop on VoIP Technology: Research and Standards for reliable applications*. IEEE Computer Society, Cannes.

Kent S 2005a IP Authentication Header RFC 4302 (Proposed Standard).

Kent S 2005b IP Encapsulating Security Payload (ESP) RFC 4303 (Proposed Standard).

Kent S and Seo K 2005 Security Architecture for the Internet Protocol RFC 4301 (Proposed Standard).

Klensin J 2001 Simple Mail Transfer Protocol RFC 2821 (Proposed Standard). Obsoleted by RFC 5321, updated by RFC 5336.

Klima V 2006 Tunnels in hash functions: Md5 collisions within a minute. *IACR Eprint Server*.

Koren T, Casner S, Geevarghese J, Thompson B and Ruddy P 2003 Enhanced Compressed RTP (CRTP) for Links with High Delay, Packet Loss and Reordering RFC 3545 (Proposed Standard).

Kornblum JD 2006 Identifying almost identical files using context triggered piecewise hashing. *Digital Investigation* **3**(Supplement 1), 91–97.

Krawczyk H, Bellare M and Canetti R 1997 HMAC: Keyed-Hashing for Message Authentication RFC 2104 (Informational).

Kristensen A 2003 JSR 116: SIP servelet API.

Lennox J 2006 Connection-oriented Media Transport over the Transport Layer Security (TLS) Protocol in the Session Description Protocol (SDP) RFC 4572 (Proposed Standard).

Lennox J, Schulzrinne H and Rosenberg J 2001 Common Gateway Interface for SIP RFC 3050 (Informational).

Lennox J, Wu X and Schulzrinne H 2004 Call Processing Language (CPL): A Language for User Control of Internet Telephony Services RFC 3880 (Proposed Standard).

Lerdorf RJ, Tatroe K, Kaehms B and McGredy R 2002 *Programming PHP*. O'Reilly & Associates, Sebastopol, CA.

Levine BN and Shields C 2002 Hordes: A Multicast Based Protocol for Anonymity. *Journal of Computer Security* **10**(3), 213–240.

Li A 2007 RTP Payload Format for Generic Forward Error Correction RFC 5109 (Proposed Standard).

Liang J and Lai X 2007 Improved collision attack on hash function md5. *Journal of Computer Science and Technology* **22**(1).

Ma M 2006 Tabu marking scheme to speedup IP traceback. *Computer Networks* **50**(18), 3536–3549.

Mahajan R, Bellovin SM, Floyd S, Ioannidis J, Paxson V and Shenker S 2002 Controlling high bandwidth aggregates in the network. *SIGCOMM Computer Communication Review* **32**(3), 62–73.

Mahy R, Gurbani V and Tate B 2008 Connection Reuse in the Session Initiation Protocol (SIP). Internet Draft draft-ietf-sip-connect-reuse-11, Internet Engineering Task Force. Work in progress.

Maiwald E 2003 *Fundamentals of Network Security*. McGraw-Hill Osborne Media, New York.

Maltz DA and Bhagwat P 1998 TCP splicing for application layer proxy performance. Research Report RC 21139, IBM.

Manral V 2007 Cryptographic Algorithm Implementation Requirements for Encapsulating Security Payload (ESP) and Authentication Header (AH) RFC 4835 (Proposed Standard).

Marshall W 2003 Private Session Initiation Protocol (SIP) Extensions for Media Authorization RFC 3313 (Informational).

Marshall W *et al.* 2002 SIP extensions for network-asserted caller identity and privacy within trusted networks. Internet Draft, draft-ietf-sip-privacy-04, Internet Engineering Task Force. Work in progress.

Mayrhofer A and Hoeneisen B 2006 ENUM Validation Architecture RFC 4725 (Informational).

McClure S, Shah S and Shah S 2002 *Web Hacking: Attacks and Defense*, chapter 17. Addison-Wesley, Reading, MA.

McClure S, Scambray J and Kurtz G 2003a *Hacking Exposed: Network Security Secrets and Solutions*, *Fourth Edition*. McGraw-Hill, New York.

McClure S, Scambray J and Kurtz G 2003b *Hacking Exposed: Network Security Secrets and Solutions*, *Fourth Edition*. McGraw-Hill, New York.

McGrew DA 2001 The Truncated Multi-Modular Hash Function (TMMH). Internet Draft draft-mcgrew-saag-tmmh-01, Internet Engineering Task Force. Expired.

McGrew DA and Fluhrer SR Datagram Transport Layer Security (DTLS) Extension to Establish Keys for Secure Real-time Transport Protocol (SRTP). Internet Draft draft-ietf-avt-dtls-srtp-05, Internet Engineering Task Force. Work in progress.

McGrew DA and Fluhrer 2000 Attacks on additive encryption of redundant plaintext and implications on internet security. *Proceedings of the 7th Annual International Workshop on Selected Areas in Cryptography*, pp. 14–28. Springer, London.

McKinney EH 1966 Generalized birthday problem. *American Mathematical Monthly* **73**, 385–387.

Mealling M and Daniel R 2000 The Naming Authority Pointer (NAPTR) DNS Resource Record RFC 2915 (Proposed Standard). Obsoleted by RFCs 3401, 3402, 3403, 3404.

Menezes AJ, van Oorschot PC and Vanstone SA 1996 *Handbook of Applied Cryptography*. CRC Press, Boca Raton, FL.

Miller BP, Fredriksen L and So B 1990 An empirical study of the reliability of unix utilities. *Communications of the ACM* **33**(12), 32–44.

Mills D 1992 Network Time Protocol (Version 3) Specification, Implementation and Analysis RFC 1305 (Draft Standard).

minisip n.d. http://www.minisip.org. Technical report.

Mirkovic J, Martin J and Reiher P 2004 A taxonomy of DDOS attacks and DDOS defense mechanisms. *ACM SIGCOMM Computer Communication Review* **34**(2), 39–53.

Mockapetris P 1987 Domain Names–Implementation and Specification RFC 1035 (Standard). Updated by RFCs 1101, 1183, 1348, 1876, 1982, 1995, 1996, 2065, 2136, 2181, 2137, 2308, 2535, 2845, 3425, 3658, 4033, 4034, 4035, 4343.

Mockapetris P 1989 DNS Encoding of Network Names and Other Types RFC 1101.

Moore D, Voelker GM and Savage S 2001 Inferring internet denial-of-service activity. *SSYM'01: Proceedings of the 10th Conference on USENIX Security Symposium*, 2. USENIX Association, Berkeley, CA.

Nakhjiri M, Chowdhury K, Lior A and Leung K 2007 Mobile IPv4 RADIUS Requirements RFC 5030 (Informational).

Niemi A, Arkko J and Torvinen V 2002 Hypertext Transfer Protocol (HTTP) Digest Authentication Using Authentication and Key Agreement (AKA) RFC 3310 (Informational).

NIST 2002 *Secure Hash Standard*. National Institute of Standards and Technology, Washington, DC. Federal Information Processing Standard 180-2.

Noldus R 2006 *CAMEL: Intelligent Networks for the GSM, GPRS and UMTS Network*. John Wiley & Sons, Chichester.

Nourse A, Liu X, Vilhuber J and Madson C 2008 Cisco Systems: Simple Certificate Enrollment Protocol. Internet Draft draft-nourse-scep-17, Internet Engineering Task Force. Work in progress.

Ohta M 2004 Simulation study of SIP signaling in an overload condition. In *Communications, Internet, and Information Technology* (ed. Hamza MH), 321–326. IASTED/ACTA Press.

Oikarinen J and Reed D 1993 Internet Relay Chat Protocol RFC 1459 (Experimental). Updated by RFCs 2810, 2811, 2812, 2813.

Ong L, Rytina I, Garcia M, Schwarzbauer H, Coene L, Lin H, Juhasz I, Holdrege M and Sharp C 1999 Framework Architecture for Signaling Transport RFC 2719 (Informational).

Oxford 2005 *Compact Oxford English Dictionary of Current English, Third Edition*. Oxford University Press, Oxford.

Pelinescu-onciul A, Janak J and Kuthan J 2003 SIP express router (SER). *IEEE Network* **17**(4), 9.

Perkins C and Westerland M 2007 Multiplexing RTP Data and Control Packets on a Single Port. Internet Draft draft-ietf-avt-rtp-and-rtcp-mux-07, Internet Engineering Task Force. Work in progress.

Perkins C, Kouvelas I, Hodson O, Hardman V, Handley M, Bolot J, Vega-Garcia A and Fosse-Parisis S 1997 RTP Payload for Redundant Audio Data RFC 2198 (Proposed Standard).

Peterson J 2002 A Privacy Mechanism for the Session Initiation Protocol (SIP) RFC 3323 (Proposed Standard).

Peterson J and Jennings C 2006 Enhancements for Authenticated Identity Management in the Session Initiation Protocol (SIP) RFC 4474 (Proposed Standard).

Peterson J, Liu H, Yu J and Campbell B 2004 Using E.164 numbers with the Session Initiation Protocol (SIP) RFC 3824 (Informational).

Peterson J, Polk J, Sicker D and Tschofenig H 2006 Trait-Based Authorization Requirements for the Session Initiation Protocol (SIP) RFC 4484 (Informational).

Postel J 1980 User Datagram Protocol RFC 768 (Standard).

Postel J 1981a Internet Control Message Protocol RFC 792 (Standard). Updated by RFCs 950, 4884.

Postel J 1981b Internet Protocol RFC 791 (Standard). Updated by RFC 1349.

Postel J 1981c Transmission Control Protocol RFC 793 (standard). Updated by RFC 3168.

Price R, Bormann C, Christoffersson J, Hannu H, Liu Z and Rosenberg J 2003 Signaling Compression (SigComp) RFC 3320 (Proposed Standard). Updated by RFC 4896.

Puthenkulam J, Lortz V, Palekar A and Simon D 2003 The Compound Authentication Binding Problem. Internet draft, draft-puthenkulam-eap-binding-04, IETF. Expired.

Rajab MA, Zarfoss J, Monrose F and Terzis A 2007 My botnet is bigger than yours (maybe, better than yours): why size estimates remain challenging. *HotBots'07: Proceedings of the First Workshop on Hot Topics in Understanding Botnets*, 5. USENIX Association, Berkeley, CA.

Ramsdell B 1999 S/MIME Version 3 Message Specification RFC 2633 (Proposed Standard). Obsoleted by RFC 3851.

Ramsdell B 2004 Secure/Multipurpose Internet Mail Extensions (S/MIME) Version 3.1 Message Specification RFC 3851 (Proposed Standard).

Reiter M and Rubin A 1998 Crowds: Anonymity for web transactions. *ACM Transactions on Information and System Security* **1**(1), 66–92.

Rescorla E 1999 Diffie–Hellman Key Agreement Method RFC 2631 (Proposed Standard).

Rescorla E 2008 Keying Material Extractors for Transport Layer Security (TLS). Internet Draft draft-ietf-tls-extractor-02, Internet Engineering Task Force. Work in progress.

Rescorla E and Modadugu N 2006 Datagram Transport Layer Security RFC 4347 (Proposed Standard).

Resnick P, Kuwabara K, Zeckhauser R and Friedman E 2000 Reputation systems. *Communications of the ACM* **43**(12), 45–48.

Richardson R 2007 2007 CSI Computer Crime and Security Survey. Technical report, Computer Security Institute.

Rigney C, Rubens A, Simpson W and Willens S 1997 Remote Authentication Dial In User Service (RADIUS) RFC 2058 (Proposed Standard). Obsoleted by RFC 2138.

Rivest R 1992 The MD5 Message-Digest Algorithm RFC 1321 (Informational).

Roach AB 2002 Session Initiation Protocol (SIP)-Specific Event Notification RFC 3265 (Proposed Standard).

Roback E and Dworkin M 1999 Conference report–first advanced encryption standard (AES) candidate conference, Ventura, CA, August 20–22, 1998. *Journal of Research of the National Institute of Standards and Technology* **104**(1), 97–105.

Rosenberg J 2002 The Session Initiation Protocol (SIP) UPDATE Method RFC 3311 (Proposed Standard).

Rosenberg J 2004a A Presence Event Package for the Session Initiation Protocol (SIP) RFC 3856 (Proposed Standard).

Rosenberg J 2004b A Session Initiation Protocol (SIP) Event Package for Registrations RFC 3680 (Proposed Standard).

Rosenberg J 2006 Requirements for management of overload in the session initiation protocol. RFC 5390 (Informational).

Rosenberg J 2007 Interactive connectivity establishment (ICE): a methodology for network address translator (NAT) traversal for the session initiation protocol (SIP). Internet Draft draft-ietf-mmusic-ice-19, Internet Engineering Task Force. Work in progress.

Rosenberg J and Jennings C 2008 The Session Initiation Protocol (SIP) and Spam RFC 5039 (Informational).

Rosenberg J and Schulzrinne H 2002a An Offer/Answer Model with Session Description Protocol (SDP) RFC 3264 (Proposed Standard).

Rosenberg J and Schulzrinne H 2002b Reliability of Provisional Responses in Session Initiation Protocol (SIP) RFC 3262 (Proposed Standard).

Rosenberg J and Schulzrinne H 2002c Session Initiation Protocol (SIP): Locating SIP Servers RFC 3263 (Proposed Standard).

Rosenberg J and Schulzrinne H 2003 An Extension to the Session Initiation Protocol (SIP) for Symmetric Response Routing RFC3581 (Proposed Standard).

Rosenberg J and Schulzrinne H 2005 Architecture and Design Principles of the Session Initiation Protocol. Internet draft, draft-rosenberg-sipping-sip-arch-00. IETF. Expired.

Rosenberg J, Salama H and Squire M 2002a Telephony Routing over IP (TRIP) RFC 3219 (Proposed Standard).

Rosenberg J, Schulzrinne H, Camarillo G, Johnston A, Peterson J, Sparks R, Handley M and Schooler E 2002b SIP: Session Initiation Protocol RFC 3261 (Proposed Standard). Updated by RFCs 3265, 3853, 4320, 4916.

Rosenberg J, Weinberger J, Huitema C and Mahy R 2003 STUN–Simple Traversal of User Datagram Protocol (UDP) Through Network Address Translators (NATs) RFC 3489 (Proposed Standard).

Rosenberg J, Peterson J, Schulzrinne H and Camarillo G 2004 Best Current Practices for Third Party Call Control (3PCC) in the Session Initiation Protocol (SIP) RFC 3725 (Best Current Practice).

Rosenberg J, Schulzrinne H and Camarillo G 2005 The Stream Control Transmission Protocol (SCTP) as a Transport for the Session Initiation Protocol (SIP) RFC 4168 (Proposed Standard).

Rosenberg J, Boulton C, Camarillo G and Audet F 2008a Best current practices for NAT traversal for client-server SIP. Internet Draft, draft-ietf-sipping-nat-scenarios-09. IETF. Work in progress.

Rosenberg J, Mahy R, Matthews P and Wing D 2008b Session traversal utilities for (NAT) (STUN). Internet draft, draft-ietf-behave-rfc3489bis-18. IETF. Work in Progress.

Rosenberg J, Mahy R, and Matthews P 2008c Traversal using relays around NAT (TURN): Relay extensions to session traversal utilities for NAT (STUN). Internet Draft draft-ietf-behave-turn-12, Internet Engineering Task Force. Work in progress.

Russell T 1995 *Signaling System #7*. McGraw-Hill, New York.

Sahami M, Dumais S, Heckerman D and Horvitz E 1998 A bayesian approach to filtering junk E-mail. *AAAI Workshop on Learning for Text Categorization*. AAAI Technical Report WS-98-05.

Saltzer J, Reed D Clark D 1984 End-to-end arguments in system design.

Savage S, Wetherall D, Karlin A and Anderson T 2001 Network support for IP traceback. *IEEE/ACM Transactions Networks* **9**(3), 226–237.

Schaad J 2005 Internet X.509 Public Key Infrastructure Certificate Request Message Format (CRMF) RFC 4211 (Proposed Standard).

Schaad J and Myers M 2008 Certificate Management over CMS (CMC): Transport Protocols RFC 5273 (Proposed Standard).

Schneier B 1996 *Applied Cryptography*. John Wiley & Sons, New York.

Schulzrinne H 2002 Dynamic Host Configuration Protocol (DHCP-for-IPv4) Option for Session Initiation Protocol (SIP) Servers RFC 3361 (Proposed Standard).

Schulzrinne H 2004 The tel URI for Telephone Numbers RFC 3966 (Proposed Standard). Updated by RFC 5341.

Schulzrinne H and Volz B 2003 Dynamic Host Configuration Protocol (DHCPv6) Options for Session Initiation Protocol (SIP) Servers RFC 3319 (Proposed Standard).

Schulzrinne H, Casner S, Frederick R and Jacobson V 1996 RTP: A Transport Protocol for Real-Time Applications RFC 1889 (Proposed Standard). Obsoleted by RFC 3550.

Schulzrinne H, Rao A and Lanphier R 1998 Real Time Streaming Protocol (RTSP) RFC 2326 (Proposed Standard).

Schulzrinne H, Casner S, Frederick R and Jacobson V 2003 RTP: A Transport Protocol for Real-Time Applications RFC 3550 (Standard).

Sengar H, Wijesekera D, Wang H and Jajodia S 2006 Voip intrusion detection through interacting protocol state machines. *DSN '06: Proceedings of the International Conference on Dependable Systems and Networks (DSN'06)*, 393–402. IEEE Computer Society, Washington, DC.

Senie D 1999 Changing the Default for Directed Broadcasts in Routers RFC 2644 (Best Current Practice).

Shacham A, Monsour B, Pereira R and Thomas M 2001 IP Payload Compression Protocol (IPComp) RFC 3173 (Proposed Standard).

Shen C and Schulzrinne H 2008 SIP server overload control: Design and evaluation. *Principles, Systems and Applications of IP Telecommunications. Services and Security for Next Generation Networks Second International Conference, IPTComm 2008*. Springer, London.

SHS 2007 FIPS 180-3: Secure Hash Standard (SHS). Federal Information Processing Standards publication, NIST.

SIPit n.d. http://bugs.sipit.net/. Technical report.

Sisalem D and Floroiu J 2009 Protecting VoIP services against DoS using overload control. *NorSec 2008*. Technical University of Denmark, Coperhagen.

Sisalem D and Kuthan J 2004 Inter-domain authentication and authorization mechanisms for roaming sip users. In *Wireless Information Systems* (ed. Mahmoud QH and Weghorn H), 89–99. INSTICC Press.

Snoeren AC 2001 Hash-based IP traceback. *SIGCOMM Computer Communication Review* **31**(4), 3–14.

Song DX and Perrig A 2001 Advanced and authenticated marking schemes for IP traceback. *INFOCOM*, 878–886.

Sparks R, Lawrence S, Hawrylyshen A and Campen B 2007 Addressing an amplification vulnerability in session initiation protocol (SIP) forking proxies. Internet Draft draft-ietf-sip-fork-loop-fix-7, Internet Engineering Task Force. Work in progress.

Stewart R 2007 Stream Control Transmission Protocol RFC 4960 (Proposed Standard).

Stone J, Stewart R and Otis D 2002 Stream Control Transmission Protocol (SCTP) Checksum Change RFC 3309 (Proposed Standard). Obsoleted by RFC 4960.

Terry DB, Painter M, Riggle DW and Zhou S 1984 The Berkeley Internet Name Domain Server. Technical Report UCB/CSD-84-182, EECS Department, University of California, Berkeley.

Tschofenig H, Hodges J, Peterson J, Polk J and Sicker D 2008 SIP SAML Profile and Binding. Internet Draft draft-ietf-sip-saml-04, Internet Engineering Task Force. Work in progress.

Tsirtsis G and Srisuresh P 2000 Network Address Translation–Protocol Translation (NAT-PT) RFC 2766 (Historic). Obsoleted by RFC 4966, updated by RFC 3152.

Turing AM 1950 Computing machinery and intelligence. *MIND: A Quarterly Review of Pyschology and Philosophy* **59**(236), 433–460.

Vemuri A and Peterson J 2002 Session Initiation Protocol for Telephones (SIP-T): Context and Architectures RFC 3372 (Best Current Practice).

VoIPSA 2005 VoIP security and privacy threat taxonomy. Report 1.0.

von Ahn L, Blum M and Langford J 2004 Telling humans and computers apart automatically. *Communications of the ACM* **47**(2), 56–60.

Wald A 1947 *Sequential Analysis*. John Wiley & Sons, New York.

Wang H, Zhang D and Shin KG 2002 Detecting syn flooding attacks *INFOCOM*, **3**, 1530–1539.

Wang X and Yu H 2005a How to break md5 and other hash functions In *EUROCRYPT* (ed. Cramer R), vol. 3494 of Lecture Notes in Computer Science, 19–35. Springer, London.

Wang X and Yu H 2005b How to break MD5 and other hash functions. *Advances in Cryptology–EUROCRYPT 2005*, 19–35. Lecture Notes in Computer Science, Vol. 3494. Springer, London.

Westin AF 1970 *Privacy and Freedom*. Atheneum, New York.

Willis D and Hoeneisen B 2002 Session Initiation Protocol (SIP) Extension Header Field for Registering Non-Adjacent Contacts RFC 3327 (Proposed Standard).

Willis D and Hoeneisen B 2003 Session Initiation Protocol (SIP) Extension Header Field for Service Route Discovery During Registration RFC 3608 (Proposed Standard).

Wing D 2007 Symmetric RTP/RTP Control Protocol (RTCP) RFC 4961 (Best Current Practice).

Wing D 2008 DTLS-SRTP Key Transport. Internet Draft draft-wing-avt-dtls-srtp-key-transport-02, Internet Engineering Task Force. Work in progress.

Wing D and Kaplan H 2008 SIP Identity using Media Path. Internet Draft draft-wing-sip-identity-media-02, Internet Engineering Task Force. Work in progress.

Wing D, Fries S, Tschofenig H and Audet F 2008 Requirements and Analysis of Media Security Management Protocols. Internet Draft draft-ietf-sip-media-security-requirements-07, Internet Engineering Task Force. Work in progress.

Wired 2000 Yahoo on trail of site hackers. *Wired.com* February.

Wolfe P, Scott C and Erwin M 2004 *Anti-Spam Tool Kit*. McGraw-Hill Osborne Media, New York.

X.509 1997 ISO/IEC 9594-8/ITU-T Recommendation X.509, "Information Technology–Open Systems Interconnection: The Directory: Authentication Framework". Technical report, ITU-T.

Yergeau F 1998 UTF-8, a transformation format of ISO 10646 RFC 2279 (Draft Standard). Obsoleted by RFC 3629.

Zeilenga K 2006 Lightweight Directory Access Protocol (LDAP) Schema Definitions for X.509 Certificates RFC 4523 (Proposed Standard).

Zfone n.d. http://www.zfoneproject.com. Technical report.

Zhang G, Ehlert S, Magedanz T and Sisalem D 2007 Denial of service attack and prevention on SIP VoIP infrastructures using DNS flooding. *IPTComm '07: Proceedings of the 1st International Conference on Principles, Systems and Applications of IP Telecommunications*, 57–66. ACM, New York.

Zimmerer E, Peterson J, Vemuri A, Ong L, Audet F, Watson M and Zonoun M 2001 MIME media types for ISUP and QSIG Objects RFC 3204 (Proposed Standard). Updated by RFC 3459.

Zimmermann P, Johnston A and Callas J 2008 ZRTP: Media Path Key Agreement for Secure RTP. Internet Draft draft-zimmermann-avt-zrtp-09, Internet Engineering Task Force. Work in progress.

Index

SIP Security   Dorgham Sisalem, John Floroiu, Jiri Kuthan, Ulrich Abend and Henning Schulzrinne
© 2009 John Wiley & Sons, Ltd